Dispersal in Plants

Dispersal in Plants

A Population Perspective

Roger Cousens, Calvin Dytham & Richard Law

OXFORD
UNIVERSITY PRESS

OXFORD
UNIVERSITY PRESS

Great Clarendon Street, Oxford OX2 6DP

Oxford University Press is a department of the University of Oxford.
It furthers the University's objective of excellence in research, scholarship,
and education by publishing worldwide in

Oxford New York

Auckland Cape Town Dar es Salaam Hong Kong Karachi
Kuala Lumpur Madrid Melbourne Mexico City Nairobi
New Delhi Shanghai Taipei Toronto

With offices in

Argentina Austria Brazil Chile Czech Republic France Greece
Guatemala Hungary Italy Japan Poland Portugal Singapore
South Korea Switzerland Thailand Turkey Ukraine Vietnam

Oxford is a registered trade mark of Oxford University Press
in the UK and in certain other countries

Published in the United States
by Oxford University Press Inc., New York

British Library Cataloguing in Publication Data

Data available

Library of Congress Cataloging in Publication Data

Data available

Typeset by Newgen Imaging Systems (P) Ltd., Chennai, India
Printed in Great Britain
on acid-free paper by
Antony Rowe Ltd., Chippenham

ISBN: 978–0–19–929911–9 (Hbk)   ISBN: 978–0–19–929912–6 (Pbk)

10 9 8 7 6 5 4 3 2 1

Preface

Dispersal is currently one of the most active areas in plant ecology. At a time of focus on climate change and plant invasions, dispersal has become a core item of the research agenda. Empirical or theoretical work on dispersal now appears in just about every edition of ecology journals.

Yet, over the years, there have been but a handful of books devoted to the subject (Beal 1989; Ridley 1930; Van der Pijl—three editions from 1969 to 1982), and these have dealt with the subject in a largely descriptive way. Most plant population biology text books, while recognizing the importance of dispersal in population dynamics, give little coverage of dispersal per se. For example, only 3% of almost 800 pages in Harper's (1977) seminal book deals with dispersal. In recent years several excellent dispersal (and frugivory) conference proceedings have been published as books, bringing together recent work by a relatively small number of ecologists (Estrada and Fleming 1986; Bunce and Howard 1990; Clobert et al. 2001; Bullock et al. 2002; Levey et al. 2002; Dennis et al. 2007). However, with so much current interest in dispersal, we feel that it is high time for a new text that brings together current knowledge of dispersal in a more systematic way.

We also feel that such a book should concentrate on dispersal from a population perspective. All three of us are interested in the population-level consequences of dispersal and it was this that brought us into this project together at the University of York in 2003. We wanted to write the dispersal chapters that have not yet found their way into general texts on plant population biology. We also wanted to show where dispersal fits into the theoretical framework of population dynamics. Theory about the role of dispersal in population and community dynamics is developing fast, but there is no systematic coverage of both the processes and the higher-level consequences of dispersal. A book which covers both data and theory should help modellers to appreciate the data and their limitations, and also guide empiricists towards the theory and the kind of data which would enlighten the subject.

In a book taking a strictly population perspective, there may be some topics about dispersal that are not covered. We have no intention that our book should provide the full story about dispersal, and the reader should not look for an authoritative review of all the relevant literature. In particular, as we wrote Chapters 6 to 8, we thought in terms of guided tours of the subjects, taking the reader through examples that illustrate the principles, rather than comprehensive reviews (how many tourists want to visit *every* chateau in the Loire?). In these chapters, we each covered the area with which we are most familiar and give our personal view of the topics.

With the focus on population dynamics, dispersal becomes a quantitative science and needs some mathematical tools. We cannot, indeed we do not think it is sensible, to hide away the mathematics completely. The formalism of the mathematical language encourages precise thinking, and in addition, the power of statistics to summarize data, allows us to become predictive. We appreciate, however, that ecologists vary in their inclination towards mathematics. Thus, for the most part, we keep the mathematics to a high school or university introductory level, putting more advanced topics into boxes or tables. We do, however, urge the reader to take the time to think about more mathematical parts of the subject because it is at the interface with the theory where we expect important advances to come about in the future.

Finally, we think it is important to stress the role of dispersal at a wide range of spatial scales. Much current research focuses on long-distance dispersal

(LLD) and could easily leave an impression that dispersal only matters at landscape and geographic scales. We take the view that distances moved by propagules are important to population processes and population dynamics at *all* scales, from a few centimetres to hundreds of kilometres. After all, dispersal at small scales is a key component of local population dynamics, able to determine whether the population can persist. We therefore stress the importance of measuring dispersal distances without an emphasis on particular scales.

There are many colleagues, some of whom we have yet to meet, and friends who played a part in the production of the book. Ian Sherman (OUP) made the book possible, provided constant encouragement and helped us through some difficult periods along the way. Jane Cousens helped enormously in the pre-submission processes; Lisa Crowfoot redrew many figures; Errol Hoffman taught RC basic fluid dynamics; Barry Hughes and Andy Rawlinson tried very hard to explain mathematical derivations (when weighty university matters already occupied most of Barry's time). Yvonne Buckley, Alex James, Mark Lewis, Andrew Robinson, Martin Schaeffer, Rob Taylor, and Kimberly With kindly read and commented on chapters (or parts thereof). Clearly, however, all errors and omissions are ours. We appreciate advice from Kimberly Holbrook, Ian Gordon, and Graham Hepworth on particular topics. Mohammad Taghizadeh generated figures from his PhD data; David Merritt revised his figures to meet our needs; David Kleinig generously hunted for trees to photograph according to our specifications; and numerous people allowed us to reproduce figures that we found on the internet (citations given in the legends). Georg Gratzer made available data for a temperate forest (Chapter 7), while George Weiblen provided data on Glacier Lily (Chapter 5). The Forest Research Institute Malaysia kindly gave us permission to use data from the Pasoh Forest Reserve (Chapter 7). Support for RL from NSF Grant NSF-DEB 0414465 is gratefully acknowledged.

Please note that for plant names we give the family name at the first mention of each genus. For animals we give only common names, unless the original source only gives genus and species.

Contents

CHAPTER 1

Introduction

1.1 Dispersal, centre stage

The science of ecology is about a century old, but through most of that century dispersal was overlooked, largely ignored, or assumed to be unimportant. Even during the plant population biology revolution of the 1970s, dispersal was seldom a focus for quantitative or experimental research. This neglect has changed and dispersal ecology has become a discipline in its own right. Dispersal is now, appropriately, seen as a central part of ecology and a clear understanding of dispersal is considered vital for addressing many important questions in pure and applied ecological research.

Two simple equations, included in most ecology text books, demonstrate why this is the case. They describe the change in the number of individuals (equation 1.1) and the number of species (equation 1.2) in an area:

$$N_{t+1} = N_t + B - D + I - E \qquad (1.1)$$

that is, N_{t+1} (the population size in the future) can be calculated by taking the population size now (N_t), adding births (B), removing deaths (D), adding immigrants (I), and removing emigrants (E). Clearly, these last two processes, immigration and emigration, are both dispersal processes.

$$S_{t+1} = S_t + C - M + X \qquad (1.2)$$

that is, S_{t+1} (the number of species in the future) can be determined from the number of species at present S_t, adding new species colonising the area (C), which is clearly a dispersal process, removing local extinctions (M), and adding any additions from speciation (X).

In this book we consider dispersal to be the phase between separation from the parent until the propagule (usually a seed) comes to final rest. We

provide a narrative that starts with consideration of single propagules and the forces that act during development and separation from the parent. Then we move on to populations of propagules arising of single parents, populations of plants, and their propagules in both expanding and saturated states, before concluding by considering the evolutionary aspects of dispersal.

This introductory chapter serves two purposes. First, we explain the format which follows in later chapters and justify the particular order we have chosen. Second, we state, as unambiguously as we can, the definitions that we will use throughout the text. Ecologists working on dispersal, use terminology that can be confusing or understood only by the specialist. Although you may disagree with the choices of terminology that we make, we hope you will be clear what we mean in using a particular word.

1.2 The framework

Our approach, taken throughout the book, builds a framework for the study of dispersal and its implications, critically appraises the approaches that have been and are being taken, distils important issues into their component parts, and indicates directions in which we believe that future research should go. We give examples to illustrate our arguments, but, because of burgeoning size of the discipline, we cannot provide a complete literature review. However, we include a lot of references to aid the newcomer to this field to engage in the literature.

What is dispersal? An internet search will show that there are many definitions, ranging in their level of detail. Common definitions refer to movement away from the origin, for example *'the movement of individuals to new locations, away from their parents'.* So how far from a parent does a seed need to land for it to have dispersed; how little movement constitutes failure to disperse? A large tree may have a canopy with a radius of 10 metres: if a seed lands under this canopy, has it dispersed? What if it lands one metre outside the canopy? Clearly one's opinion on this depends on the scale of interest. If we are considering invasions at a landscape or geographic scale, and we are interested in long-distance dispersal (LDD, the topic of much current research), we may have little interest in seeds landing within a few metres of their parent. However, if we are considering the formation of pattern within populations and the resulting competition between neighbouring plants, then (in the case of annuals) distances of a few centimetres may be critical. Our view is that dispersal is important at a wide range of scales, all of which are important for a variety of ecological processes. No seed, or other plant propagule, can land at the same position as its parent: the parent, alive or dead, occupies that position at the time of seed release. All propagules move away to some extent and we can summarize this by their distances relative to the centre of the parent (i.e. the point dispersed to in the previous generation). Rather than attempt to pronounce on which propagules have and have not dispersed, we consider that *all* have dispersed to a greater or lesser extent and in principle the dispersal distance can be measured for every individual. The ecological context in which the data are viewed will then determine whether or not short or long distances are most important.

As plants grow, they develop seeds (perhaps in various enclosing structures) or vegetative structures that are capable of survival and propagation when moved to another location. Each of these individual units occupies a unique position on the parent; it develops according to the environment at that particular point in space and according to the resources that it receives from the plant. When the particular threshold combination of developmental stage and applied force (by the plant itself or by some external agent) co-occur, the unit moves (or is moved) away from its point of origin. The trajectory followed and the final resting place will depend on the attributes of the unit, the particular agents acting on it and the sequence in which they act, the way in which their impacts change over time, and the attributes of the surrounding habitat. The act of dispersal thus involves an interaction of a number of processes acting on each individual unit. In Part A (Chapters 2, 3, and 4) we consider these various components. Chapter 2 considers the development of the propagule and the contributions of the parent to dispersal, Chapter 3 the traits that will determine movement away from the parent, and Chapter 4 the forces that influence the unique path of a single propagule after it has separated from its parent. The potential complexity of this apparently simple process is shown in Fig. 1.1.

Each plant, if it grows successfully, produces a number of dispersal units: as few as one or two or as many as several million in a single season. As a result of variation in the trajectories of individual units, the precise pattern in which they come

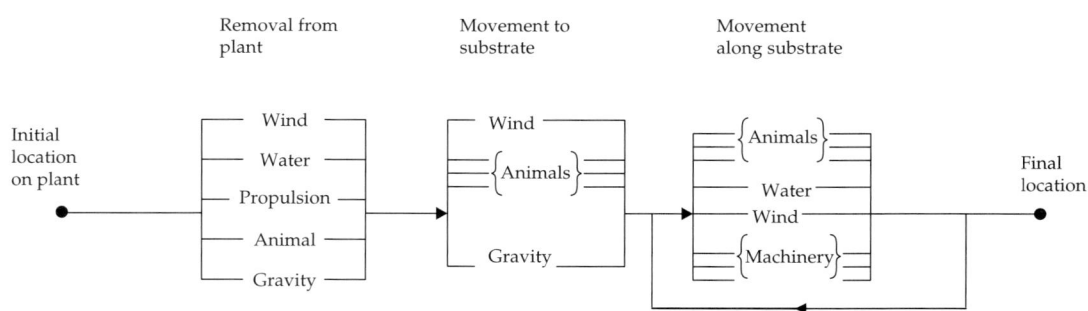

Figure 1.1 Vectors involved at different stages of the dispersal process. Different vectors may act in sequence (in series), or on different propagules from the same source (in parallel).

to rest will be unique to each parent plant and to each set of seasonal conditions. However, although each pattern may be unique at a fine scale, certain attributes might be shared, or differ, among plants at a coarser resolution. We therefore need to be able to describe dispersal patterns quantitatively, to summarize these higher-level properties. This will enable us to compare dispersal from parent plants, for example among species and environments. In addition, functions and parameter values can be generated for input into population and community models. Part B (Chapter 5) explores various ways of describing dispersal patterns, considers issues related to sampling designs, statistical analyses and equations, and summarizes what we know about patterns of dispersal around individual plants. This is the pivotal section of the book as it allows the processes acting at the level of an individual plant to be summarized to allow us to consider how individuals act together at the much bigger scales of Part C.

Having determined the immediate outcome of dispersal, that is, the pattern of dispersed units on the ground, we can then consider the consequences of dispersal for a population, a species, or a community over a time scale of generations. At some point, a new genotype or species arrives at a location within a landscape. It must first reproduce, but having done so successfully, dispersal occurs inevitably (according to our definition, since no unit can occupy the space already occupied by the base of the parent). Survival, reproduction, and dispersal then drive the dynamics of the population, together determining its success or failure, its rate of spread, or its decline towards extinction. Thus we arrive at some of the key questions for population ecologists. Why did trees spread so much more rapidly after the last ice age than we might expect from looking at dispersal from individual trees? Will species in the future be able to spread fast enough to track human-mediated climate change? What management options will either accelerate or, in the case of invasive plants, slow the rate of spread? What are the consequences of habitat fragmentation, such as through farming, for the conservation of rare species? Over what distances can colonization of unoccupied patches, or islands, occur? Do we need to enhance reproduction, dispersal, or both for a species to persist? Why are agricultural weeds, and many other plants,

so patchy: even in apparently homogeneous environments, when they can apparently be dispersed throughout a field? Chapter 6 sets out to examine the ways in which ecologists have sought to answer some of these questions. Because of the long time scale of population dynamics, most of our insight has come from theoretical modelling. This area cannot be covered without citing equations, but we will keep them to a minimum and hope that the non-mathematician can still follow the arguments.

Populations of one species do not, however, occur in isolation from other species. Each may be responding differently to the environment and may be patchy within the landscape. Some will be at the same trophic level, thus competing for resources, while others may be herbivores or pathogens. Although much of the early work on modelling species interactions assumed implicitly that spatial distribution was unimportant, recent work has predicted that a range of outcomes may occur when spatial dimensions and dispersal are incorporated. Intraspecific competition will be high close to the parent plant, as may predation. Therefore, there is an advantage to individuals that are able to disperse effectively (the so-called Janzen–Connell hypothesis). Chapter 7 follows the development of models of species occurring within a community and the predictions that they make for different scenarios.

In Chapter 8 we review the evolutionary forces acting on dispersal. First the processes acting to mitigate against dispersal, such as habitat heterogeneity and the risks of travel, then those forces acting to promote dispersal, such as kin competition, escape from pathogens, and the undoubted benefits of expansion. We then consider some of the many other processes that impact the evolution of dispersal, such as the effect of interspecific interactions and the problems of phylogenetic independence in comparative studies.

1.3 The terminology

Accurate use of terminology is always important in science. The introduction of terms with strict definitions helps us to communicate ideas and concepts unambiguously. On the negative side, a proliferation of jargon will decrease our ability to communicate with people from outside the immediate research area. Here we will define basic terms that will be used throughout the book and explain why we have avoided some of the usual terminology applied to the study of dispersal.

Propagule. Nearly all the dispersal events we consider in this book will be movements by seeds. However, plants may move through space in a number of different forms, so it would be incorrect to use seed as our generic term. They may move as individual seeds, as seeds contained within fruits, as whole plants, or as units (such as bulbs or stem/shoot units) that are genetically identical to the parent. Various terms have been used to refer to dispersal units in general, without regard to their biological form. In a great many parts of this book, it is irrelevant whether the unit is a seed, a group of seeds, or a vegetative ramet. So, which term should be used? A *diaspore* (according to van der Pijl, 1982, from *diaspeiro*, meaning 'I broadcast') may be defined as any propagative part of a plant, especially one that is easily dispersed. *Disseminule* refers to a plant part, although its etymology implies a seed, that is capable of growth in its new location; *propagule* has a similar definition but without the emphasis on seeds. Potentially all of these words could be used for our purposes. Rather than vary throughout the text, we have chosen to standardize on *propagule*.

Dispersal. This is the process of a propagule moving through space, whether this is active or passive (Begon et al. 1996a), although van der Pijl (1982) notes his preference for terms that distinguish between active and passive movement. There is a very similar term, *dispersion*, which tends to be used by ecologists as the pattern that results. Inevitably, confusion of the two sometimes results, in which they are treated as synonymous. For this reason, Pielou (1977) regarded *diffusion* as preferable to dispersal, although this perhaps implies a directional process (diffusion in chemistry involves movement from high to low concentrations). Krebs (1985) considers different types of dispersal, with *diffusion* as a gradual process through a landscape, whereas *jump dispersal* is over a shorter time period and across large distances, such as in new invasions. We do not see any need to use distinguishing terminology for these two ends of a continuum: essentially they are the same (movement of propagules) and

merely differ in spatial or temporal scale. Van der Pijl (1982) notes that *dissemination* implies movement of seeds and is therefore insufficiently general for studies of population biology; he also mentions the rarer terms *germule, migrule,* and *chore.* Because of its dominance of usage, we will use the term *dispersal* throughout.

Chorology. Van der Pijl (1982) preferred the term *chorology* for the entire field of study of dispersal (from *chorein*, meaning 'to wander'). He quotes Dammer (1892) as using the suffix '*chory*' after a prefix that denotes the vector of dispersal, and adopts this throughout his book. He argues that it is simpler to refer to plants as myrmecochorous than to say that its seeds are regularly dispersed by ants. However, we argue the opposite. We believe that it is less confusing to use plain English than for the non-specialist to have to look up a term in a book just to see what it means! 'Dispersal by ants' is not that cumbersome. Furthermore, some species may be telechores, endozoochores, ornithochores, mammaliochores, chiropterochores, i.e. polychores, all at the same time, as seeds from the same plant may be dispersed through being eaten by birds, mammals, or bats. Moreover, the suffix is both used for the vector and in some cases for the scale over which dispersal takes place (e.g. telechory). We refrain from using the 'chory' terminology in the rest of this book.

Anatomy and morphology. The particular structural adaptations that have evolved to aid dispersal are the subject matter for anatomists and are (arguably) not directly relevant to population biology. It is sufficient in most cases to consider that the seeds are surrounded by a tasty flesh, or attached to a parachute-like structure, rather than being concerned with the part of the plant from which the flesh or parachute evolved. The terminal velocity of a propagule is more important to its dispersal by wind than the way that it has achieved a reduction in its rate of fall. We are aware that the detail of the adaptation may constrain the distance that the propagule may be moved, but we need to constrain the scope of the book to some extent and prefer to leave this area to others.

Effective dispersal. Many ecologists draw attention to the fact that what matters to a population is those propagules that disperse *and* establish themselves at a new location. The distinction is made between dispersal and *effective* or *realized* dispersal (i.e. the locations of the offspring). We have no argument with this philosophy although there are potential pitfalls with using, for example, realized dispersal to describe seeds that have successfully germinated, as it could be argued that they have not actually dispersed effectively until they, in turn, have produced progeny. Also, when we look at dispersal mechanistically, it is apparent that the term dispersal refers only to the process of movement, whereas realized dispersal is the complex result of movement *and* habitat suitability, competition, predation, pathology, and so on. In fact, it brings together many aspects of population biology that we seek to understand individually, as well as their interactions. Although it may be easier to study the seedlings that emerge, rather than hunting for tiny seeds themselves, we do not know what remains in the seed bank to come up in later years and which may vary spatially. Modellers may find it simpler to deal with the overall outcome of all processes than to have to parameterize all possible mechanisms, and this may be appropriate for understanding population dynamics at a superficial level (hence the proposed restriction of the term, dispersal kernel, to realized dispersal—Higgins et al. 2003a). Ecologists who seek to quantify dispersal through the use of genetic markers similarly only study relatedness of the successful dispersers.

Dormancy. It has often been noted often that dormancy is a mechanism for dispersing through time (e.g. Mayhew 2006). Although this is an interesting concept, we consider dormancy is beyond the scope of this book.

Vectors of dispersal. Dispersal is often mediated by forces external to the plant, provided by such things as wind, water, and animals (see Chapters 3 and 4). In the older literature, these are referred to as dispersal *agents*. In pathology, the term *vector* has been used to refer to the host for a disease and hence the cause of its spread. This terminology has now entered the plant ecology literature and it is common to refer to dispersal vectors. In mathematics, a vector is defined as having both magnitude and direction, and it is these aspects of the dispersal agent that we seek to understand. We will therefore adopt the term vector in this book.

Seed rain, curves, and kernels. The term 'seed rain', or perhaps 'propagule rain', is used as an analogy. Like raindrops, seeds fall on the ground. We appreciate that the number of landing seeds will vary spatially and we seek to quantify this. For plant propagules, we also expect the amount to change with distance from the source. Some authors refer to 'dispersal curves', with some measure of the amount of dispersal on the y-axis and distance from the source on the x-axis. But, as we discuss in detail in Chapter 5, the y-axis is sometimes the number per unit area (density) and sometimes the frequency falling within a particular range of distances. These are two quite different things, although mathematically inter-convertible (see p. 83). Adding to this confusion, the y-axis on a graph will often be labelled simply as 'number'. We consider that the term 'dispersal curve' is best avoided. Nathan and Muller-Landau (2000) refer to the *distance distribution* (for a y-axis of the frequency or probability of landing at a given distance) and the *dispersal kernel* (where the y-axis is the number landing per unit area, such as per trap, i.e. density). It is true that consistent use of this nomenclature would make the literature less ambiguous. We discuss these problems in greater detail in Chapter 5. Until then, we avoid any ambiguity by making it clear what variables we are referring to.

Modellers, predicting the dynamics of populations over generations, have often used the terms *redistribution kernel* or *dispersal kernel*. Their models invariably consist of a population growth element (the outcome of births and deaths) and a dispersal kernel (determining immigration and emigration). As with the term *curve*, *kernel* is not used consistently with respect to density and probability. Some models are written with respect to density (mostly cellular models), while others are based on the probability of dispersing a given distance (mostly continuous space models). However, functional forms for the models are often taken without regard for the nature of the data.

Dispersal of individual propagules

Dispersal is just one step in a complex series of events that make up a plant's life cycle. A plant grows, develops, and then liberates its propagules into the environment which, with luck, start a new generation. Consider a seed that has recently germinated. At first, all tissues produced by the young plant are vegetative: meristems become stems, leaves, and roots, producing both the structural framework or 'skeleton' of the plant and the means to obtain and transport resources. Vegetative propagules may result from the severing or breakdown of sections of this framework. However, at some point, often associated with day length, low temperatures, or perhaps merely time, meristems begin to produce reproductive structures. The components of the flowers: petals, sepals, anthers, stigma, and ovary, are differentiated. Ovules form, attached by short stalks (funicles) to the ovary wall. Pollen reaches the flower, brought by wind or animal. Some of the pollen grains that land on the stigma send tubes down inside the style and fertilize the ovules. The ovules develop into seeds, each one containing an embryo and the food supply necessary for them to produce a new plant. The ovary wall and other structures develop further, perhaps forming more food reserves or protective structures around the seeds. As the seed matures, fruits, seed coats, or other structures may develop appendages to aid dispersal, either directly or by attracting animals. Cell division in a band within the funicle and the peduncle (at the base of the fruit), produces an abscission zone. Cells in this zone begin to break down, often leaving the vascular bundles as the only means of attachment, until finally being severed by the action of an external force and/or due to forces building up inside the developing fruit. It is at this point that the propagule leaves the parent plant and dispersal begins.

Clearly, the sequence of events leading up to dispersal will have a profound influence over the distance that a propagule moves and its ability to establish a new plant when dispersal ends. The architecture of the maternal plant determines where the dispersal trajectories of propagules begin, as well as the forces to which a propagule will be exposed. The physiology of the maternal plant will set the timing of maturity in relation to the availability of dispersal agents. In some species it may also determine the velocity with which the propagule leaves the plant. The maternal structures around the seed influence the external agencies that will be involved in dispersal (through their desirability to animals or their ability to glide through the air), the force required to liberate the propagules, and the food reserves available to establish the new plant.

After the propagule has left its parent, the external environment has the greatest influence on dispersal. Movement away from a parent is affected by a range of physical and biological mechanisms, several of them often acting in sequence. For example, propagules might be blown by wind, washed by water, ingested then defaecated by animals, or stuck with mud on to animals. While in transit, they might hit an obstacle, or be brushed off by an animal. Even after reaching the ground they may resume their motion, perhaps by different mechanisms, which might take them further away from or back towards their parent. The pathway taken by each propagule will thus be unique and subject to considerable uncertainties, depending on chance encounters with animals, weather patterns, and so on.

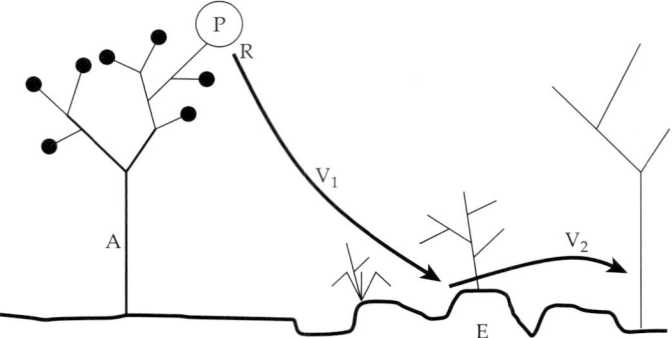

Figure A.1 Components affecting the dispersal trajectory of a propagule: the architecture of the plant (A), the traits of the propagule P (such as mass, shape, appendages), the mechanisms of release (R), the nature of the dispersal vectors (V_1, V_2, etc.), and the environment (E) into which the propagule disperses.

This uncertainty means that it is impossible to predict the dispersal of an individual propagule precisely. However, we can make predictions about the scale of dispersal distances that are likely to result for different types of propagule, different types of dispersal agent, and different types of receptor environment. By considering each of the component processes, we can predict how dispersal will change from one situation to another. We can also explain the sources of variation that we see from a whole plant (see Chapter 5). The uncertainty associated with dispersal means that, rather than making deterministic calculations for a single propagule trajectory, we must consider probability distributions. These statistical distributions represent the range of dispersal outcomes and the probabilities that given outcomes will occur. Therefore we are as interested in the variance and skewness of predicted dispersal distances as we are in the mean value (i.e. both shape and scale of the probability distributions are important).

The *components of dispersal* that together determine the trajectory, and thus the final resting-place, of an individual propagule (Fig. A.1) can be summarized as:

1. Contribution of the maternal parent plant to dispersal (Chapter 2).
   - initial location of propagule (P) in space, determined by plant architecture (A);
   - timing of maturity;
   - force required for abscission;
   - launching mechanism (if any) (R).
2. Propagule traits that determine the influence of external vectors (Chapter 3).
3. Movement of the propagule during dispersal (Chapter 4).
   - effectiveness and direction of any vectors (V) in moving propagules;
   - environment (E) into which the propagule disperses;
   - number and sequence of vectors acting.

CHAPTER 2

Contribution of the parent plant to dispersal

2.1 Introduction

The act of dispersal begins with the propagule leaving its parent plant. However, factors contributing to the dispersal trajectory occur very much earlier than this moment. Indeed, we could argue that dispersal is being influenced right from the germination of the parent plant. Growth and development throughout life determine the architecture of the parent and hence the locations of the propagule release points in space, as well as the number of propagules. The phenology of the parent plant determines the time at which the reproductive structures mature, and hence the weather conditions and the availability of dispersal agents at the time of release. The force required for the propagule to separate from the parent depends on the abscission tissues laid down by the parent, while in some species parents provide a mechanism by which propagules are given an initial force to launch them. The morphological structures of the propagule itself that aid dispersal are also provided by the maternal plant, but we consider these in the next chapter.

A thorough consideration of dispersal must, therefore, start with the parent plant. In this chapter we discuss the various components of dispersal contributed by the parent.

2.2 Initial locations of propagules

No two plants are identical: each structure is uniquely determined by genetics, providing the general pattern of growth, by adjustments of that pattern determined by its environment, and by chance events. From the point of view of dispersal, each plant is a unique population of propagule release sites, which can defined by a frequency distribution of horizontal (x, y) and height (h) locations. Here we consider the causes of variation in these distributions.

The size of plants range from a few millimetres in height to around one hundred metres, and there is also an enormous variety of plant structural 'formulae' that can be classified according to their branching patterns (see e.g. Hallé et al. 1978, for tropical trees). Some growth forms are simple, such as the coconut palm (*Cocos nucifera*: Arecaceae), where all fruits are at roughly the same height and held close to the trunk (Fig. 2.1). For such species we can adequately describe the population of release sites by the mean height of the propagules. Slightly more complex is the cylindrical spike of *Xanthorrhoea* spp. (Xanthorrhoeaceae), where seeds are spread evenly over a clearly defined height range. Many terrestrial orchids produce one or a few capsules at discrete heights, each constituting almost a *point source* for the release of many thousands of seeds. For prostrate species, such as *Portulaca oleracea* (Portulacaceae) or *Juniperus communis* ssp. *nana* (Cupressaceae), there is little variation in propagule height, but propagules may be located well away from the base of the plant because of lateral growth.

For large, spreading trees the range of heights at which propagules are located within the tree may be considerable, as may the range of distances away from the central trunk. For propagules with very short potential dispersal distances, this lateral growth of the parent plant may move them much further than they are capable of moving from their point of release. A large oak tree growing in the

Figure 2.1 Examples of differing plant architectures, with propagule sources being: (a) at a single height (coconut palm, *Cocos nucifera*, reproduced courtesy of Michael Alvard), (b) along a narrow cylinder (grass tree, *Xanthorrhoea australis*, reproduced courtesy of Jill Kellow), (c) widely located laterally and vertically (horse chestnut, *Aesculus hippocastanum*), (d) confined to a horizontal plane (pigface, *Portulaca oleracea*).

those near the ground may be within the reach of ground-dwelling animals. Propagules released at the top of the tree will have a greater distance to fall than those released close to the ground, so wind will be able to move them further before they hit the ground.

Species differ in their strategies of propagule placement. Horse chestnut (*Aesculus hippocastanum*: Hippocastanaceae) flowers are held at the ends of branches, thus tending to confine seeds close to the canopy surface. Although flowers of most *Eucalyptus* spp. (Myrtaceae) are produced close to the canopy surface, in some species their seeds may take two to three years to mature, by which time the canopy may have extended some distance outwards, leaving the fruits beneath the canopy. The small shrub *Cistus ladanifer* (Cistaceae) produces fruits in a zone from the edge of its canopy inwards to about half its radius (Acosta et al. 1997), while cluster fig (*Ficus racemosa*: Moraceae) trees produce their fruits on the branches and trunk, spread throughout the internal volume of the canopy. In indeterminate herbs such as *Sinapis arvensis* (Brassicaceae), seed pods are produced on stems that continue to grow, so that by the time the plant dies (perhaps due to drought or frost) seeds have been dispersed from positions throughout much of the aerial volume occupied dynamically by the growing plant. For a grass, even though seeds may be held together in a tight inflorescence on each stem, there may be stems with a range of sizes and heights and, as a further complication, both the height above ground and the distance of the seeds away from the base of the parent vary enormously as stems sway in the breeze.

Although it is relatively easy to discuss canopies subjectively, quantitative data on the locations of propagules on a plant are uncommon. Few researchers have sought to answer questions that require such information. Most dispersal models, for example, have assumed that plants are point sources, at least with respect to the dispersal vectors being considered, and have therefore not required such data. Perhaps researchers have baulked at the task of trying to characterize an entire canopy. There are a few examples of studies where quantitative data on propagule locations on plants have been assembled.

open may have seeds located from, say, 1 m above the ground up to 20 m, and extending 10 m or more horizontally from the central trunk. Such a tree is a *distributed* population of point sources. Propagules dispersing from different points on the tree may be subject to different 'strengths' of dispersal vector, or even types of vector. Seeds near the top will be subject to the greatest wind speeds, while

In many coniferous trees, female flowers may be produced only on the uppermost branches (Wareing 1958; Greene and Johnson 1996), thus restricting sites of seed release vertically within the tree. Hard (1964) found that the mode of the height distribution of *Pinus resinosa* (Pinaceae) cones was just above the middle of the tree and that most were in the top two thirds. Nathan et al. (2001) found that the vertical distribution of *Pinus halepensis* cones was well described by a normal distribution, while Greene and Johnson (1996) assumed in their model of wind dispersal that seeds are distributed normally in the upper half of the canopy and with a mode 25% from the top.

The small vine, *Amphicarpaea bracteata* (Fabaceae) produces both subterranean and aerial cleistogamous pods, as well as aerial chasmogamous pods. Aerial pods release their seeds explosively. Subterranean seeds are, on average, 16 times heavier than those produced in the canopy and are borne in the top few centimetres of the soil on runners up to 3 m in length. In a study by Trapp (1988), both types of aerial pod were produced up to about 1.5 m above the ground (Fig. 2.2).

While the data on *Amphicarpaea bracteata* were obtained using very simple methods, advances in technology mean that precise data on the spatial coordinates of plant structures and fruit positions can be easily, though often laboriously, captured for plants up to the size of small trees. Fig. 2.3 shows the structure of a plant of the annual weed *Raphanus raphanistrum* (Brassicaceae) produced by a three-dimensional digitizer, along with the frequency distribution of heights of seed pods extracted from it. Despite the pods being arranged almost linearly along stems, the frequency distance of pod heights on the plant as a whole follows a fairly smooth, somewhat asymmetric distribution. Fruit positions on a walnut tree have been captured using the same type of device (Sinoquet and Rivet 1997), while a theodolite was used to produce three-dimensional images of the spatial positions of kiwifruits (*Actinidia deliciosa*: Actinidiaceae) grown on trellises (Smith et al. 1992). Propagule height frequency data are, however, so uncommon that it is difficult to make any generalizations about their patterns.

Simulation modelling is one way to improve our understanding of how plant growth and structural development lead to particular patterns of pre-dispersal propagule locations. 'Virtual' plants can be generated using simple empirical construction rules, that encapsulate the architecture of a given species (Oppenheimer 1986), together with computer algorithms known as L-systems (e.g. Prusinkiewicz and Hanan 1990). In this way, single plants, dense stands of trees, and clonal herbs have been simulated (Mech and Prusinkiewicz 1996; Renton et al. 2005). Although most attention so far has been given to modelling the positions of branches (to understand

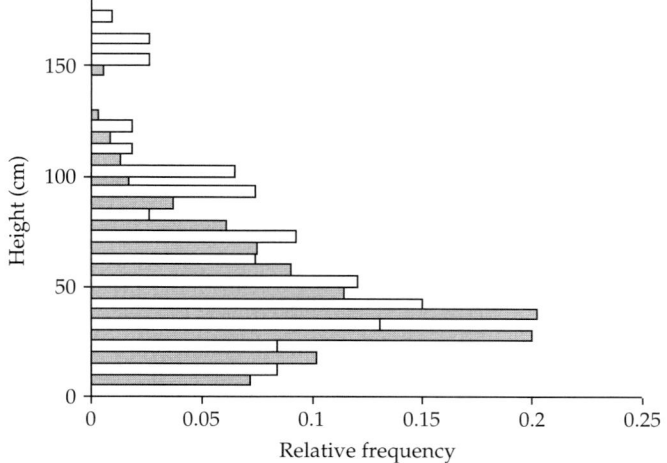

Figure 2.2 Frequency distributions of the heights of aerial pods in *Amphicarpaea bracteata* (redrawn from Trapp 1988, reproduced courtesy of the Botanical Society of America): white bars are chasmogamous pods, grey bars are cleistogamous pods.

(a)

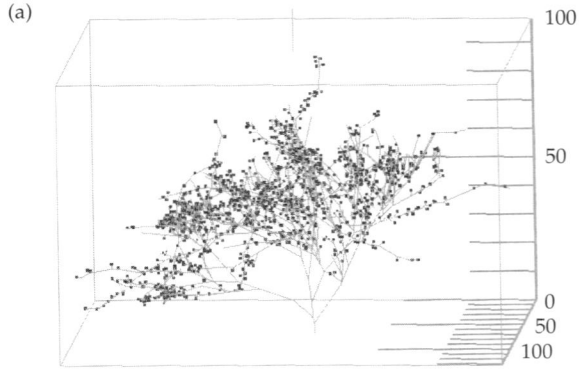

(b)

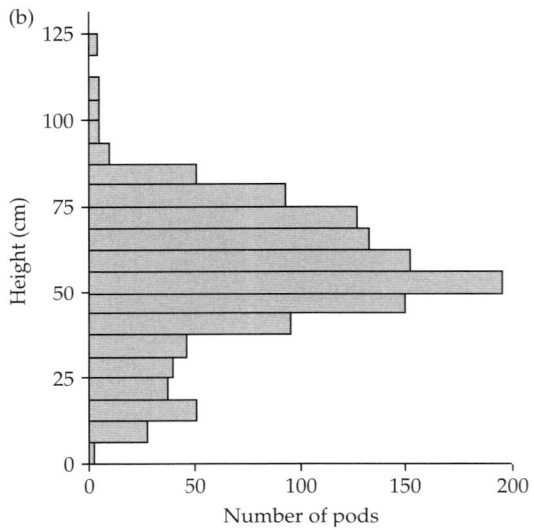

Figure 2.3 Digitized plant, with scale in centimetres (a) and frequency distribution of pod heights (b) of *Raphanus raphanistrum* (Taghizadeh 2007). Plant was growing in a low density wheat crop. Dots show the locations of bases of pods (silicules). In this species, whole pods usually fall from the plant, later breaking up into single-seeded segments.

plant 'form') and leaves (to predict canopy photosynthesis), there is clearly the potential to simulate flower and fruit positions (Prusinkiewicz and Lindenmayer 1996), and hence to predict the statistics of propagule positions for different types of plant architecture. If the models are made more mechanistic, by linking the empirical branching rules to physiological models of vegetative growth and developmental switches, future research will be able to predict variation in plant structures caused by environmental conditions (Cici et al. 2006).

While the overall growth form of a plant is set genetically, any factor that affects plant growth or removes plant material has the potential to affect both the number and positions of propagules being released. There are often simple allometric relationships between seed number and plant mass (Watkinson and White 1985; Niklas 1994); thus conversion from biomass to propagule number can be straightforward. The effects of growing conditions on seed positions, however, have seldom been reported and may not be as easy to estimate. A few conclusions about heights of release can be reached on purely logical grounds from our knowledge of plant growth. Generally, older plants will produce propagules at greater heights and spread over a larger projected horizontal area than when they were younger (at least until they start to senesce). Propagules on emergent trees in a dense forest will be exposed to windier conditions (see Chapter 4) and

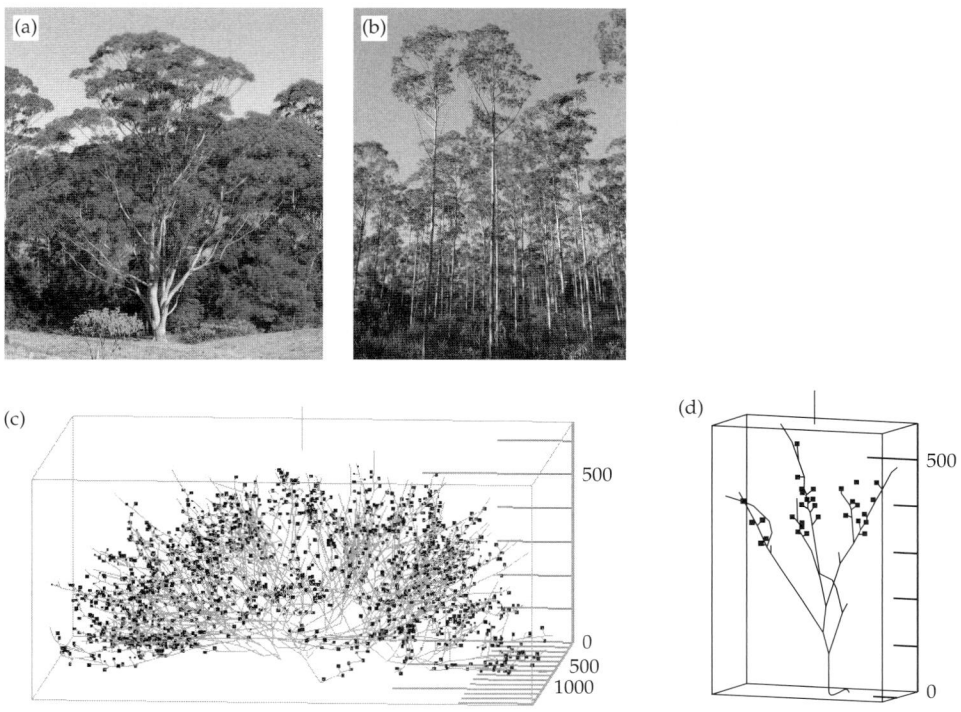

Figure 2.4 Illustration of the difference in morphology between a plant grown in the open (a, c) and a plant that has grown within a dense stand (b, d). If propagules are located throughout the canopy, it is easy to see that they will be starting out from very different locations in the two cases. a, b: *Eucalyptus grandis* (reproduced courtesy of David Kleinig); c, d: *Raphanus raphanistrum*, digitized plants with scales in millimetres, showing locations of seed pods; (c) was growing without competition, (d) was growing in a dense wheat crop (Taghizadeh 2007).

will be more visible to aerial frugivores than when they were young. Trees subjected to browsing may, as a result of shoot damage, have a canopy (and therefore seeds) that begins well above the ground (such as in savannahs: Archibald and Bond 2003), while grazed herbs may result in a growth form producing seeds very close to the ground. Long-day annuals will flower very quickly if plants emerge in summer, whereas plants emerging in autumn or winter may grow vegetatively into large plants before they flower. As a result, summer-emerging plants will have fewer seeds (e.g. Cheam 1986) and these will be held, on average, closer to the ground than winter-emerging plants. Shorter plant stems resulting from a lack of resources will reduce both the mean and maximum heights of propagule release.

Plant density has a strong influence on plant form as well as on size. Under intense competition, herbs

and trees produce fewer and shorter side branches (e.g. Geber 1989; Donohue 1998, 1999) (Fig. 2.4). This results in plants that produce a greater proportion of their seeds at the top of the canopy and concentrated closer to the main stem (i.e. a smaller range of x, y, h dimensions). In contrast, plants at low density will have more branches, and more seeds, many of which are produced low down (i.e. a large range of x, y, h). Moreover, very dense stands contain a skewed frequency distribution of plant size, with a small proportion of large plants and many smaller plants. At low density, stands tend to be more uniform, with a normal distribution of plant sizes and a greater mean plant mass (Koyama and Kira 1956). Hence, in dense stands there will be many plants producing a few propagules each and a smaller number of plants, with more branches, producing larger quantities of propagules (Weiner et al. 1990). It is likely that propagules on the smaller individuals will have a narrower spread of locations in the x, y plane. Dominant plants are often taller under competition, perhaps etiolating in order to reach light (Weiner et al. 1990; Donohue 1998), while herbs growing in the absence of competition may sometimes be reduced in height as a result of bending under the weight of fruits that they produce. However, some studies report shorter plants at high densities (e.g. Wender et al. 2005).

Dense stands of plants, such as forests, inevitably have edges, and plants growing at an edge may have a different morphology from plants within the stand. Edge plants experience high competition on one side and low competition on the other. As a result, they produce most of their branches, leaves, and fruits on the open side, resulting in asymmetry (termed the 'riverbank effect' by Hallé et al. 1978). They will also be the most fecund individuals in the stand, making a disproportionate contribution to the seed rain. On the side open to the wind, propagules may be produced all the way to the ground, while on the inside they will only be produced at the very top of the canopy.

So far we have been assuming that plants arise from the ground as a single stem, or a tight cluster of stems (as in grasses) that can be considered together as a single unit. Clonal perennials, however, produce their branches on, or below, the soil or water surface. The shoot system spreads its canopy horizontally, producing modules referred to by ecologists as *ramets*, from which inflorescences may (or may not) be produced. The entire system of shoots, arising from one original individual, is called a *genet*. Unlike trees, this branching process can carry on almost indefinitely, potentially resulting in dispersal of ramets over hundreds of metres, merely through vegetative growth. In a well-developed clone, the horizontal distance over which inflorescences are located may be orders of magnitude greater than the maximum height of propagule release. As the shoot system grows, connections between ramets may decay or fracture, and they will then live independently, with some perhaps being carried to distant locations. Thus, there can be dispersal of both vegetative units and sexual propagules. Like trees and shrubs, clonal species may have characteristic branching angles (examples are *Alpinia speciosa* [Zingiberaceae] and the seagrass *Cymodocea nodosa*

[Cymodoceaceae]) which again enable the growth of clones to be modelled (Bell 1979; Sintes et al. 2005).

Clonal species tend to control their own module density via feedback through the shoots (Hutchings and Bradbury 1986). Also like other plants, growth rates and branching structures of clonal perennials respond to their environment. Stolons and rhizomes can 'forage' for resources, responding to localized concentrations by changing the number of branches produced and internode length; they can also redistribute resources from one module to another (Slade and Hutchings 1987; Hutchings and Wijesinghe 1997). Clones will also respond to competition from other plants. For example, the density and maximum distance dispersed by ramets of *Elymus repens* (Poaceae) is affected by both resource availability and competition with other grass species (Marshall 1990).

There can be variation in the height of release of sexual propagules in clonal species, between, as well as within, ramets. For example, flowering shoots of *Solidago canadensis* (Asteraceae) may vary in height from 0.4 m to over 1 m (Bradbury 1981). *Hieracium aurantiacum* (Asteraceae) stems towards the edge of a patch may have taller stems and more flowers per stem than those in the patch centre (Stergios 1976). This will clearly have important consequences for the pattern of dispersal of propagules from the genet as a whole.

So far we have considered only the distribution of propagules within the plant as a whole. The architecture of individual branches, however, can also influence the effectiveness of different dispersal vectors. For foraging birds, positions of fruits relative to branches will affect their ability to eat fruits while perched in a tree. Some birds have great difficulty in reaching fruits that hang down below the branches (Moermond et al. 1986). Hanging fruits can still be taken, but this must be achieved through flight, involving a greater expenditure of energy. Animal-dispersed propagules also have to be displayed so that they can be seen from a distance: many species thus have bright fruits located at the ends of branches. Spines on stems can also determine which animals are able to access fruits.

The behaviour of the peduncle after fertilization has important consequences for dispersal in some herbaceous plants. The leaves of *Taraxacum* spp.

(Asteraceae) form a rosette and produce flowers that first open close to the ground. However, the hollow peduncle extends upwards and by the time the achenes in the inflorescence are ready to disperse, the capitulum is held above the surrounding vegetation in a position where wind can blow them away more effectively. Other species exhibit peduncle behaviour that actually prevents dispersal. In *Romulea rosea* (Iridaceae) and *Trifolium subterraneum* (Asteraceae), the peduncle bends over, pushing the fruits into the ground or under other vegetation. This ensures that they are placed in a microsite suitable for germination, but they are not exposed to most dispersal vectors.

2.3 Timing of maturity

The timing of propagule maturity is important because there is seasonality in the availability and strength of many dispersal vectors. If propagule maturity and the vector coincide, then clearly dispersal will be greater than if they do not. In most regions, including the tropics, there are distinctive seasons, dictating the availability of rain and strong winds. The timing of maturity in plants adapted to wind dispersal, for example, may coincide with the windiest season (although there are also many species for which there is no such correspondence: van Schaik et al. 1993) or when deciduous forests are mostly leafless and consequently wind speeds and eddies within the canopy are greater (Nathan and Katul 2005; see Chapter 4). Peaks in fruiting may coincide with visits by migratory birds (Herrera 1984). This may not necessarily mean that seeds are dispersed over very long distances, since passage through the gut is quite rapid (see p. 61) and the migrants may stay within the area for extended periods before leaving. However, dispersal via the gut of a bird will be greater than if the fruits simply fall on the ground beneath the plant.

It is thus plausible to deduce that there may be a causative relationship in some instances, that is, certain patterns of plant phenology have evolved specifically because they enhance plant dispersal. Though it is easy to speculate at length on adaptive significance, it is impossible to prove what the selective pressure leading to a particular trait actually was. A great many life history traits influence the plant phenotype and the one that we may have identified as being closely related to fruiting time may not be the primary selective influence. Correlation does not prove causation. For example, the windiest season may also be the wettest: selection for a particular phenology may have been because of the suitability of conditions for seed germination and seedling establishment, rather than for dispersal distance.

In the case of dispersal of plant propagules by migratory animals, fruits produced in the season when the animals arrive will achieve enhanced dispersal, while animals visiting at that time find ample supplies of food. Both organisms benefit, and therefore perhaps their phenologies have co-evolved (though the original driver may still have been unconnected with plant dispersal, such as seasonal changes in temperature or water availability). There may, however, be negative consequences of species fruiting synchronously. Although an abundance of fruits may ensure that there are many frugivores in the general vicinity (e.g. Carlo 2005), an over-supply of food may mean that the animals feed wastefully, scattering many fruits directly under the parent plant and leaving some fruits untouched, thus reducing mean dispersal distance. It will also be more likely that seeds from several species will be mixed within animal faeces, resulting in interspecific competition between emerging plants in the next generation (Stiles and White 1986). If generalist feeders (as most frugivores are) concentrate on the most productive or most attractive species, other less productive or conspicuous species may be ignored completely. It can be argued that being the only plant fruiting at another time of year will attract a large proportion of the animal dispersal vectors, which may be experiencing periods of food shortage. On the negative side, such species will also draw the attention of seed predators: so synchronous fruiting may be a means to glut predators (thus increasing seed survival) rather than increasing dispersal distance. However, perhaps it is best to avoid such speculation altogether and to concentrate on the factual data.

A great deal of information on fruiting phenology has come from tropical rainforests. Data are from studies focusing directly on dispersal, as well as from studies of fruits as sources of food for animals (van Schaik et al. 1993). Individual species may produce fruits for a range of durations, from just a few weeks in a particular season, to continuously for several months (spanning more than one season), or sometimes in repeated bursts throughout the year (Frankie et al. 1974; Hilty 1980). The number of species fruiting at any one time in a tropical forest varies throughout the year, often with one distinct peak (Peres 1994) or sometimes two less distinct peaks (Hilty 1980; Dinerstein 1986). The species in the peak may vary between years, as in some years a species may not reproduce at all (Medway 1972; Wheelwright 1986). The main peak in the number of species fruiting is often towards the end of the dry season (Fig. 2.5a), but it may also be at other times of the year (van Schaik 1986), depending on the region and climate. The resident animals which eat and thereby disperse these fruits may time their breeding to coincide with the annual peak in fruit abundance (e.g. Medway 1972; Estrada and Coates-Estrada 2001). For part of the year, there may be only a few species fruiting, but they may do so prolifically. This has led to the concept of 'keystone plant resources', without which the community of larger animals within the system could not survive (Terborgh 1986; van Schaik et al. 1993). There has been considerable debate about what environmental triggers cause reproduction in tropical regions, where there is only minor variation in temperature and day length (Borchert 1983; Borchert et al. 2005).

Peaks in the number of species fruiting, as well as in the quantity of fruits available to animals, are characteristic of arctic and sub-arctic communities, where the cold winters severely limit the period in which growth can occur and in which migratory animals visit. In a Mediterranean shrub community, most species produce ripe fruits in autumn and late summer, regardless of when they flower (Fig. 2.5b; fruiting of tropical species may also be more seasonal than flowering—Hilty 1980). In wet sclerophyll forests in southeastern Australia, where most frugivores are residents, there is a peak in the number of species fruiting in late summer and early winter (French 1992). In other regions hot, dry seasons may restrict plant and fruit growth. Fruit production in species growing in seasonally flooded habitats may coincide with peak flood conditions,

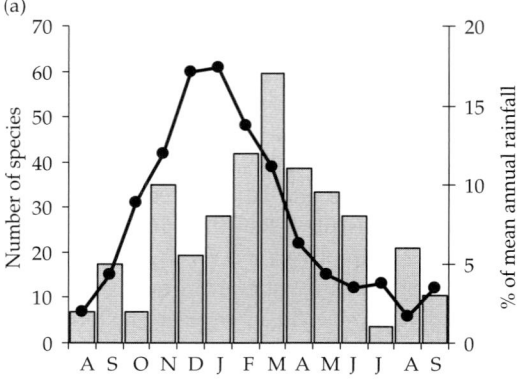

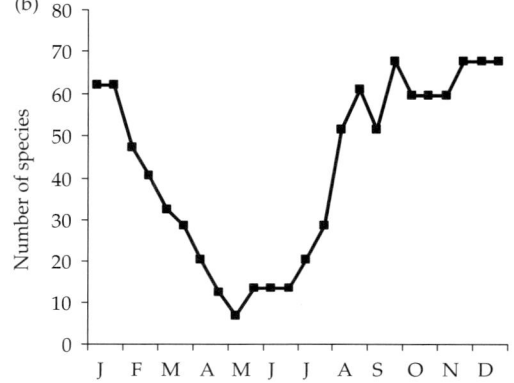

Figure 2.5 Seasonal variation in number of fruiting species: (a) overstorey species in an Amazonian forest (redrawn from Peres 1994, reproduced courtesy of Blackwell Publishing): histogram columns show the percent of mean annual rainfall (3256 mm); (b) Mediterranean scrubland (redrawn from Herrera 1984, reproduced courtesy of Ecological Society of America). Note the different scales for a and b.

such as spring melt-water in cold regions and monsoon floods in the tropics, maximizing dispersal by water and fish (Williamson et al. 1999; Merritt and Wohl 2002). For example, Kubitzki and Ziburski (1994) found that the plant community in a white-water area of Brazil fruits mainly in the early parts of the flooding season, whereas the fruiting periods of species in a blackwater area (where water is high in tannins) are spread throughout the flooding season.

In many temperate and Mediterranean species, reproductive development is initiated by day length or low temperatures (vernalization), though once reproduction has been triggered the rate of growth of floral structures is then positively correlated with temperature. The physiology of these mechanisms has been studied in great detail in crops, and it is reasonable to assume that flowering and propagule production of wild species is controlled in a similar manner. Several annual grasses differ in time to flowering (and hence time to maturity) in relation to latitude and length of growing season (Whalley and Burfitt 1972; Gill et al. 1996). Considerable geographic variation is found in responses of flowering to vernalization and day length within species of *Avena* spp. (Poaceae) (Loskutov 2001). As was discussed on p. 13, temperate annuals in which flowering is triggered by long days may take several months to flower if they emerge in autumn or winter, but they flower within a few weeks if they emerge in summer. The times of reproduction of different cohorts within a population will thus be much closer together than their seedling emergence dates. For example, a 121 day difference in emergence of *Raphanus raphanistrum* resulted in only a 41 day difference in flowering dates (Cheam 1986).

Many temperate weeds have a very similar phenology to the crops in which they grow. Harvesting operations, and especially modern machinery, provide the opportunity for agricultural weeds to disperse within the same farm, as well as to other cropping regions and to other countries (Cousens and Mortimer 1995), so a similar maturity date will enhance this means of dispersal. Such weeds may also have a similar morphology to the crop and are sometimes referred to as 'crop mimics'. Baker (1974) argues that for species whose seeds are likely to be cleaned easily from harvested grain, due to their small size or their shape, there would be selection for a maturity date either before or after crop harvest. Thus their dispersal will be severely limited. However, it can also be argued that there is strong selection to produce seeds of similar size or shape to those of the crop: this, along with a similar phenology, would ensure widespread dispersal. Timing of floral development, along with non-shattering inflorescences at maturity, will also be important for species growing in fields to be cut for hay. In addition to the obvious implications for dispersal of agricultural weeds, hay bales are often used for land reclamation works in National Parks and may initiate weed infestations into endemic communities. The recognition of the potential for hay bales to cause weed dispersal has led some authorities to insist on strict standards of hygiene in the source fields and inspections prior to purchase (I. Pulsford, pers. comm.).

Individuals in many species produce all their mature propagules at roughly the same time, although there is variation within and between populations (Howe 1982). Plants on slopes of different aspect may take different lengths of time to mature, as a result of the different temperatures experienced. In Alaska, a distance of 100 to 200 kilometres or an increase in altitude from 100 to 950 metres can mean a one month shift in flowering and seed production in trees (Zasada 1986). There may also be variation in time of maturity of propagules within a plant. For example, fruits within each raceme on *Rubus* spp. (Rosaceae) mature over a period of several weeks. Annuals with indeterminate growth forms may have seed/fruit production spread over several weeks or even months. In determinate annual grasses (Poaceae), such as *Avena fatua*, *Lolium rigidum* and *Phalaris canariensis*, seeds from different parts of the inflorescence mature and fall from their spikelets at different times, with the dehiscence period broadly coinciding with maturity of the cereal crops in which they grow. An early harvest may mean that a significant proportion of seeds are taken up into the harvesting machinery, whereas a late harvest may result in the majority falling to the ground. The persistence of some seeds on the plant may thus be the difference between *some* long distance dispersal and none at all (Wilson 1970; Barroso et al. 2006).

2.4 Separation from the parent

During growth and development, propagules must be connected securely to the parent by tissues which supply them with resources. Eventually, however, propagules and parent part company. An abscission layer usually forms between the maternal tissues and seeds and fruits, providing a line of weakness at which the propagule can separate. The threshold force required for release from the parent thus decreases as the propagules become more mature. Propagules must not be released prematurely: they must be adequately developed so that they can survive and grow once separated from the parent. For

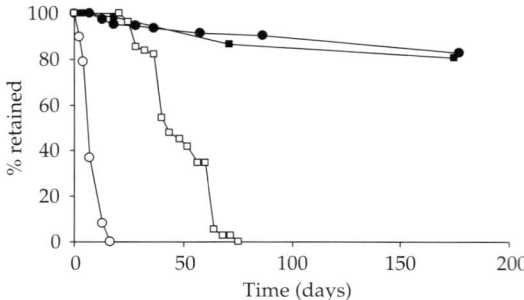

Figure 2.6 Retention of seeds of four Australian species of Acacia in the canopy (redrawn from O'Dowd and Gill 1986, reproduced courtesy of Elsevier). *A. myrtifolia* (○); *A. pycnantha* (▫); *A. melanoxylon* (●); *A. cyclops* (▪).

a wind-dispersed species, longer dispersal distances will be achieved if only the strongest winds are able to liberate them (Greene and Johnson 1995) so easy release of propagules may reduce the distance dispersed. Hooked propagules need to remain at a height where they can be brushed by passing animals and a strong separation force will make it more likely that a sturdy animal, rather than gusts of wind or swaying vegetation, will dislodge them (Graae 2002). Fruits that rely on arboreal dispersers need to remain available within the canopy; however, they must not be so firmly connected that animals cannot detach them. Among Australian Acacias, those species with fruits dispersed by birds hold their seeds on the plant for as long as 18 months, whereas those dispersed by ants readily release their seeds (ballistically in some cases) to fall to the ground (Fig 2.6; O'Dowd and Gill 1986). Thus, a reasonably strong abscission layer will be to the advantage of the propagules of many species.

Eventually, the abscission zone may become so weak that even the slightest force will cause the propagules to fall from the parent, thus ensuring that they reach conditions under which they can potentially grow. Release at some point is better than no release at all, even if this means dropping to the base of the parent plant. Regardless of the dispersal adaptation of the propagule, the final breakage may be caused by wind, rain or hail, a knock from a neighbouring plant (likely to be also caused by wind), or from an animal brushing against it.

Because propagules ripen in a particular season, rather than throughout the year, there will be only a small window within the year when release forces are at a minimum. For those species in which the release force is provided by wind, the first gusts of sufficient velocity will result in dispersal. However, since there is some variation in maturity date among propagules, as well as variation in the threshold separation force at maturity, the dispersal period for a plant population may last for a few weeks or perhaps months. A strong wind following shortly after previous windy days may cause only a few tightly-held propagules to disperse, while a moderate wind following a period of still weather may release a great many propagules. The temporal patterns of numbers dispersing can be charted by collecting propagules landing in traps on the ground (i.e. by measuring seed 'rain'). Many temperate forest species release most of their propagules during the autumn or spring (Nathan and Katul 2005), although for some species releases may occur over many months. A period of calm weather may delay the start of dispersal, while a period of very windy weather may curtail dispersal by causing a high proportion of propagules to become separated from their plant. In countries with cold winters, timing of the release of propagules when snow is on the ground can result in considerable dispersal along the ground surface, because the surface provides less friction than vegetation (Matlack 1989). In the Rockies, willow dispersal coincides with peak river flow (Fig. 2.7): while the propagules would usually be classified as wind dispersed, they are very effectively dispersed by water (Chapter 4).

The force required to separate propagules from their parent has rarely been recorded. Donohue (1998) measured the force required to detach pod segments of *Cakile edentula* (Brassicaceae) from dried plants using a tensiometer. In this species, pods contain seeds in two sections: the distal pod section abscises readily and disperses widely, whereas the proximal section often remains permanently attached to the plant (Keddy 1980). For the distal segment, the force required was related to the diameter of the abscission layer; there was no such relationship for proximal segments. Rather than measure the force required directly, in the case of

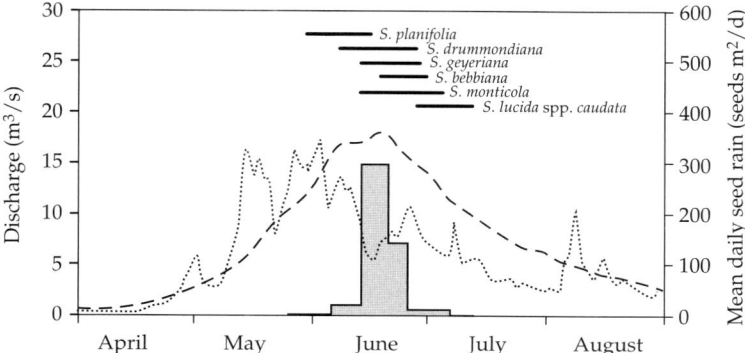

Figure 2.7 Periodicity of propagule release in willows (*Salix* spp.) in the Rocky Mountains, Colorado in relation to stream flow (Gage and Cooper 2005, reproduced courtesy of National Research Council of Canada). Horizontal bars show the period over which seeds were dispersing; histogram shows mean daily seed rain; the mean daily flow of water in 2001 and the average over 1946–1998 (dotted and dashed lines respectively).

wind dispersal we can measure the strength of the vector required to separate propagules from their parents. This makes clear sense, since meteorological data and models of dispersal by wind are based on wind velocity. Measurements of the minimum wind speed for propagule release can be made by placing inflorescences in a wind tunnel and steadily increasing the air flow (e.g. Maier et al. 1999).

There are a number of difficulties associated with measurements of the force or wind speed required for propagule detachment. Firstly, it is difficult to define the stage of maturity: high forces required for abscission may result from propagule immaturity rather than being an inherent characteristic of the species. Stages of maturity may be indicated by changes in colour; readiness for abscission may also be indicated if other propagules have already been released. However, the fact that dispersal has begun may lead to bias in the estimation of the minimum release force, as those propagules remaining on a parent (and which would be those collected by a researcher) may be less mature than those already detached, since the propagules on a plant will vary in their ripening times; or they may be a biased sample consisting of those requiring greater release forces. Secondly, the dynamics of wind can be complex, with rapid changes in horizontal and vertical components of velocity (wind gusts). Wind tunnels or fans in a laboratory provide a simplified form of wind: even if the wind speed is varied, imitation

gusts are unlikely to have the same properties as real wind.

Temperate trees and grassland herbs have been the focus of much research on wind dispersal. For example, Greene and Johnson (1992) found that few mature *Acer saccharum* (Aceraceae) samaras, whose peduncles were held rigidly, were released below $1\,\mathrm{m\,s}^{-1}$, but the number increased rapidly above that speed. Fifty percent abscized at wind speeds of $3\,\mathrm{m\,s}^{-1}$, while 77% abscised at wind speeds of $5\,\mathrm{m\,s}^{-1}$. Less wind was required if it was in an upwards, rather than a horizontal, direction. Not surprisingly, there are large differences among species. For two herbs of the *Taraxacum* genus, Ford (1985) found that the wind speed required for release of achenes differed considerably: $3.5\,\mathrm{m\,s}^{-1}$ vs $5.6\,\mathrm{m\,s}^{-1}$. Minimum wind speed for seed release in two species of *Phyteuma* (Campanulaceae) was $2\,\mathrm{m\,s}^{-1}$, while in two other species of the genus $5\,\mathrm{m\,s}^{-1}$ was needed (Maier et al. 1999). At wind speeds of $13\,\mathrm{m\,s}^{-1}$, only 10–25% had been released. Clearly, failure to allow for the release force in models of wind dispersal will bias predictions of dispersal and overestimate the number of propagules reaching short distances (see Greene and Johnson 1996).

The threshold abscission force is likely to be negatively correlated with humidity, since tissues will be less pliable and more likely to fracture if they are dry. In *Taraxacum*, the bracts around the inflorescence may close up under high humidity (Sheldon and Burrows 1973) and thus the wind does not reach the achenes. Greene and Johnson (1992) noted that air humidity is usually least in the early afternoon, coinciding with higher strength winds. How many children have blown on the achenes of dandelion (*Taraxacum officinale*) to supposedly 'tell the time'? Mature capsules of some *Mesembryanthemum* spp. (Aizoaceae) open only when there is rain (Ridley 1930). Humidity is also important in many species where the seeds are enclosed in a capsule or some other structure. Seeds separate from the funicle but may be physically unable to disperse because there is no exit route from the maternal structure. Release only occurs when external environmental conditions cause the structures to open. For example, *Betula lenta* (Betulaceae) catkins in the laboratory released their seeds under low humidity, regardless of the temperature (Matlack 1989), although it was not possible to find such a relationship in the field. Seeds of *Pinus* spp. are released as their cones dry and open (Allen and Wardrop 1964; Harlow et al. 1964). A regression equation based on temperature and humidity was used by Nathan et al. (2001) to predict the number of propagules ready to disperse from *Pinus halepensis*. Sensitivity analysis of their simulation model led these researchers to conclude that greater increases in dispersal would occur from synchrony of seed release with windy seasons than from any (additional) morphological adaptations to increase aerodynamic properties of seeds or to change the release position on the tree.

In some cases, such as fire-adapted species, seed release may require the influence of heat. For example, in some species of *Hakea* and *Banksia* (Proteaceae), seeds can only escape from their enclosing woody structures after heat from a fire causes the follicles to open (a phenomenon known as serotiny).

2.5 Launching mechanisms

Rather than relying entirely on external vectors, some plants supply the release force themselves. Furthermore, by providing a force well in excess of that required for mere separation, they can achieve significant lateral movement. The potential distance travelled through still air by a projectile from a given height of release will depend on its initial speed, angle of release, mass, and aerodynamic properties (Beer and Swaine 1977). Variation in each of these physical parameters will contribute to variation in distance propelled. The parameters that concern us here are initial speed and angle; post release movement will be dealt with on page 56.

Explosive, or 'ballistic', release of seeds (Fig. 2.8) occurs in many species, from a wide taxonomic range (Ridley 1930). Tension builds up in the fruit wall as it matures and the tissues dry out. When the structure finally fails, the fruit wall springs back explosively, flinging the seeds outwards and upwards into the air. There may be structures within the fruits to direct the seeds (referred to as 'jaculators' in *Ruellia brittoniana* (Acanthaceae): Witztum and Schulgasser 1995a). Dispersal distances of up to 45 m from the parent have been recorded for an 11.2 m high, 9 m radius specimen of the tropical tree *Hura crepitans* (Euphorbiaceae; Swaine and Beer 1977). Most explosive annual herbs, however, achieve maximum dispersal distances of only 5 m or less. Accounts of ballistic dispersal, like those of other dispersal mechanisms, often stress maximum distances.

Studies of ballistic propagule release are usually laboratory-based. A fruit is taken from a plant and secured by a stem or by its pedicel to a rigid stand, at an angle that corresponds (at least approximately) to the angle that a fruit hangs on the parent plant. Methods for the measurement of initial release speed and angle have varied, from extremely simple to the use of advanced technology. For example, Stamp and Lucas (1983) measured angles of release by suspending a surface covered in a sticky substance above the exploding fruits of seven herbs. Mean projection angles, measured above the horizontal, ranged from 46° for *Geranium molle* (Geraniaceae) to 76° for *Phlox drummondii* (Polemoniaceae). Garrison et al. (2000) used dual-angle high-speed stroboscopic photography for their studies of *Vicia sativa* (Fabaceae) and *Croton capitatus* (Euphorbiaceae). The mean initial speeds were $4.64 \, \mathrm{m \, s^{-1}}$ and $4.71 \, \mathrm{m \, s^{-1}}$ respectively and the release angles were 25° and 31°. Further examples are summarized in Table 2.1.

Although the release parameters for a given fruit may be determined with great precision, there can be considerable variation between fruits: for *V. sativa*

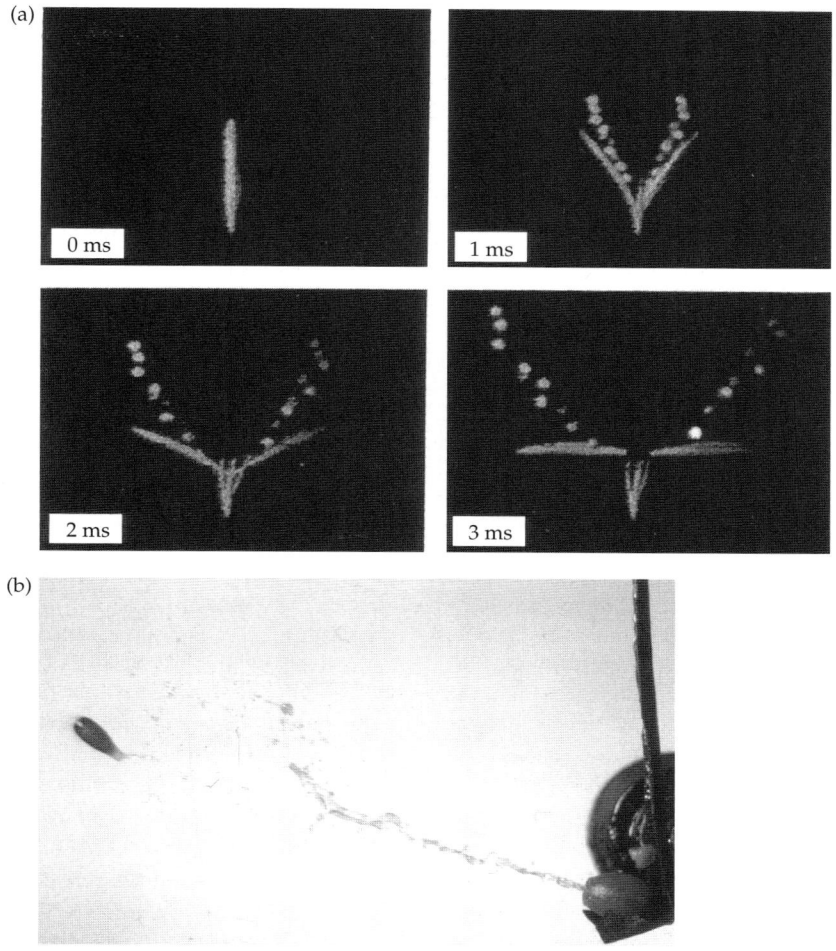

Figure 2.8 Examples of explosive release of seeds: (a) *Ruellia brittoniana* (Witztum and Schulgasser 1995a, reproduced courtesy of Elsevier); (b) *Arceuthobium vaginatum* (T. E. Hinds, reproduced courtesy of USDA Forest Service). Seeds in (a) are about 2 mm in diameter and in (b) are around 3 mm in length.

the range of launch speeds was 0.83 to $9.03\,\mathrm{m\,s}^{-1}$ while the range for launch angle was $-50°$ to $+69°$ (Garrison et al. 2000). Angles at which fruits were held on the parent plants ranged from $-40°$ to $+90°$ and varied with stage of development. One complication is that rigidly held pedicels in such experiments have no recoil, resulting in a greater estimate of launch speed than if a fruit is freely swinging on a real plant. Therefore, even though it is possible to measure the appropriate launch parameters in the laboratory, it must be appreciated that the prediction of the frequency distribution of

ballistic dispersal distances for a real plant will be complex.

There are various other launching mechanisms that could be mentioned, but these have received little quantitative attention. In the squirting cucumber *Ecballium elaterium* (Cucurbitaceae), for example, a high pressure builds up within the viscous fluid inside the fruit, and its sides stretch elastically. When ripe, the fruit detaches suddenly at its base, causing the fluid and seeds to gush out through the resulting hole propelling the fruit away from its parent. Ridley (1930) states that seeds travel around 2 m and

Table 2.1 Variation in physical parameters for ballistic propagule release. All are measured under laboratory conditions. Sources are: Hinds and Hawksworth (1965); Garrison et al. (2000); Swain and Beer (1977); Stamp and Lucas (1983); Witztum and Schulgasser (1995a, b).

	Propagule mass (mg)		Angle of projection above horizontal (degrees)*		Launch speed ($m s^{-1}$)	
	mean	CV (%)	mean	CV (%)	mean	CV (%)
Hura crepitans	1020	17	34	23	43	40
Geranium maculatum	6.0	15	47	21	–	–
Geranium carolinianum	3.5	9	46	7	–	–
Geranium molle	1.1	9	61	3	–	–
Phlox drummondii	1.7	18	76	12	–	–
Viola eriocarpa (Violaceae)	3.0	13	68	21	–	–
Arceuthobium spp. (Viscaceae)	0.9–2.3	–	–	–	21–26	9–19
Ruellia brittoniana	1.8	–	≈40	–	12	–
Blepharis ciliaris (Acanthaceae)	10	–	49	–	5	–
Vicia sativa	12.8	12	25	96	4.64	47
Croton capitatus	23.3	13	31	81	4.71	38

*Some references give angles only relative to the position of the fruit as it is held in the laboratory, while others convert this to the angle ejected from a fruit in its typical attitude on a plant. Original papers should be consulted for detail.

that the direction of the squirt is highly unpredictable.

Dynamic release may also be achieved through an interaction between an external dispersal vector and the parental structures, rather than as a direct developmental process. Wind causes plants to sway, especially in annual plants lacking secondary thickening. This movement can provide propagules with forward momentum, as well as assisting in the act of separation. Poppies (*Papaver* spp.: Papaveraceae) appear to have adapted specifically to this opportunity. The seeds are in capsules and are situated below the openings that provide the opportunity for seeds to escape. They can only be released if the stem bends over far enough (and as a result, away from the parent) or stops abruptly, with the momentum of the seeds propelling them out of the capsule with a net forward motion. Seeds in the bottom of the capsule will only be dispersed in the strongest of winds. To examine the effect of this release mechanism, Salisbury (1942) clamped stems of three poppy species in the laboratory at their typical heights and used a fan to simulate a gusty wind. All species had frequency distributions of dispersal distance (i.e. the combined result of release mechanism *and* post-release movement by wind and by rolling) that were peaked. The species with the longest capsules (*Papaver dubium*) had the sharpest peak, while the species with the most spherical capsules (*P. hybridum*) had the greatest proportion of seeds landing close to the release point. The vector responsible for the launch of propagules may also be water rather than wind. The capsules of some plants open out, presenting the seeds on a flat or cupped surface (e.g. *Mitella* spp. (Saxifragaceae) and *Mesembryanthemum spp.*): rain drops hitting the surface transfer their energy to the seeds, causing them to be propelled into the air (Burrows 1986).

2.6 Other parental traits

Any categorization of maternal plant attributes must, to some extent, be artificial and there are some things affecting dispersal that do not fit readily into the previous sections. We conclude the chapter by discussing a range of parental traits that affect the dispersal of propagules by humans.

There are many reasons why plants have been deliberately dispersed by humans (for reasons other than the structures that have evolved to aid dispersal by non-human vectors). We use plants directly for our nutrition, to treat ailments, to smoke, to feed our animals, to produce paper and other wood products, to tan leather, to produce material for clothing and other products, for land reclamation, and for a range of other reasons. We also disperse plants simply because we find them attractive and because they remind immigrants of home (Mack and Lonsdale 2001).

Chicory (*Cichorium intybus*: Asteraceae) and salsify (*Tragopogon porrifolius*: Asteraceae), for example, have escaped in some countries after their introduction as vegetable crops for their edible roots. Wormwood (*Artemisia absinthium*: Asteraceae), Mullein (*Verbascum thapsus*: Scrophulariaceae), and yarrow (*Achillea millefolium*: Asteraceae) have long histories of herbal and medicinal uses and may have been spread deliberately for this reason. White clover (*Trifolium repens*) has naturalized in most temperate regions of the world following its introduction as a pasture species. *Cannabis sativa* (Cannabaceae) has been spread for the use of its stem fibres for ropes and canvas, for medicinal and herbal uses of its leaves, as well as being cultivated illegally as a drug. *Panicum miliaceum* and *Sorghum halepense* (both Poaceae) have been dispersed in North America after their introduction as forage crops. In these cases the maternal plant attributes leading to dispersal as animal feed include palatability, digestibility, and nutritional value of stems and leaves. Hay bales often contain viable weed seeds which are dispersed with it (Thomas et al. 1984). Dispersal in hay may be between fields on the same farm, between islands, states, or even between continents.

Many trees have been spread for commercial purposes or to be grown in gardens. For example, several *Pinus* spp. have been introduced thousands of miles from their native range for timber production and have then naturalized (Rejmánek and Richardson 1996). *Hakea* and *Acacia* (Fabaceae) species have been introduced from Australia into South Africa for their tannins (used in leather manufacture) and are now spreading through the fynbos. Some unusual uses may result in significant spread: single sex clones of willows (*Salix* spp: Salicaceae) in Tasmania, originally introduced as ornamentals, have apparently been spread by anglers who use forked twigs as rod supports: these sometimes take root and produce a new tree. A range of plant species have been spread because their roots, stolons, or rhizomes aid revegetation (e.g. marram grass (*Ammophila arenaria*: Poaceae) in sand dunes). Finally, aesthetic attractiveness to humans has been an extremely important maternal plant attribute: it has been estimated that 65% of plants naturalized in Australia between 1971 and 1995 were introduced as ornamentals (Groves 1998). Important traits resulting in dispersal for ornamental planting include bright colours, large flowers, and colour, texture, and shape of foliage and stems.

2.7 Conclusions

In this chapter we have considered various aspects of the maternal parent which contribute to the frequency distributions of dispersal distance. Genotype, the abiotic environment, competition, and herbivory all help to determine the plant phenotype and hence where the trajectories of propagules will begin. Plant phenology determines when the force required to separate the propagule from its parent are at a minimum and therefore when a trajectory is likely to begin. In some cases the parent itself provides a force greater than that required merely for separation, launching the propagule away from the parent. Some of these parental contributions are widely recognized: as we show in Chapter 4, the importance of release height has been investigated in considerable detail, while both experimental and theoretical studies have explored ballistic dispersal. The timing of fruit maturity can be critical for the survival of animal vectors as well as for movement of seeds contained in fruits. However, the implications of plant form are less well appreciated. As we will show in Chapter 5, both the horizontal and vertical distributions of propagule positions on the parent plant can be critical in determining the frequency distributions of dispersal distance in species with relatively limited dispersal ability. This will, in turn, have consequences for competition among offspring and between parent and offspring, with consequent implications for population dynamics (Chapter 7) and evolution (Chapter 8). We therefore expect increased attention in future studies to the role of the maternal plant in dispersal.

CHAPTER 3

Attributes of propagules that aid dispersal

3.1 Introduction

Natural historians have long been fascinated by the various structural adaptations that aid the dispersal of seeds. Every ecologist interested in dispersal should browse through Ridley's (1930) classic dispersal text at some time, to be fascinated by the many ways in which evolution has enhanced dispersal. These traits have been thoroughly described and classified (both anatomically and functionally) and their evolutionary paths speculated upon. They include an astonishing array of fruit colours, sizes, and chemical compositions to attract animals, appendages to attach to animal hair, to catch the wind or to attract ants, sticky coatings to attach to animals, flat surfaces that cause propagules to glide or to spin through the air, waxy surfaces and air pockets to enable them to float, and hard, thick surfaces to protect them when passing through an animal's gut.

To be transported effectively by a single dispersal vector, a propagule often needs to possess several traits, not just one. Long flight distances, for example, are achieved by having a thin appendage that acts as a wing: but the seed also needs to be located in the correct position and its mass must be within a particular size range (Augspurger and Franson 1987). To be dispersed by ants, seeds need both an external lipid body and a size that makes it possible for the ant to carry them. A group of traits related to dispersal by a particular vector is often referred to by ecologists as a dispersal 'syndrome'. For example, we can identify wind dispersal syndromes, water dispersal syndromes, and animal dispersal syndromes. Within the range of colours, sizes, chemical, and structural traits that enable dispersal by animal ingestion, it is further possible to identify an ant dispersal syndrome, mammal dispersal syndrome, bat dispersal syndrome, and so on, although there may be considerable overlap in the fruits eaten by each type of animal. Groups of plant species possessing common sets of dispersal traits may also be referred to as dispersal 'functional groups' or 'guilds'. Species sharing the same syndrome may, however, have evolved in quite different ways. Hooks, spines, and barbs all enable propagules to attach to animal hair, and each of these can be derived from a number of alternative floral or reproductive structures (Sorensen 1986). Similarly, various structures may provide the carbohydrate or lipid rewards sought by animals.

An unfortunate outcome of the classification of dispersal adaptations into syndromes is a tendency to view dispersal as one plant species, one dispersal vector. If there is a clear adaptation that aids dispersal by wind, for example, then we tend only to examine dispersal by wind, even though the same propagules may be moved considerable distances by water. In a given year or a given location, propagules from one plant may be exposed to a number of vectors: some act in parallel on separate proportions of the population at the same time, or they may act on the same propagules sequentially (i.e. in series). For example, only some propagules with barbs or hooks encounter an animal (often a very small proportion): many fall from the plant and stay where they land, a few may be transported with mud on farm machinery, and others may fall into a river and be spread downstream. Propagules may move as far, if not further, by alternative vectors than the vector to which they are apparently adapted. Indeed, a trait that has arisen through selection by one particular vector may pre-dispose the propagule to dispersal by another vector: for example, adaptations enhancing wind dispersal often result in seeds that also float better on water. For some vectors, the trait giving the greatest potential for dispersal may not have evolved in relation to dispersal at all: an impervious seed coat might have evolved due to its ability to regulate the time of year at which germination occurs, but this also increases its buoyancy and hence will aid dispersal by water. When considering the spatial dynamics of a population of propagules, it is thus essential to consider the influences of *all* potential dispersal mechanisms and not just the ones that appear to have selected directly for a particular trait.

If we are to predict the dispersal trajectory of an individual propagule (see Chapter 4), we need to determine some measurable physical, chemical, or biological characteristics of the propagule that are relevant to movement by a given vector: if we can estimate both the 'strength' and direction of the vector over the duration of the dispersal event, then, at least in principle, we could predict the trajectory resulting from the action of that vector, even if the propagule is not obviously adapted to it. For example, if we measure the time that any given propagule remains buoyant and its falling velocity when it sinks, we could predict the dispersal distance of the propagule in a given set of water flow conditions. In this chapter, we discuss traits that determine dispersal by different vectors and that can be determined for any propagule. We pay special attention to dispersal by air, water, animals, and humans, since these vectors have received the greatest attention in the scientific literature.

3.2 Aerodynamic traits

There are a great many structural modifications of propagules that clearly enhance dispersal through the air. It is traditional to emphasize the importance of air *movement* by referring to these aerodynamic features as adaptations to dispersal *by wind*, even though they may increase dispersal when there is no wind at all (Burrows 1986). Certainly, strong winds can disperse some types of seeds over very great distances and wind has no doubt played a central role in their evolution. However, a significant proportion

of propagules may be released on relatively calm days. There are also types of propagule which lack obvious aerodynamic adaptations but which, nevertheless, pass through the air on their way to the ground. These may be propelled by force from the parent plant, or simply fall under the influence of gravity. All of them are affected by the fluid medium (the air) that they are passing through, whether it is still or in motion.

The forces acting on a particle as it moves through the air are well-known and have been characterized in the science of fluid dynamics. It is these forces that determine a propagule's speed and direction, and hence its trajectory and final location, under any given air flow. Fluid dynamics also tells us the properties to measure if we want to assess a propagule's *potential* for movement in air. A brief review of the relevant theory and important definitions, such as lift, drag, and terminal velocity, is given in Box 3.1. More detailed accounts can be obtained elsewhere (Burrows 1986; Minami and Azuma 2003).

In purely physical terms, fluid dynamics tells us that to increase the dispersal distance of a propagule carried passively by air, a morphological adaptation must increase the drag or lift coefficients or decrease the ratio of mass to area. One type of morphological adaptation may, of course, result in beneficial changes in more than one of the forces on the propagule. When a propagule is propelled into the air with additional force supplied by the plant (see p. 34), its horizontal velocity may well be greater than the air. Drag thus acts to *decrease* its forward speed. Therefore, to increase the distance travelled by a ballistically projected propagule, mass per unit area must be increased or the coefficient of drag must be decreased.

Adaptations for dispersal by air are most abundant on grassy plains, open heaths, sand dunes, and deserts (van Rheede van Oudtshoorn and van Rooyen 1999), all habitats in which the parents (and hence propagule release sites) are easily exposed to the effects of wind. Aerodynamic traits are also abundant amongst the emergent species in forests, but less abundant in the understorey where there is less air movement (van der Pijl 1982). Morphologies of propagules can be classified by the aerodynamics that they display. For example, Burrows (1986) used

Box 3.1 A brief overview of fluid dynamics

For convenience, we assume only two dimensions, horizontal and vertical, parallel to the propagule's trajectory. Note also that at this point, we make no assumptions about specific shapes of propagules. Consider the velocities of both the propagule and the air at an instant in time (note that a velocity denotes both a speed and direction). It is usual to resolve air velocity into horizontal (V_{AH}) and vertical (V_{AV}) components of movement; the net velocity is the resultant of these two vector quantities, V_{AR} (Fig. 3.1). The trajectory of the propagule at any point in time is the outcome of all the forces acting sequentially on the propagule up until that point, which have resulted in its current velocity V_P. The velocity of the propagule *relative to the medium* can be calculated as the resultant of $V_P - V_{AR}$. In order to determine the velocity of the propagule in the next instant of time, we need to consider the forces acting on an object travelling through a fluid medium. The principal forces will be gravity, drag (in a direction opposite to $V_P - V_{AR}$ and acting through the centre of pressure) and lift (at 90° to the drag and acting through the centre of lift). There is also a buoyancy force acting as a result of the displacement of the fluid by the propagule, but this is negligible in air for almost all plant propagules. If these forces completely balance in all directions, the propagule will travel at constant velocity along the same trajectory; if the forces do not balance, the propagule will accelerate (or decelerate) in a direction governed by the resultant of the three force vectors. The force due to gravity is *mg*, where *m* is mass and *g* is gravitational acceleration (9.81 m s^{-2}). The size of the drag depends on the characteristics of the air and of the object passing through it, as well as the velocity of the propagule relative to the air.

Drag has a complex relationship with velocity, determined by the size of the object and the properties of the fluid as summarized by the dimensionless Reynolds number (R_e):

$$R_e = \frac{\rho A^{0.5} V}{\mu} \tag{B3.1.1}$$

where V is the velocity of the propagule relative to the air ($V_P - V_{AR}$), A is the cross-sectional area in the direction of travel, while ρ and μ are the density and viscosity of air respectively. R_e is seldom calculated: it is sufficient to know simply its order of magnitude: whether it is low (viscous forces dominate) or high (inertial forces dominate). Most plant propagules moving through air have R_e greater than 1. For example, a particle of diameter 5 mm (5×10^{-3} m) travelling at 0.5 m s^{-1} through air (density 1.09 kg m^3; viscosity 1.84×10^{-5} N s m^{-2}) has a Reynolds number of

about 150. Under conditions resulting in a high R_e,

$$Drag = 0.5 \rho A C_D V^2 \tag{B3.1.2}$$

where C_D is the coefficient of drag. For very small 'dust' seeds, spores, and pollen falling very slowly through the air, R_e may be less than 1, in which case Stokes' Law (accounting for the friction a small particle experiences moving through air) applies and

$$Drag = k \mu A^{0.5} V \tag{B3.1.3}$$

For most propagules, however, we need only consider equation 3.2.

Lift results from differential pressures created on the surfaces of a propagule parallel to the direction of drag. If the pressure on the upper surface is less than that on the lower surface, a force will be generated in an upwards direction. In the absence of vertical air flow, the lift can never overcome the effect of gravity, but it will be able to both slow the rate of descent and increase the horizontal component of velocity. It is defined in a similar way to drag:

$$Lift = 0.5 \rho A C_L V^2 \tag{B3.1.4}$$

Now let us consider the consequences of lift and drag for a dispersing propagule. Firstly, what happens when a propagule separates from its parent into a strong flow of air (without any additional force provided by the plant)? Its velocity is initially zero, while its velocity relative to the air is thus high and negative. Drag is therefore also high in the direction of the wind and causes the propagule to accelerate in that direction: a high drag coefficient results in a greater acceleration. If the propagule's mass is low, it will almost instantly reach approximately the same horizontal speed as the air ($V_{PH} = V_{AH}$), at which point there is no drag in the direction of flow. For many light propagules, it is thus reasonable to assume that the horizontal velocity is the same as that of the wind.

Now, let a propagule be released into still air ($V_{AR} = V_{HR} = 0$). If there is no appreciable sideways movement generated by lift, the propagule will accelerate downwards until the force due to gravity exactly equals the drag force. Hence, from equation 3.2,

$$mg = 0.5 \rho A C_D V^2 \tag{B3.1.5}$$

continues

Box 3.1 continued

Rearranging this equation, we obtain the *terminal velocity* (or *falling velocity*) as

$$V_{term} = \sqrt{\frac{2mg}{\rho A C_D}} \qquad (B3.1.6)$$

The ratio of mass to area is often referred to as the wing loading. From equation 3.6 we can see that the terminal velocity is proportional to the square root of wing loading, an observation that will be discussed later in the text.

Terminal velocity is commonly used as an index of dispersability by wind: the longer it takes a propagule to fall to the ground, the further it can be blown horizontally away from the parent. Terminal velocity is thus an input into many dispersal models (p. 53). However, although it may be a reasonable predictor of rate of descent in a horizontal air flow, the terminal velocity will overestimate the rate of vertical descent if the air flow or the shape of the propagule generates lift (Hensen and Müller 1997). There are two ways to measure terminal velocity in the laboratory. The most common is to drop a propagule in still air: after a short distance sufficient for it to have accelerated to the terminal velocity, its rate of descent is measured by the distance moved in a given time interval (using a stop-watch, stroboscopic photography (Matlack 1987) or photo-detection (Askew et al. 1997)). Alternatively, the air speed in a vertical wind tunnel can be adjusted until the propagule floats at a constant height (Jongejans and Schippers 1999).

When the propagule reaches the ground, there can clearly be no further downwards motion (unless it later encounters a soil crack or other physical feature). A horizontal air flow can still act on the propagule, although there will be an additional force due to friction with the ground surface (μmg, where μ is the coefficient of friction). A propagule will begin to move when the drag caused by air flow can overcome the initial static friction

$$\mu_s mg < 0.5\rho V_{AH}^2 A C_D^* \qquad (B3.1.7)$$

i.e.

$$V_{AH} > \sqrt{\frac{2\mu_s mg}{\rho A C_D^*}} \qquad (B3.1.8)$$

where μ_s is the coefficient of static friction, and will reach a maximum velocity when

$$V_{AH} - V_P = \sqrt{\frac{2\mu_k mg}{\rho A C_D^*}} \qquad (B3.1.9)$$

where μ_k is the coefficient of kinetic friction ($\mu_k < \mu_s$); note that the coefficient of drag, C_D^*, will have a different value from C_D for movement through the air. Again, velocity is proportional to the square root of wing loading (Johnson and Fryer 1992). The larger the surface area per unit mass, the lower the velocity required to overcome static friction and the lower the difference between air velocity and propagule velocity (and hence the greater the propagule velocity).

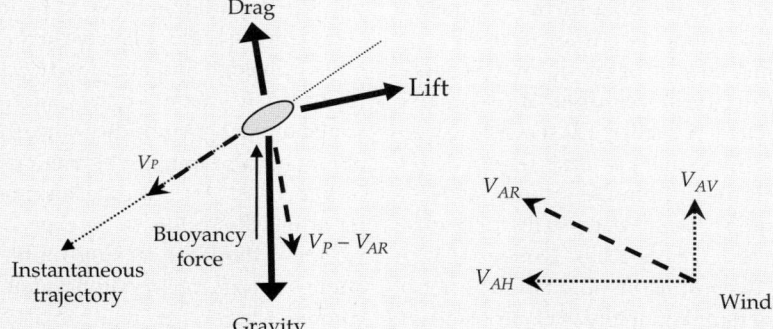

Figure 3.1 Forces acting on a propagule travelling through the air. The length of an arrow Indicates strength: solid arrows are forces, dashed arrows are velocities. Wind velocity has horizontal (V_{AH}) and vertical (V_{AV}) components, giving a resultant V_{AR}. The velocity of the propagule is V_P, giving a resultant velocity of the propagule relative to the air of $V_P - V_{AR}$.

the following classification:

- dust seeds and spores (and pollen);
- plumed and woolly propagules;
- plane winged propagules with a central or more-or-less central concentration of mass;
- winged propagules which rotate when falling;
- seed-carrying tumbleweeds;
- aerodynamically unimportant propagules.

(Note that authors vary in their classifications and terminology.) We would add to this list seeds expelled ballistically from the parent, since their aerodynamic properties are important to their dispersal. In the following sections, we review some of the relevant published data on the aerodynamics of these types of propagule.

3.2.1 Dust propagules

Some plants have evolved tiny seeds, with a mass of as little as 0.3 μg in *Schomburgkia undulata* (Orchidaceae: Arditti and Ghani 2000). These seeds may also contain air pockets that increase their surface area and (from equation 3.2) thus increase drag and decrease rate of fall. The small mass is at the expense of the energy supply for the germinating seed and young seedling. In orchids, there has been the coincident development of symbioses with mycorrhizal fungi, from which they obtain the carbohydrates needed in the early stages of growth. At these very small propagule sizes, viscous forces in air are important in slowing the rate of fall (R_e may be less than 1), such that terminal velocities of less than $0.1\,\mathrm{m\,s^{-1}}$ are achieved (Gregory 1973). Only very slight upwards air movement is thus required to lift these dust seeds away from the ground, allowing winds to disperse them potentially to great distances (Arditti and Ghani 2000). Ridley (1930) illustrates the effectiveness of this adaptation by noting that orchids are found on many oceanic islands, hundreds of kilometres from the most likely sources.

3.2.2 Plumed propagules

Plumes and woolly appendages attached to propagules (Fig. 3.2a) are highly effective in increasing drag, since they greatly increase the surface area for a given mass (and therefore decrease wing loading—equation 3.6). This reduces their rate of descent. In many species of the Asteraceae (exemplified by *Taraxacum officinale*), for example, there is a pappus on top of an achene: the descending propagule is reminiscent of a parachute and acts in the same way. An additional feature of some *Asteraceae* propagules is that the seed is located well below the plume, resulting in a centre of gravity away from the centre of lift. Although there may be phylogenetic constraints that have led to this arrangement, the outcome is that it reduces any tendency to rotate in the vertical plane, thus increasing stability during flight.

There have been more studies of the flight of plumed propagules than any other adaptation for air dispersal. Terminal velocities for most plumed species range between 0.1 and $1\,\mathrm{m\,s^{-1}}$ (Tackenberg et al. 2003). A negative curvilinear relationship between terminal velocity and the ratio of pappus diameter:achene diameter has been found for data representing 17 species (Fig. 3.3a). Although a single terminal velocity is often assumed to be representative of a species, its value may vary considerably between populations, within the progeny of a single plant and across experimental conditions. In 15 studies of *T. officinale*, the mean terminal velocity ranged from 0.10 to $0.66\,\mathrm{m\,s^{-1}}$ (Tackenberg et al. 2003). Most researchers assume standard 'laboratory conditions', but these conditions vary from day to day. For example, if it is raining outside the building, the humidity is likely to be high inside, while some climates naturally have higher humidities than others. Sheldon and Burrows (1973) found that drag increased in *Senecio vulgaris* (Asteraceae) propagules stored under low humidity, since the pappus opened out more, resulting in a slower rate of descent.

A focus on the average terminal velocity when reporting the results of an experiment implies that variation about the mean is unimportant. The variation between observations may be largely measurement error, or real variation of no ecological significance. Indeed, sensitivity analyses of models have shown that variation in terminal velocity may be much less important in determining dispersal distance than variation in height or timing of release (Nathan et al. 2001). It has also been concluded from field studies that long-distance dispersal occurs because some seeds are

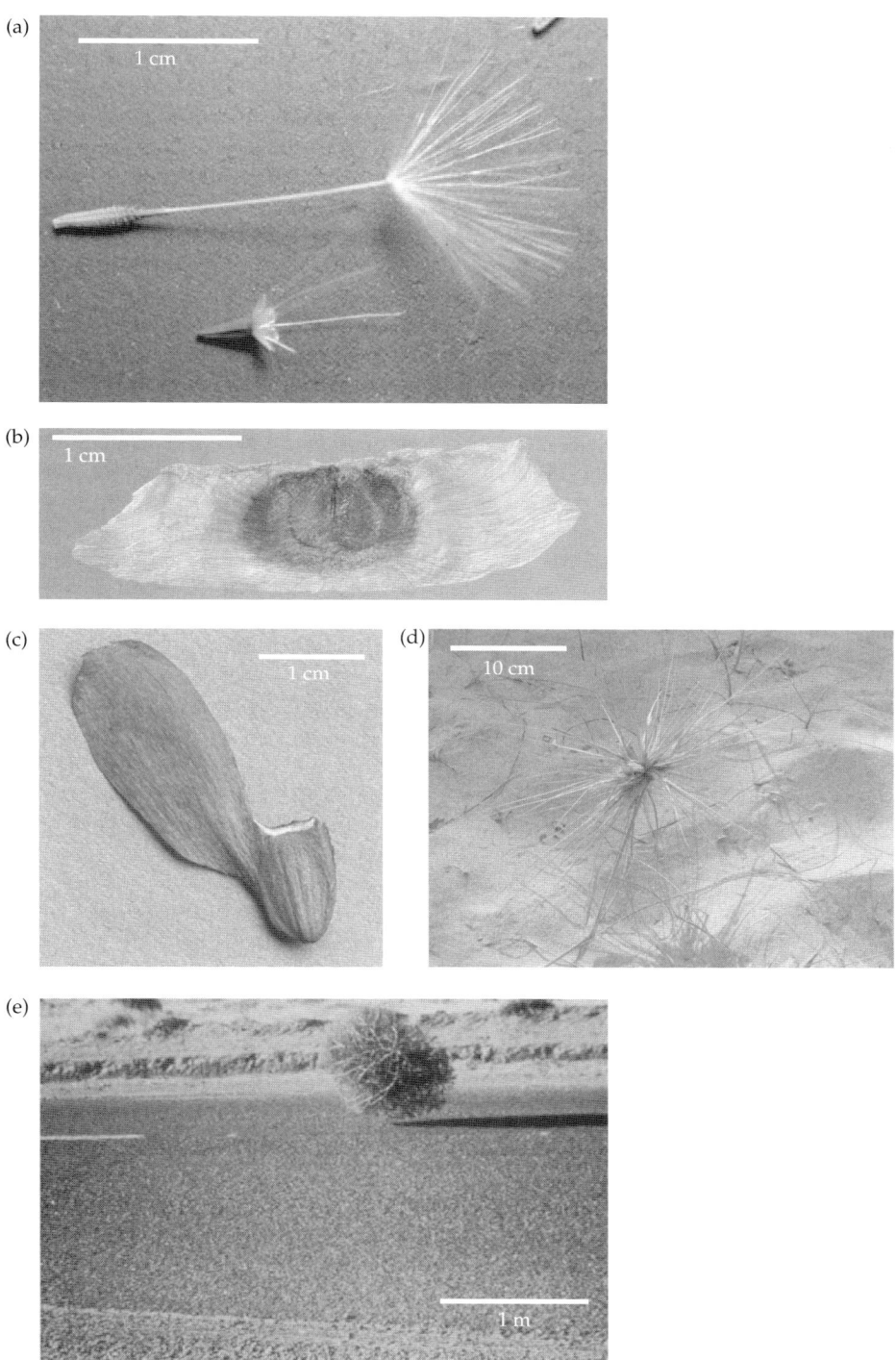

Figure 3.2 The range of structures enhancing the movement of propagules by wind: (a) *Taraxacum officinale* and *Krigia virginica* (Asteraceae) (picture courtesy of Wayne Hughes), (b) *Campsis radicans* (Bignoniaceae) (reproduced courtesy of Steve Hurst @ USDA-NRCS PLANTS Database), (c) *Agathis robusta*, (d) *Spinifex sericeus*, (e) *Kochia scoparia* (reproduced courtesy of Carol Mallory-Smith). Scales are approximate.

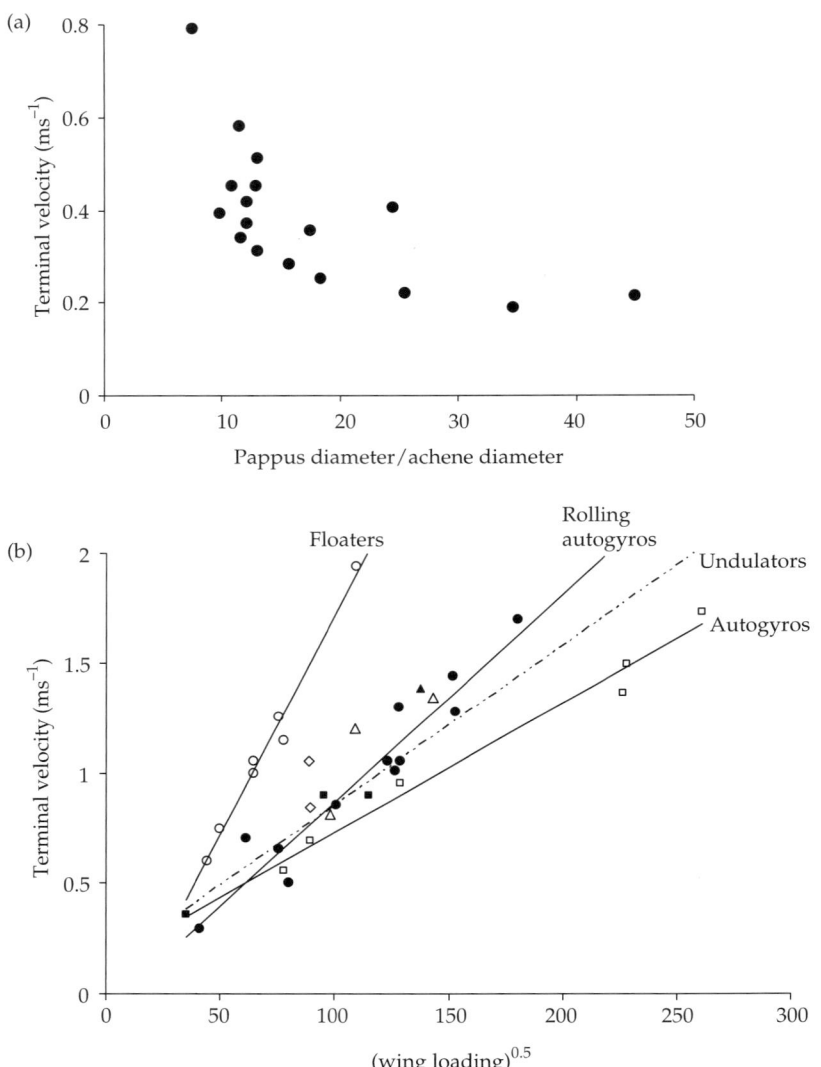

Figure 3.3 Relationships between rate of descent of propagules and morphology: (a) redrawn from Sheldon and Burrows (1973, reproduced courtesy of Blackwell Publishing); (b) redrawn from Augspurger (1986, reproduced courtesy of Botanical Society of America). Wing loading is the ratio of mass:maximum cross section. Symbols in (b) are △ helicopters; ○ floaters; □ autogyros; ◇ unclassified; ■ undulators; ● rolling autogyros; ▲ tumblers.

uplifted by abnormally strong winds, not because some exceptionally mobile propagules move further during average wind conditions (Horn et al. 2001). However, variation within propagules from the same parent plant can be considerable and this *will* have an effect on the trajectory of an individual propagule under a given set of wind conditions.

Variation among propagules is particularly apparent in the Asteraceae, where seeds from a single capitulum of some species have very different seed weights and pappus dimensions; indeed, some seeds may have no pappus at all. In *Senecio jacobaea* and *Crepis sancta*, for example, inner florets on the capitulum produce achenes with a pappus, while the outer florets produce heavier achenes with no pappus (McEvoy and Cox 1987; Imbert and Ronce 2001). Over 80% of *Carduus pycnocephalus* and *Carduus tenuiflorus* achenes may be of the pappus-bearing type (Olivieri et al. 1983). In *Heterotheca latifolia*, ray achenes produce no pappus, but are of similar mass to disc achenes (Venable and Levin 1985). Populations of the same species may differ in their proportions of different propagule types. For example, the proportion of pappus-bearing achenes of *Carduus pycnocephalus* and *C. tenuiflorus* has been found to decrease along a successional gradient (Olivieri and Gouyon 1985). A reduction in the ratio of pappus size to achene size was found in island populations of *Hypochaeris radicata* compared with mainland populations, while lower proportions of pappused achenes were found in older populations of *Lactuca muralis* compared with newer colonies (Cody and Overton 1996). It is unclear in either of these studies whether the observations were due to genetic differences or phenotypic plasticity; however, the existence of variation between populations has been established.

The proportions of plume-bearing propagules may also vary with growth conditions. For example, lower nutrient levels produced a greater proportion of pappus-bearing achenes in plants of *Crepis sancta* (Imbert and Ronce 2001). In another study, the number of disc achenes of *Heterotheca latifolia* was greater on average than the number of ray achenes, but for plants below the median height the proportion of ray achenes increased (Venable and Levin 1985). In *Hypochaeris glabra*, the ratio of two achene types (light and heavy) may be affected by plant density and time of year (Fig. 3.4). The proportion of the lighter achenes (mean terminal velocity in still air 0.45 m s^{-1}) from the centre of the capitulum was greatest in the middle of the year; this propagule type was also more abundant than the heavier achenes (terminal velocity 0.69 m s^{-1}) at lower plant densities throughout the year. Along with two types of aerial achenes (and intermediates), *Catananche lutea* (Asteraceae) produces two types of amphicarpic (self-burying) achenes. The proportion of amphicarpic capitula, and therefore the number of non-dispersing seeds, has been found to increase in proportion as plant density increases (Ruiz de Clavijo and Jiménez 1998). Within the aerial achenes,

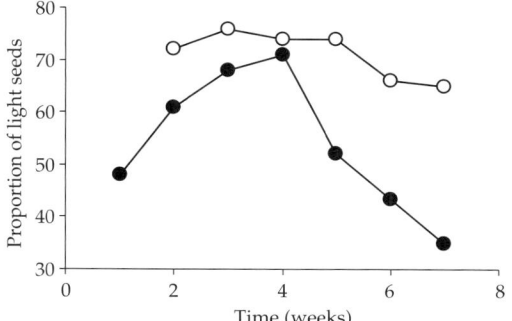

Figure 3.4 The ratio of light:heavy seeds in *Hypochaeris glabra* (redrawn from Baker and O'Dowd 1982, reproduced courtesy of Blackwell Publishing), as affected by plant density (● high and ○ low) and time of year.

the proportion of fully-pappused seeds fell only slightly with density.

While the preceding studies have mostly examined variation in propagule morphology, indicating that variation in propagule aerodynamics will probably occur, few studies have measured the actual variation in terminal velocity within a random sample. A three-fold range of terminal velocities, however, has been recorded for propagules of *Asclepias syriaca* (Asclepiadaceae) in two years (Fig. 3.5). There was considerable variation in rate of descent among seeds from the same pod and from the same clone in both years; a greater number of 'slow' seeds in 1982 was mainly due to some very light seeds from a single pod. Mean seed mass varied 20-fold (heaviest:lightest) within the sample in 1982, and slightly under 10-fold in 1983. Coefficients of variation within six *Solidago* species were found to range from 15 to 35% for terminal velocity, from 15 to 85% for propagule mass, and from 7 to 10% for plumule radius (Werner and Platt 1976). A two-fold range of terminal velocity has been reported in propagules of three Asteraceae and one Dipsacaceae (Soons and Heil 2002), with rate of descent being well correlated with the ratio of the square root of the achene mass to the maximum diameter of the structure (for propagules with a pappus) or with the square root of seed mass (for those without a pappus).

Although terminal velocity is a useful descriptor of aerodynamic properties, it is not the only parameter that determines dispersal distance. The taller the plant, the further the propagule will fall,

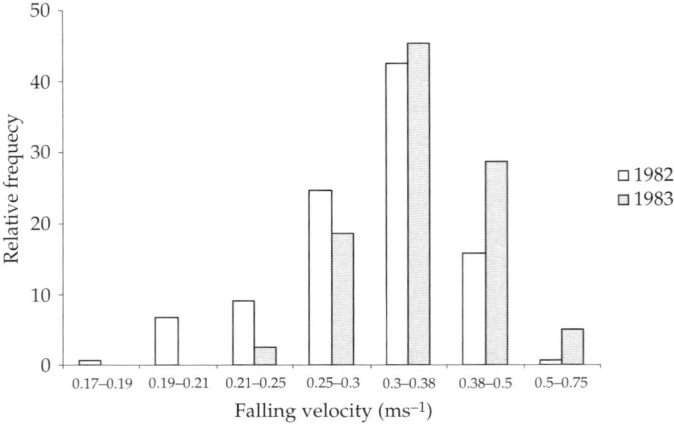

Figure 3.5 Variation in falling velocity of propagules of *Asclepias syriaca* in two years (redrawn from Morse and Schmitt 1985).

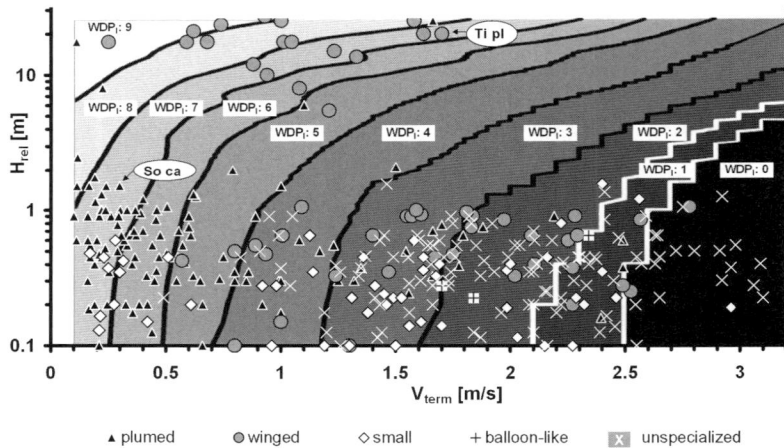

Figure 3.6 Categories of 'wind dispersal potential' in relation to release height (H) and terminal velocity (V) (Tackenberg et al. 2003, reproduced courtesy of Ecological Society of America). Symbols indicate data for different species, categorized according to their morphology. See text for details.

the greater the duration of flight, and the further it can be blown horizontally (see Chapter 4). Tackenberg et al. (2003) have therefore proposed a Weed Dispersal Potential (WDP) index, based on the predicted proportion of propagules of a given terminal velocity (V) traveling a distance of 100 m from a release height (H, a typical fruiting height for the species) under a reference set of weather conditions (Fig. 3.6). Plumed propagules have lower terminal velocities than most other types. The authors commented that terminal velocity is the

more influential parameter for long-distance dispersal, while height of release becomes more important under conditions of shorter dispersal distances.

3.2.3 Plane-winged propagules

Many propagules have an extended surface which is bilaterally symmetrical about the seed (Fig. 3.2b). This generates lift, causing the propagule to glide as it falls. A smooth flight to achieve maximum dispersal distance also requires a condition known in aeronautics as *stable trim*. This is dependent on the location of the centre of mass on the wing relative to the centre of lift: the wrong combination can cause rotation, leading to unstable flight and faster rates of descent. Some gliding motions include periodic changes in flight path, with alternation of diving and stalling: this is referred to as fugoid oscillation (Burrows 1986). Other possible trajectories are spiral and 'rocking' glide paths, which may arise from asymmetry of the propagule (Minami and Azuma 2003).

The larger the ratio of lift to drag, the slower the rate of sink in the glide (Burrows 1986). Lift:drag ratios have been measured at between 3 and 4 for *Alsomitra macrocarpa* (Cucurbitaceae, with a wingspan of about 13 cm), resulting in rates of vertical descent of 0.3–$0.7 \, \mathrm{m \, s^{-1}}$ (Azuma and Okuno 1987), 30 times slower than a spherical particle of the same mass (McCartney 1990). For twelve species from five families with propagules undergoing gliding and 'straying' flight, Minami and Azuma (2003) measured rates of descent of 0.33 to $1.19 \, \mathrm{m \, s^{-1}}$. Augspurger and Franson (1987) showed experimentally how mass, area, and shape can all affect dispersal distance in this type of propagule, using artificial models designed to mimic *Tachigalia versicolor* (Fabaceae). Mean rate of descent of the real propagule was $1.76 \, \mathrm{m \, s^{-1}}$, but in the models this could be decreased to $1.57 \, \mathrm{m \, s^{-1}}$ and increased up to $3.68 \, \mathrm{m \, s^{-1}}$ depending on surface area and mass (if we consider only the bilaterally symmetrical shapes). If the wing loading was held constant, by changing mass and area at the same time, terminal velocities varied only from $1.79 \, \mathrm{m \, s^{-1}}$ to $2.01 \, \mathrm{m \, s^{-1}}$.

The threshold wind velocity required to move winged propagules along a wooden surface in a wind tunnel has been shown to be approximately linearly related to the square root of their wing loading (as predicted by equation 3.8) in an analysis of mean data from ten species (Greene and Johnson 1997). However the relationship did not hold up well when examined within a species.

3.2.4 Rotating propagules

Some propagules are characterized by having their centre of gravity in a very asymmetric position with respect to the wing surface (e.g. *Acer*

platanoides; *Agathis robusta* (Araucariaceae); Fig. 3.2c). This arrangement causes the propagule to rotate about a horizontal circle: lift is generated by the action of the wing slicing through the air as it spins (reminiscent of a helicopter rotor blade). Some propagules roll about their longest axis (e.g. *Fraxinus excelsior*: Oleaceae): this also produces lift, but the rates of descent are faster than for 'helicopter' propagules (Fig. 3.3b). In both cases, however, relatively large seed masses are able to achieve dispersal away from the parent plant with the aid of the wind. Rates of descent of *Pinus luchuensis*, *Cedrus deodara* (Pinaceae) and *Fraxinus griffithii* have been measured as 0.78, 0.84, and $1.12 \, \mathrm{m \, s^{-1}}$ respectively, while their corresponding rates of spin were 1798, 1012, and 1158 rpm (Minami and Azuma 2003). Tackenberg et al. (2003) give values of terminal velocity for winged propagules ranging from about 0.3 to $2.8 \, \mathrm{m \, s^{-1}}$. Peroni (1994) recorded variation in wing area, mass, and wing loadings among ten populations of *Acer rubrum* propagules. Although differences were not large, they indicated reduced dispersal in late succession populations compared to those in early succession habitats. For samples of *Liriodendron tulipifera* (Magnoliaceae) seeds collected after dispersal, Horn et al. (2001) measured terminal velocities ranging from 0.95 to over $1.6 \, \mathrm{m \, s^{-1}}$.

Matlack (1987) altered wing loading experimentally by either removing tissues from the embryo and/or adding lead shot to the hole thus created: hence propagule shape was maintained while wing loading was altered. It was found that the linear relationship between rate of fall and the square root of wing loading was maintained in both rotating (*Acer platanoides*) and plumed (*Asclepias syriaca*) propagules across 300-fold and 20-fold variation in propagule mass respectively.

3.2.5 Tumbling structures

In some species, the unit blown along by the wind may be an entire (dead) plant or just an inflorescence, as in the spiny balls of spikelets of *Spinifex sericeus* (Poaceae, Fig. 3.2d). When tumble-plants die, they break off at ground level; being roughly spherical in shape, they are readily rolled along by the wind (Fig. 3.2e). Seeds fall off as they travel, resulting in dispersal along a long, narrow band.

On open grasslands or in deserts, where there is little to impede them, they can roll for many kilometres: distances of up to 4.07 kilometres over 6 weeks have been recorded for *Salsola iberica* (Chenopodiaceae: Stallings et al. 1995). Other examples of such species are *Oenothera deltoides* (Onagraceae) in California and the tumbleweeds (e.g. *Kochia scoparia*: Chenopodiaceae) familiar to us from movie westerns, where they are inevitably shown rolling through ghost towns.

3.2.6 Ballistic propagules

Propagules thrown (or squirted) away from their parent have a significant forward momentum and will commonly be travelling faster than the air around them. A dense, oblate spheroid with a smooth surface will have the greatest horizontal travel in non-windy conditions, since these traits will minimize drag. Flat, plumed, or woolly propagules have high drag and would rapidly decrease the speed. A rounded shape (e.g. seeds of *Geranium* spp.) also allows ballistic propagules to roll when they land. The size of ballistic propagules will be related to the maternal investment in the launching mechanism, which will in turn be related to the parent's morphology: in annuals and small shrubs propagule mass ranges from about 1 to 25 mg, whereas in at least one tree (*Hura crepitans*) they may reach 1 g (Table 2.1). In the case of *Arceuthobium* spp., the seeds are oblate spheroids: during flight, they become aligned with their major axis along the flight trajectory (Hinds et al. 1963). Not all ballistically-released propagules are rounded: seeds of *H. crepitans* are discs, while propagules of *Erodium* spp. (Geraniaceae) have part of the capsule still attached (as an aid to seed burial when on the ground: see p. 48). *Erodium* seeds are sharply pointed: although the appendage increases drag, it will also act as a rudder during flight, keeping the point of the seed aligned with the direction of travel. In the Fabaceae, propagules of many species are released forcefully, but they are also adapted to secondary dispersal by ants by having a lipid appendage (*elaiosome*) on the end of the seed. The appendage has only a slight effect on the surface area and its effect on the aerodynamics of the seed will probably be slight.

3.2.7 Un-adapted propagules

The force required for liberating propagules will often be provided by the wind, either directly or indirectly via swaying foliage, even in propagules with no obvious adaptations for wind dispersal. Once they begin their descent, it is highly likely that the wind will still be blowing and that it will exert at least some horizontal force on 'un-adapted' propagules, especially if they are falling from a tall tree (Higgins et al. 2003b). It is also possible that some un-adapted propagules will produce lift. For example, some florets or spikelets of grasses (e.g. *Bromus inermis*: Poaceae) are flat. Rabinowitz and Rapp (1981) noted that grass seeds dropped in still air do not all land directly beneath the point of release (though it is almost impossible to be certain that absolutely no sideways force is applied when releasing an object). Whether the resulting sideways movement is of ecological significance is unclear: most such propagules will still probably land beneath the parental canopy. Some, perhaps those produced at the very edge of the canopy, may be moved just enough to escape the zone of influence of the parent. Their surface area and mass will, necessarily, determine the trajectory that each propagule follows and, in theory, this could be predicted.

3.2.8 Comparisons of aerodynamic propagule types

How do the terminal velocities of different types of propagule compare? Fig. 3.6 shows that there is considerable overlap in values for the different propagule types in temperate European species. However, plumed propagules are consistently among the slowest in their descent, with most species having terminal velocities below $1\,\mathrm{m\,s^{-1}}$. Winged propagules have values spread over an order of magnitude, but with almost all having a terminal velocity greater than $0.5\,\mathrm{m\,s^{-1}}$. Rates of descent of propagules in the tropical forest of Barro Colorado Island, Panama also show considerable overlap between morphological types (Fig. 3.3b). In this study, Augspurger (1986) used slightly different terminology to Burrows (1973), classifying the species as floaters (various shapes generating little sideways motion), undulators (plane-winged; gliding and undulating in flight), autogyros (a single blade, spinning around the seed), rolling autogyros (plane-winged, rotating about their longest axis), helicopters (spinning propagules with multiple blades), and tumblers (with wings causing them to tumble as they fall). (Analogies for dispersal adaptations are often drawn with human inventions, though the reality may be that the 'inventions' were originally copied from nature!). For each propagule type, their rate of fall in still air was linearly related to the square root of wing loading. For a given wing loading, autogyros tended to have the slowest rate of descent, while floaters (mostly from the Bombacaceae) were the fastest (Fig. 3.3b). On theoretical grounds, Minami and Azuma (2003) have argued that the exponent of wing loading relationships should be 3/4 for pappose propagules and 2/3 for winged propagules. Based on a selection of species from a range of geographic origins, they also concluded that parachuting propagules have a slower rate of descent than gliding, rocking, and spinning propagules.

3.3 Traits enabling flotation

The ability to float is common among plant propagules, even in species that we would not readily identify as having morphological adaptations enhancing dispersal by water. For many species, as their propagules mature on aerial stems their tissues dry and their density becomes less than that of water: as we will show, this means that most will float, at least initially. Forty percent of fruits and seeds in the British flora float for more than a week, 25% for more than a month and 15% for over 6 months (for *all* species, not just those adapted to aquatic environments: H. B. Guppy, quoted by Ridley 1930). Most freshwater aquatic or riparian species hold their mature fruits above the water and when they fall from the parent plant, they float. Flotation may be aided by the presence of air chambers within a propagule: for example, in *Swartzia polyphylla* (Fabaceae) there is an air pocket between the cotyledons (Williamson et al. 1999), while for *Cocos nucifera* there are air spaces between the fibres of the husk as well as in the central chamber. Other adaptations for floating include a spongy mesocarp and a cork-like pericarp (Ridley 1930; van der Pijl

1982). Many seeds also contain oils and lipids, whose specific density is less than water. These serve both to increase flotation and to prevent waterlogging, which would otherwise decrease flotation over time. Propagules may float only temporarily or for a considerable period of time: they will drift along with any water currents until they are washed up on land or they take on water and sink.

The forces acting on propagules submerged in water are the same as those acting in air, but the buoyancy force (Fig. 3.1) will now be significant. This force (F_N), which will act in the opposite direction to gravity, is equal to the weight of fluid displaced (Archimedes' Principle), given by

$$F_N = k\rho_{fluid}v_{prop}g \tag{3.1}$$

where ρ_{fluid} is the density of the fluid, v_{prop} is the volume of the propagule, and k is the proportion that is submerged. The gravitational force can be expressed in similar terms

$$mg = \rho_{prop}v_{prop}g \tag{3.2}$$

From these equations, it is easy to show that the ability of a propagule to float is related to the relative magnitudes of the densities of propagule and water. Consider a propagule that is released just below the surface of the water, so that initially $k = 1$. If the density of the propagule is greater than that of the water ($\rho_{prop}/\rho_{fluid} > 1$), $mg > F_N$, and it will sink (and drag will result as it does so). If the density of the propagule is less than that of water ($\rho_{prop}/\rho_{fluid} < 1$), $F_N > mg$ and the top of the propagule will rise above the surface until the two forces balance, at which point $k = \rho_{prop}/\rho_{fluid}$. This ratio, the density of an object relative to water, is an index of flotation ability and is termed its *specific gravity* (or *specific weight*). The density of water will vary, depending on whether it is fresh, brackish or salty, and this must be taken into account by using the appropriate density of the water in any calculations.

Specific gravities are usually presented as species averages (for smaller seeds it is easier to measure volume of many seeds together than to do so for individual seeds). For newly matured propagules of tree species in Amazonian flood plains, for example, Kubitzki and Ziburski (1994) recorded 77% of species as having specific gravities less than 1.0

(either measured directly or by observing that they floated when placed in water). The lowest specific gravity was 0.67 for *Aldina latifolia* (Fabaceae), while the largest value was 1.17 for *Astrocaryum jauary* (Arecaceae).

While the specific gravity of newly dispersed propagules determines whether they will float initially, this does not necessarily correlate with the length of time that they will stay afloat. For 6 tree species from a seasonally flooded forest in Panama, Lopez (2001) recorded specific gravities of 0.78 to 0.98, while for 6 species from a drier community the range was 0.20 to 1.62. There was no significant correlation between the specific gravity of fresh seeds and the percentage still floating after 15 days. The only way to assess the ability of propagules to stay afloat for long periods, and thus have the capacity to disperse over long distances, is direct observation: by putting them into water and seeing how long they float.

While extreme flotation times are often quoted, such as several years for *Sophora microphylla* (Fabaceae) (Murray 1986), these may not adequately represent the floating capacity of a population as a whole. Only a proportion of seeds of a given species may float at the outset and, of these, individuals progressively sink over time; some may even re-float later on. In a study of the Amazonian floodplain species *Swartzia polyphylla*, specific gravities of seeds ranged from 0.95 to 1.09 (Williamson et al. 1999). For a sample of 27 seeds, 45% of them never floated, 22% sank but later floated, while the remainder always floated over the duration of the study. In the Panamanian study by Lopez (2001), the proportions of seeds initially floating for different tree species ranged from 0 to 100%. Regardless of the initial proportion floating, the number still afloat declined rapidly for all six species from a non-flooding habitat, but declined much more slowly in species from a seasonally flooded forest (Fig. 3.7). Hence, rather than categorizing species as either floaters or sinkers, it is better to quantify their floating duration as a mean, median, or some other percentile (ter Steege 1994, refers to the median floating time as the 'floating half-life').

Median floating times have been used by Boedeltje et al. (2003) to assess the potential for species to be dispersed by water in the Netherlands. Emergent aquatic species had median floating times that were considerably longer, on average, than semi-aquatic, terrestrial, and submerged aquatic species. However, all categories had a high frequency of species with median flotation times of less than 20 days (Fig. 3.8). A similar study in Sweden defined long-floating seeds as those having a maximum floating time of more than two days (Nilsson et al. 2002).There were more species with long-floating seeds near or in still or slow-running water than along fast-running water courses. This does not mean that rivers are less important for seed dispersal: because of turbulence, rivers will disperse seeds regardless of whether they float or sink, whereas to be dispersed by slow-moving water a propagule needs to float for a considerable length of time. Turbulence, particularly that following heavy rainfall, can erode riverbanks and move sediments containing seeds. Whole rafts of floating vegetation may be seen travelling down tropical rivers in flood. Kelley and Bruns (1975) collected seeds of 84 species in samples of irrigation water, many of which would not have been classified as being adapted for water dispersal. Propagules also do not have to be in or near waterways to be dispersed by water. Newly shed seeds (as well as dormant seeds from the seed bank) can be washed along in surface flow of water across the ground after heavy rainfall. For example, unidirectional movement of *Helianthus annuus* (Asteraceae) seeds by 80 m down a slope (gradient 1.3 m in 100 m) has been recorded in a maize field (Burton 2000): the most plausible explanation was due to flow of excess rainwater along furrows.

So far we have considered dispersal of sexual propagules by water. Clonal propagation, however, is also important in many aquatic species. Separation into floating units of one or more ramets allows fully functioning plants to colonize new areas: this has been referred to as 'falling apart as a lifestyle' (Room 1983). The traits that allow plants to float on the surface of water also enable them to disperse effectively. For example, the highly successful invader, *Eichhornia crassipes* (Pontederiaceae), has air-filled inflated petioles that enable the plants to float. Although it was originally introduced outside its native range by humans as an ornamental species, *E. crassipes* has spread within tropical and sub-tropical countries, almost exclusively by water-borne dispersal

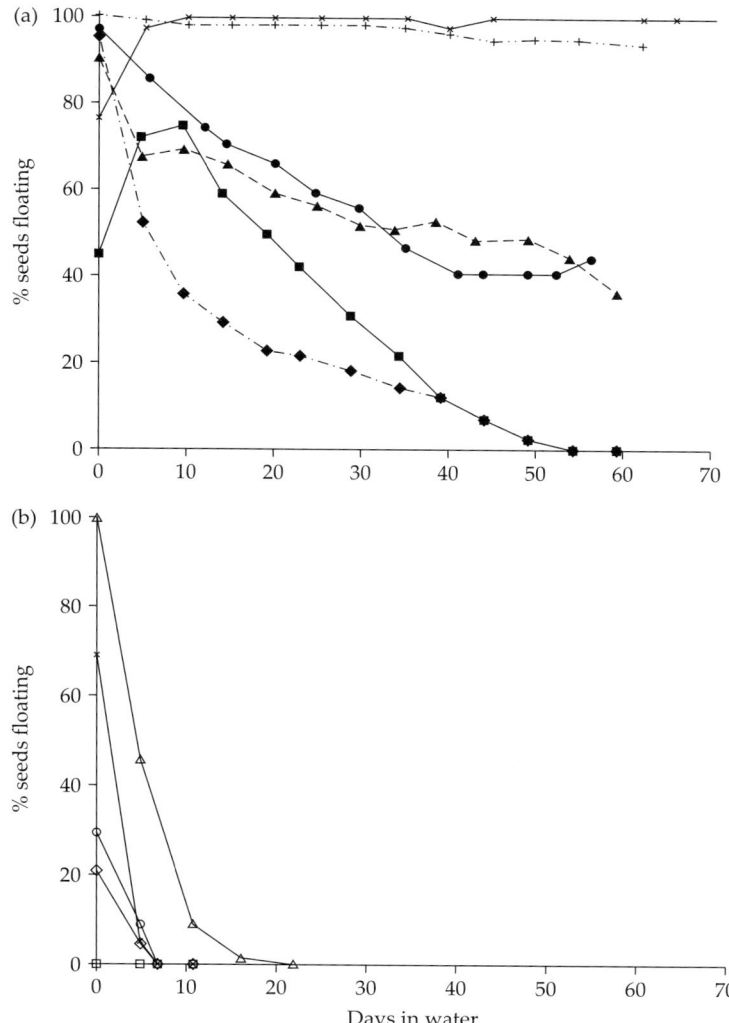

Figure 3.7 Duration of floating in water for tropical tree propagules from (a) a seasonally flooded forest and (b) a non-flooding area ('*terra firme*') in Panama (redrawn from Lopez 2001, reproduced courtesy of Blackwell Publishing). Species in (a) are: + *Pterocarpus officinalis* (Fabaceae); × *Pachira aquatica* (Bombacaceae); ● *Carapa guianensis* (Meliaceae); ▲ *Prioria copaifera* (Fabaceae); ■ *Pentaclethra macroloba* (Mimosaceae); ◆ *Pterocarpus* sp. Species in (b) are: × *Virola surinamensis*; △ *Tabebuia rosea* (Bignoniaceae); ○ *Anacardium excelsum* (Anacardiaceae); □ *Gustavia superba* (Lecythidaceae) and *Dipteryx panamensis* (Fabaceae); ◇ *Calophyllum longifolium* (Clusiaceae).

of its ramets (Ridley 1930); the same is true for *Salvinia molesta* (Salviniaceae), a sterile pentaploid (Room 1983). Sampling of the surface water in a waterway in the Netherlands found that 90% of propagules were vegetative units and only 10% were seeds (Boedeltje et al. 2004). Dispersal of vegetative units by water is not restricted to floating plant

modules: sea ice in sheltered bays in eastern Canada can lift and relocate entire blocks of soil from salt marshes, containing rhizomes of *Spartina alterniflora* (Poaceae).

As we have discussed above, at some point (if not immediately) most propagules floating on or submerged in water will eventually sink. The movement

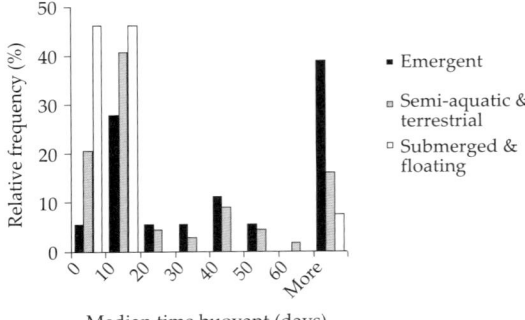

Figure 3.8 Relative frequencies of the median durations of flotation for species along a waterway in The Netherlands (based on data from Boedeltje et al. 2004).

of these sinking seeds can be explored through the same fluid dynamics equations that we used for dispersal in air (p. 27): only the parameters differ. We can therefore also use the equations to determine traits that affect their rate of descent through the water and hence the distance that they are dispersed by horizontal water movement (though the distances thus moved will be very much less than for floating propagules). For propagules sinking in water, the Reynolds number can still be relatively high, since both density and viscosity are greater for water than for air (the ratio of density to viscosity at 20°C is 17 times higher in water than in air). The terminal velocity of an object sinking in water, however, will be very much less than in air. This is because of both the greater buoyancy force and the greater drag (equation 3.2 includes the density of the medium, which for water is three orders of magnitude greater than air). Published values of terminal velocity of seeds in water are uncommon. For the seeds of two seagrasses, *Enhalus acoroides* (Hydrocharitaceae) and *Thalasia hemprichii* (Hydrocharitaceae), terminal velocities have been measured at $0.1 \, \mathrm{m \, s^{-1}}$ in seawater (Lacap *et al.* 2002), similar to the lowest rates of descent for plumed propagules in air and despite weighing 249 and 102 mg respectively (a plumed seed with a $0.1 \, \mathrm{m \, s^{-1}}$ terminal velocity in air would weigh considerably less than 1 mg). The terminal velocity of the seeds of *Zostera marina* (Zosteraceae) has been measured at $0.06 \, \mathrm{m \, s^{-1}}$ (Orth et al. 1994); the 'erosion threshold' water velocity required in the

laboratory to cause seeds to move along a sandpaper substrate was $0.007 \, \mathrm{m \, s^{-1}}$.

We end this section with a note of caution. It is easy to leap to the conclusion that since aquatic plants are intimately associated with water, that water will also be their main dispersal vector and that they will have evolved propagules suited to this mechanism. However, this is very often not the case, and aquatic plants are commonly dispersed by other means. Indeed, without alternative methods of dispersal riparian plants would not be able to spread upstream from their maternal parent. If a plant is growing in or near a very still lake, it can be argued that much greater dispersal distances will be achieved by vectors other than by water. Small propagules, for example, may be transported on the feet and plumage of water fowl (Darwin 1859). Some species strongly associated with water, such as *Populus* spp. (Salicaceae), *Salix* spp. and *Typha* spp. (Typhaceae), have plumed propagules and can disperse well through the air. These structures aiding dispersal by wind have the added advantage that they also enhance flotation on water: Geertsema (2002) found that about 50% of *Linaria vulgaris* (Scrophulariaceae), a wind-dispersed species, floated for a month, while a similar proportion of two other wind-adapted species floated for ten days. For very fluffy propagules, surface tension may, at least initially, keep most of their surface area above the water. It is, perhaps, more appropriate to consider such propagules as dispersing in the air but subject to considerable friction with the water.

3.4 Traits enabling dispersal by animals

We can predict the effects of wind and water on propagules with considerable confidence, since they vary predictably in their characteristics. They obey physical laws and, as we have seen, conform to particular equations. In contrast, animal vectors display a vast range of morphologies, physiologies, and behaviours. They range in size from ants to elephants; they inhabit aquatic, terrestrial, and aerial environments; they move by swimming, walking, climbing, or flying; and they differ in their preferences for different food sources. Although we are steadily learning to understand the ways that they disperse plant propagules, through observation and experimentation, the generalizations that we can make are based on statistical correlations and associations, perhaps with attendant qualifiers such as 'often' and 'sometimes'.

There are three main ways by which animals move propagules away from the parent plant:

- immediate ingestion at the source and subsequent transport via the gut;
- deliberate removal to another site, with the intention of ingesting of all or part of the propagule later;
- inadvertent removal on the outside of their bodies.

In the following sections, we will discuss plant morphological traits that enhance dispersal by each of these methods.

3.4.1 Traits encouraging immediate ingestion

Many plant species, in many families, provide animals with a source of food: by ingesting entire propagules, or parts of propagules containing at least one seed, animals inadvertently disperse the seeds via their digestive systems. For this strategy to be successful for dispersal, a proportion of the seeds must still be viable after defaecation or regurgitation. The risk of attracting animals is that they may not all be potential dispersers: some may be efficient consumers of seeds, destroying them as they bite or chew or as they pass through the gut. However, since the strategy of food provision is so common, enough seeds must survive in the long run.

All ripened ovaries are technically fruits. The fleshy 'fruits' that we regard as having evolved to assist dispersal by animals consist of any structure surrounding the seeds or, in some species (e.g. *Fragaria* spp., Rosaceae), a structure on which the seeds are positioned. The flesh can be formed from a range of tissues: the ovary wall, the receptacle, extensions of the funicle or seed coat (in the case of an aril), or trichomes within the ovary (as in *Citrus* spp.: Rutaceae). In some ecosystems, a large proportion of plant species produce fleshy fruits. For example, it has been estimated that 51–93% of canopy trees and 77–98% of sub-canopy trees and shrubs in neotropical forests have fleshy fruits adapted for animal dispersal (Howe and Smallwood 1982). Fifty-two

percent of the southern African tree flora produces drupes and berries, of which almost half are dispersed primarily by birds (Knight and Siegfried 1983), while about one third of the tree species in a Cameroon forest benefit from dispersal of fruits by primates (Poulsen et al. 2001).

The largest component of the flesh is usually water. In a Mediterranean community, water contents of pulp ranged from 19 to 91% across species, with an average of 71% (Herrera 1987). In arid ecosystems, the water content itself may be an important aspect of the reward to animals (Wolf et al. 2002). Other constituents, providing a nutritive reward, may be sugars (and other non-structural carbohydrates), lipids, oils, and proteins. Carbohydrates usually contribute a high percentage of the non-water components. It is difficult to generalize on their levels, because of the range of analytical methods used (Corlett 1996), but the concentration of soluble carbohydrates commonly exceeds 50% of the total dry weight. Fruits usually contain glucose, sucrose, and (with a few exceptions) fructose, their proportions varying among species (Ko et al. 1998). Where the flesh is derived from an aril, the lipid content may be as much as 85% of the dry weight of the fruit tissue (e.g. Corlett 1996). Protein levels are usually below 10% (Debussche et al. 1987), but in rare instances can reach almost 30% (Herrera 1987). There is usually a broad negative correlation between lipid and carbohydrate content (Herrera 1987), while in some communities fruits produced in certain seasons may be higher in lipids and proteins than in others (Stiles 1980). In southern France, a study by Debussche et al. (1987) found that winter fruiting plants had lower carbohydrate and higher lipid contents than summer fruiting species, while summer fruiting species had higher protein contents than those fruiting in autumn and winter. Herbaceous plants had more watery fruits and higher protein and carbohydrate levels than woody plants. There can also be considerable variation in fruit quality among plants within a population: for example, Howe (1986) reported a two-fold range of lipid and protein levels among different trees of *Virola surinamensis* (Myristicaceae).

Fruit size is an important trait for dispersal by animals, since there is a physical maximum limit to the size of an object that can be fitted into the mouth. If a fruit is smaller than the gape width of the animal, it can be swallowed intact: the flesh is then removed and digested as fruits pass through the gut. For example, in a Mediterranean scrubland community, most seeds smaller than the gape width of birds were removed, whereas less than 20% of larger fruits were taken (Herrera 1984). If the fruit is too large, then the flesh may be removed and the seeds dropped straight to the ground without ingestion. However, if fruits are too small they may be ignored as the investment in harvesting may not be worthwhile. In species with compound fruits or single fruits containing multiple seeds, a fruit can be eaten either whole by larger animals or it can be broken into smaller portions by smaller animals. Hence, it does not necessarily follow that if a fruit that is too large to be swallowed whole it will not have its seeds dispersed via the gut.

There is a large range of fruit sizes within most communities, with smaller sizes contributing the greatest proportions (Fig. 3.9). The fruit size distributions within a regional flora have been published on several occasions. In Herrera's (1987) Mediterranean study, for example, fruit widths of species range from 4 mm to 31 mm, with a median of 7.9 mm. For a Peruvian tropical forest, fruits had a larger maximum: several species had a diameter of over 50 mm, while the median was 14 mm (Janson 1983). Different sizes of animals are able to ingest fruits of different suites of plant species within a plant

community. In southern Africa, the fruits eaten by mammals have a mean size almost three times as great as fruits eaten by birds (Knight and Siegfried 1983). Wheelwright (1985) also found that plant species with large seeds or fruits in montane forests in Costa Rica have a reduced number of bird species that attempt to ingest their propagules.

Within some communities, there are fruits that are too large even for the biggest frugivores to ingest and it has been proposed that these species may have co-evolved with frugivores that are no longer present. For example, in South America most of the megafauna has become extinct since the Pleistocene (Janzen and Martin 1982; Hallwachs 1986): these extinct species may have been the primary dispersers of the large fruits. The consequences for a plant community of a reduction in abundance of animal species or their complete disappearance will be considerable: some plants will no longer be dispersed effectively (McConkey and Drake 2002). Over the course of generations, this will result in some plants becoming restricted to isolated clumps and they may decrease significantly in overall abundance.

While fruit size varies between species, it also varies within and between populations of a single species and this may govern which fruits are removed from a plant. In Wheelright's (1985) study, even amongst those birds targeting a particular plant species, bulky fruits were often simply plucked from the trees and dropped (or after some attempt at ingestion) directly to the ground. For *Prunus mahaleb* (Rosaceae), the average mass per fruit remaining at the end of the season was smaller than when recorded earlier in the season, suggesting that larger seeds are taken preferentially (Jordano 1995). Small fruits of *Santalum album* (Santalaceae) in India are preferred by birds (Hegde et al. 1991). The smaller fruits tend to be passed in faeces, whereas the larger ones tend to be regurgitated. In contrast, Sallabanks (1993) found that American Robins preferentially ingested larger seeds of *Crataegus monogyna* (Rosaceae) and artificial fruits. Larger numbers of smaller fruits of *Prunus virginiana* (Rosaceae) were produced in riparian habitats than on drier slopes: birds preferentially consumed the smaller fruits (Parciak 2002). In Hegde's et al.'s (1991) study of *Santalum*, individual seeds from some trees were less than half the mass of those on other trees.

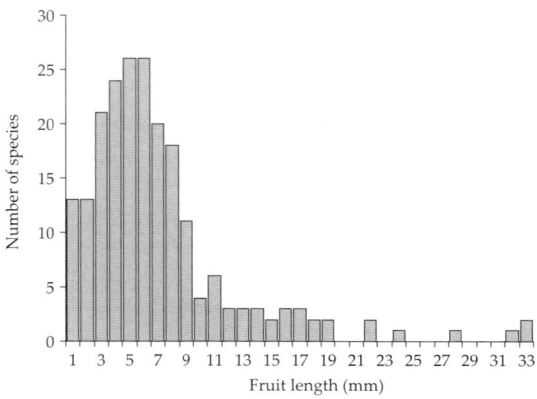

Figure 3.9 Frequency distribution of lengths of fleshy fruits within the flora of New Zealand (redrawn from Lord et al. 2002, reproduced courtesy of CABI).

Other fruit attributes that can be easily quantified are the number of seeds per fruit, the quantity of flesh, and the ratio of flesh to seed. The tightness of the bond between seeds and flesh and the toughness of any skin can also affect feeding behaviour (Gautier-Hion et al. 1985), but these are difficult to quantify. While there may be many species within a flora whose fruits contain only a single seed, the maximum number of seeds may be very large. In the tropical species *Cecropia obtusifolia* (Moraceae), for example, a maximum of 2792 seeds has been recorded in a single fruit (Estrada and Coates-Estrada 1986). In the Iberian Peninsula over 40% of the species have a single seed per fruit, a further 35% have between 2 and 5 seeds, while just a few have over 100 seeds (Herrera 1987). Pulp mass generally increases with fruit size, but small propagules have proportionally more pulp (in terms of dry mass). When adjusted for seed mass, fruits of trees and vines tend to have more pulp per seed than shrubs or herbs. Once again, there will be genetic and environmental variation in these traits: *Frangula alnus* (Rhamnaceae) fruits from southern Iberia, for example, are larger and contain more, larger seeds than fruits from a central European population, resulting in a lower pulp:seed ratio (Hampe and Bairlein 2000).

Merely containing a reward and being the right size may still not be sufficient to ensure a high level of dispersal. Colour and smell often guide animals to ripe fruits. Unripe fruits are commonly green and do not stand out from the surrounding foliage. In many species they change colour as they ripen, with the result that seeds are seen by potential dispersers and are ingested at the right time for maximum seed viability. Ripe tropical fruits may be red, black, orange, yellow, brown, white, blue, purple, or mixtures of colours (Janson 1983). Red and black are particularly common fruit colours in most floras. However, some experimental studies have found little specific preference by frugivores for these colours. It would appear that it is the level of contrast with the background colour that is important to species locating food by sight, rather than the colour per se (Schaefer and Schmidt 2004). However, not all fruits are of highly contrasting colours and some fruits even remain green at maturity. Strong smell will attract animals to fruits that are widely spaced

within a forest, without the need for colour signals. Perhaps the best examples of this are the durians of south-east Asia (*Durio* spp.: Bombacaceae), which are renowned for the overpowering smell they emit when ripe.

While smell and colour may attract animals, poisonous compounds may act as a deterrent. Some fruits contain toxic levels of alkaloids, glycosides, or other compounds. Clearly, deterring all animals would be counter-productive for a plant: this strategy for dispersal relies on the fact that some animals are immune to the chemical (either pre-adapted or as a result of co-evolution), so that only these species disperse the seeds. For example, in a Mediterranean community, Debussche and Isenmann (1989) found that mammals tended not to eat species with high (greater than 5%) protein fruits and noted that six out of seven of the plant species avoided contained toxins. These fruits were, however, eaten readily by birds. It has been suggested that there is an inverse relationship between nutrient content of fruits and the concentration of secondary compounds, resulting in the more rapid removal of nutritious fruits (Cipollini and Levey 1997a, b); supporting evidence for this has been obtained from plant communities in different parts of the world (Schaefer et al. 2003; Tang et al. 2005).

The proportions of seeds of a given plant species surviving ingestion/defaecation will vary among animal species. For example, Murray (1988) found that seeds of three plant species passed with 100% success through three bird species, but were completely destroyed on passage through a fourth species. A hard seed coat will protect seeds passing through birds that have a gizzard, but a hard coat is not necessary to pass intact through some bird species. Most grazing animals defaecate viable seeds of at least some plants (Salisbury 1961), but the proportion that remains viable after passage through the gut varies from zero up to at least 90%. For example, smaller seeds and those with an impermeable seed coat pass more successfully through horses than those with larger or softer seeds (St. John-Sweeting and Morris 1990).

It should be noted that it is not just fleshy fruits that are dispersed via ingestion. Many seeds, particularly in grasslands, may be swallowed by animals that are browsing vegetation in general, not just reproductive tissues. Janzen (1984) suggested that for large herbivores the foliage is functionally equivalent to the flesh of fruits, in that it attracts animals and leads them to ingest seeds. Any seeds that are able to survive this ingestion will be dispersed when the animals defaecate: seedlings can be seen sprouting from the dung of most herbivores, indicating that a proportion of seeds ingested in an indiscriminate way survive passage through the gut (Salisbury 1961). For seeds to retain their viability after being grazed by an animal, they must escape mastication and then be able to withstand the digestive enzymes of the gut for the period of time taken for them to pass through the animal (the maximum passage time may be up to a month: see p. 61). Traits likely to enhance seed survival under grazing pressure are small size and a hard, impermeable seed coat. If a high proportion of the seeds in a grassland is ingested, we would expect the evolution of such traits in response to grazing animals (Janzen 1984). It is clear that many plant species of grasslands do, indeed, have one or other of these traits (e.g. Quinn et al. 1994). Gardener et al. (1993) found that the rate of passage through cattle by ten legume species was correlated with the specific gravity of seeds, the proportion of 'hard' (impermeable) seeds and seed size (Gardener et al. 1993). However, in other studies such correlations are weak or non-existent (Bruun and Poschlod 2006).

As we have seen, there is considerable variation in fruit traits among plant species: size, water content, sugar, protein and other chemical components, seed number, texture, smell, colour, and so on. Can we identify traits or groups of traits that attract different animal groups? There are, in fact, few instances where a single plant species attracts a single animal species: most animals are generalist feeders to some extent, while most species of plant attract more than one family of animals. Despite this, it is possible to identify general dietary preferences and thus to relate attributes of fruits to the types of disperser that *tend* to be attracted to them. These dispersal traits are referred to as syndromes (the term *syndrome* is also applied to characters related to dispersal by wind or water). Within a tropical African rainforest, for example, Gautier-Hion et al. (1985) identified seven fruit syndromes using multivariate analysis (Fig. 3.10).

It may be difficult to establish animal dispersal syndromes by examining a single floral community, because of small sample sizes of some animal types (such as megafauna) or plant phylogenies. A broad literature review of many communities, or a more formal metadata analysis, may therefore be applied (though reviews are often biased by the over-representation of tropical studies). For example, Howe (1986) suggested the following syndromes based on a review of the literature:

- Bird fruits.
  - Obligate frugivores: large arillate seeds or drupes; seeds larger than 10 mm diameter; scentless; rich in lipids or protein; black, blue, green, purple, or red.
  - Opportunistic frugivores: small or medium arillate seeds, berries, or drupes; seeds less than 10 mm; scentless; rich in lipids, protein, sugar, or starch.
- Mammal fruits.
  - Arboreal frugivores: large arillate or compound fruits; aromatic; rich in protein, sugar or starch; brown, green, orange, yellow, or white.
  - Aerial frugivores (bats): Odourless or musky; often rich in starch or lipids; green, yellow, white or whitish; often pendant.
- Megafaunal fruits: large, tough and indehiscent; often oily; fibrous pulp; dull colours; resistant seeds.

Rather than focusing on an entire flora, information on syndromes can sometimes be obtained by examining variation among fruits within a single genus or family. Because they share a phylogenetic background, the variation among such species is limited to fewer traits and relationships with disperser animals may thus be easier to detect. For example, there is considerable variation among the elaiosomes, lipid-rich appendages, within the genus *Acacia* in Australia. Bright colouration, bicoloured display, large aril mass, and high lipid content have been found to be typical of *Acacia* species dispersed by birds, while the opposite attributes can be found in ant-dispersed species (O'Dowd and Gill 1986). The mass of the lipid body in bird-dispersed species is on average three times as large as in ant-dispersed acacia species of a similar seed mass.

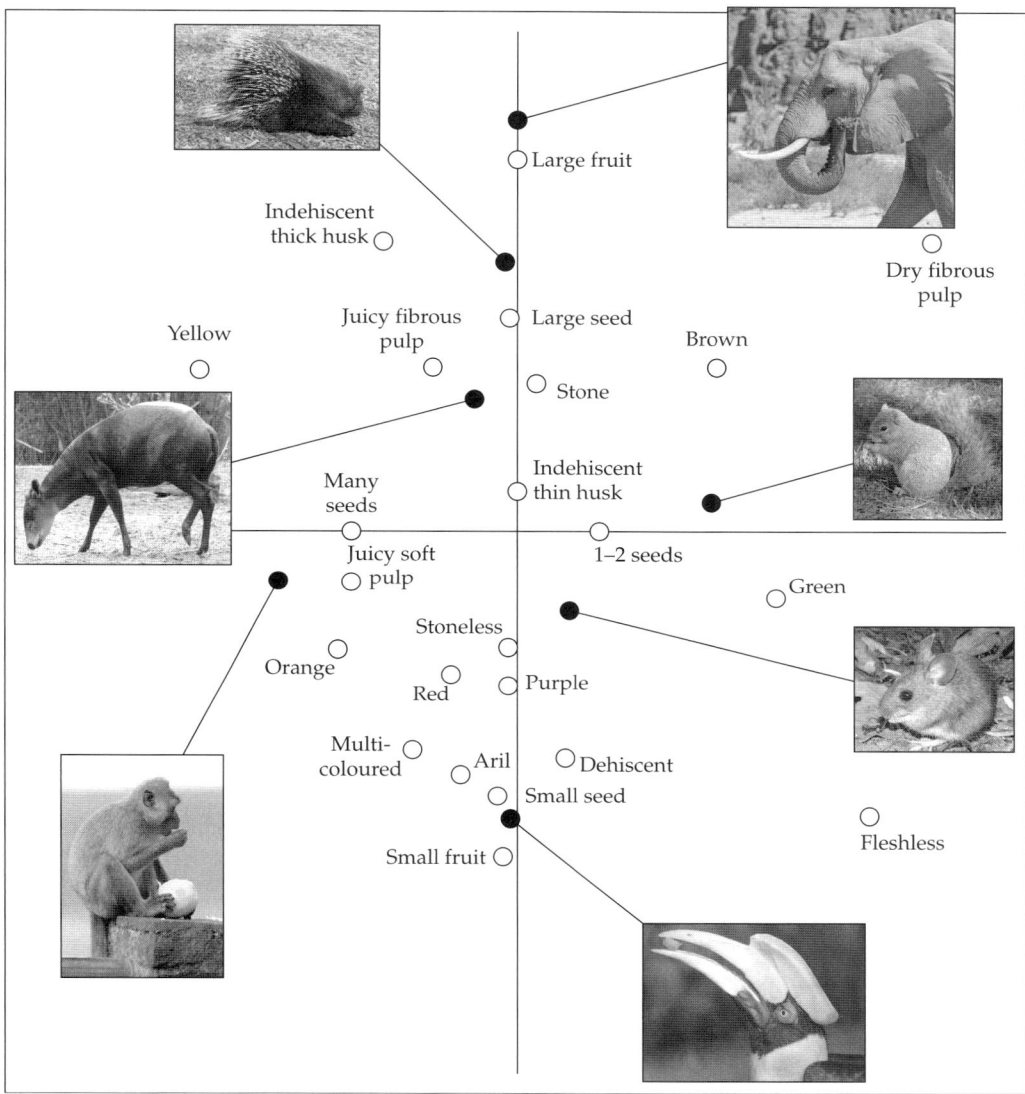

Figure 3.10 Ordination showing fruit traits and animal disperser associations for a tropical forest in Gabon (adapted from Gautier-Hion et al. 1985).

Natural polymorphisms, that is, variation within a plant species, may also be used to indicate animal preferences for selection/ingestion of fruits. If plants with different fruit types (such as colour) occur side-by-side and a greater proportion of one type is eaten consistently, it is reasonable to assume that animals have a preference for that trait (e.g. Whitney 2005). However, it is still difficult to express this preference quantitatively: removal rates may depend on confounding variables, such as relative abundance of the morphs, population sizes (both plant and animal), and local community composition. Whitney and Lister (2005) found clinal variation in the frequencies of plants of *Acacia ligulata* with red, orange, and yellow elaiosomes along a gradient of temperature and rainfall. These colour-morphs did not differ in seed size, aril mass, fatty acids, or flavonoids. Populations were dominated by the red morph (as might be expected for bird-dispersed species), seed predatory birds removed seeds of yellow and orange morphs more often than would be expected at random, while the preferences of seed-dispersing bird varied between populations and years. There may be correlation between fruit traits that are not immediately obvious, such as colour and chemical composition. Choice tests to determine animal preferences are often conducted under laboratory or cage conditions, thus controlling many of the confounding variables but having the disadvantage of being highly artificial. For example, Willson (1994) demonstrated preferences of American Robins for particular colours of *Rubus spectabilis* and *Sambucus racemosa* (Sambucaceae), though there was variation among individual birds. Choice tests have been used to compare selection of fruit size, species, and ripeness (e.g. Moermond and Denslow 1983); artificial fruits have been used to compare bird responses to size, colour, and content of both desirable and toxic compounds (Willson 1994; Levey and Cipollini 1998; Schmidt et al. 2004). Despite all these studies, there are probably few generalizations that can be made, other than that animals are often selective, their preferences can vary considerably between individual animals, and that preferences may be expressed for some fruit traits and not others.

3.4.2 Traits encouraging deliberate removal

The plant traits encouraging deliberate fruit removal are mostly the same as those discussed in the previous section. However, the propagules must also be large enough not to be consumed more easily at the source. In this section, we will consider propagule traits related to two particular types of animal take-away: colonial animals, where scavenger individuals remove propagules to a communal nest where feeding occurs (e.g. ants); and animals which store items elsewhere for consumption at a later date (some rodents and birds).

Many species in a wide range of families, but especially the Fabaceae, have a lipid-rich body (the elaiosome) attractive to ants attached to one end of their seeds. Ants transport the seeds to their nest,

where they eat the elaiosome and discard the seed, either within their nest or around its entrance. It has been shown experimentally that seeds without an elaiosome are taken less often to the nest, while those with larger elaiosomes (within the same species as well as among species and in artificial propagules) are preferred (Hughes and Westoby 1992; Mark and Olesen 1996). Amongst the Australian Acacias harvested by ants, mean seed mass ranges up to about 60 mg, while the eliaosome may be as large as 4 mg and contains an average of 36.5% lipid (O'Dowd and Gill 1986). Within a species, the relative sizes of the eliaosomes can also vary considerably: for example, in *Hepatica nobilis* (Ranunculaceae) the width of the eliaosome has been found to range from 0.7 to 1.7 mm while the achene width varies from 1.4 to 2.1 mm (Mark and Olesen 1996). Species also vary considerably in the lipid and protein concentrations of the eliaosome (Lanza et al. 1992; O'Dowd and Gill 1986). Elaiosomes may also contain diglycerides which stimulate carrying behaviour in ants (Skidmore and Heithaus 1988).

In nuts, the food sought by the animal is the seed. There is a thus a real danger that the dispersal event will in fact be fatal: a consumption event of no benefit to the plant. The success of nuts as dispersal units relies on the inability of animals to consume the entire crop. Some nuts may be removed by animals directly from the tree, while many drop to the ground, and are found and removed from there. At peak periods of production, excess nuts may be taken away to be stored (cached) for later consumption, either singly (by 'scatter-hoarding' animals) or in large numbers (by 'larder-hoarders'). The benefit to the plant from caching occurs because not all seeds are recovered: either more seeds are buried than are later needed, or there is inefficient recovery. If the cache is in the ground (often not the case for larder-hoards), non-recovered seeds achieve both dispersal and burial, so that they can later germinate in situ. It has been suggested that 'mast' seeding, where there is extraordinarily high production in just some years, may be an adaptation to ensure that consumers are overwhelmed by seeds, so that far more will be overlooked or buried than will later be needed (e.g. Boucher 1981). For example, in one study only 3% of beech seeds (*Fagus sylvatica*: Fagaceae) were consumed in a mast year, compared to 50% or greater in years of low seed production (Wauters and Lens 1995, analysed by Steele et al. 2005).

Nut size is an important trait: if a nut is too large, it will not fit into the beak, mouth, or cheek pouches (or fewer will fit in at the same time). There has been less research on the importance of size in nuts than in fleshy fruits. In *Astrocaryum mexicanum* (Arecaceae) nut size varies considerably between trees (Brewer 2001). Larger nuts are removed at a greater frequency than smaller nuts, but both sizes are taken to burrows or buried: slightly greater proportions of small seeds survive.

A hard or leathery shell makes it both difficult for animals to consume the kernel rapidly, and aids the longevity of the nuts in hoards. If nuts are easy to extract, animals are more likely to consume them immediately, whereas a longer processing time will make it more likely that a nut is hoarded. This has been tested by offering grey squirrels hazelnuts (*Corylus spp*: Betulaceae) with shell either left intact or removed (Jacobs 1992). Laboratory preference studies have shown that whole acorns (*Quercus* spp.: Fagaceae) are preferred to whole black walnuts (*Juglans nigra*: Juglandaceae) or hickory nuts (*Carya ovata*: Juglandaceae) by squirrels, even though they prefer to eat walnut and hickory kernels (Smith and Follmer 1972). Although acorns have a lower calorific value than the other species, squirrels obtain more energy per unit time from them, since the shells are thin and easy to penetrate. Both acorns and walnuts are cached.

There is considerable variation in nut content between species. Among northern hemisphere nuts, lipid content has been found to range from 2 to 74% of the kernel dry weight, while protein varies from 4 to 33% (Vander Wall 2001). As in the case of fruits, there is a negative correlation between lipid and carbohydrate content. Many nuts, such as acorns, are no more nutritious than seeds dispersed by non-animal means and merely constitute a large package of food (Vander Wall 2001). Tannin levels may be high in some species, for example constituting 6–11% in black oaks (*Quercus* spp.). In some species, essential oil content may provide olfactory cues as to fruit location and ripeness (Mikich et al. 2003).

Dormancy may also differ among nut species, affecting the ability of seeds to survive caching by animals (see p. 68). Most nuts in temperate forests have little dormancy: if there is a high probability of being recovered to be eaten, there is a strong selection pressure to germinate immediately. Thus, white oak acorns germinate in the same autumn as they are harvested, while red oaks, with higher tannin levels, germinate in the spring (Hadj-Chikh et al. 1996); squirrels prefer intact nuts over sprouted seedlings (Fox 1982). Some squirrels will kill many white oak acorns prior to burial, by excising the embryo (Fox 1982) and thus preventing their autumn germination and resulting in predation rather than dispersal. There may also be a preference for caching red oak acorns, which are dormant until spring and thus are less perishable as a food store (Smallwood et al. 2001), while preferentially consuming white oak acorns in the autumn (Hadj-Chikh et al. 1996).

3.4.3 Traits leading to dispersal on the outside of animals

Hooks and spines on propagules (Fig. 3.11) have evolved in many plant families and are most easily explained as mechanisms for dispersal on the outside of animals. These traits are present in less than 6% of species in most floras (Sorensen 1986). They are, however, conspicuous in the ground-flora of temperate forests, in the grasslands of Africa and Australia, and in the weeds of Mexican maize crops (Vibrans 1999). Most species with hooks and spines grow less than 2 m in height (Sorensen 1986; Hughes et al. 1994): if they were taller, the likelihood of contact with animals would be reduced. Hooks, acting like the artificial equivalent Velcro, become entwined with the wool and hairs of the coats of animals (as well as on human clothes). Most dog owners are also familiar with just how firmly they can become attached! As a result, propagules may be dispersed for long distances, only becoming dislodged when the animals groom, brush against bushes, or when they moult. In an extreme example, it is believed that *Xanthium spinosum* (Asteraceae) was transported to Australia in this way, on the tails of horses brought from Chile (Hocking and Liddle 1986). The distance travelled by propagules with these adaptations, even without the aid of humans, can be considerable.

Many grass propagules have outwards-pointing bristles (e.g. Rabinowitz and Rapp 1981) towards

Figure 3.11 Examples of propagules that attach to animals through hooks and spines: (a) *Agrimonia parviflora* (Rosaceae: reproduced courtesy of Kenneth Robertson), (b) *Bidens aristosa* (reproduced courtesy of Kenneth Robertson), (c) *Tribulus terrestris* (Zygopyllaceae: reproduced courtesy of Steve Hurst @ USDA-NRCS PLANTS Database). Scales are approximate.

their bases. If they are dislodged from the plant and pushed into the animal's coat, they can become entrapped, working their way through to the bases of the animal hairs and even penetrating the skin. Fischer et al. (1996), for example, recovered 8,511 propagules of 85 species from one half of a sheep's fleece; the majority of these were of grasses with bristles.

Some propagules may adhere to animals because of the presence of sticky hairs (Ridley 1930), or because they are coated in mucilage. It has been suggested that the rapid expulsion of sticky seeds by the squirting cucumber (*Ecballium elaterium*) causes them to spray on to the passing animal that may have caused the fruit to detach. Seeds of the mistletoe *Viscum album* (Viscaceae) stick to the beaks of birds as they are eating; the birds wipe them off on to the bark of trees, hence ensuring dispersal into an appropriate habitat for establishment (e.g. Lang 1987). Mistletoe seeds ingested by the Australian mistletoebird are sticky. When they are excreted they stick to the rump of the bird, causing the animal to wipe them off on a perch, in an ideal habitat for establishment. Other ways of attaching to animals include spines, which can penetrate skin (as well as the soles of shoes), and structures that clasp on to hooves and feet.

Very small seed size may aid transport on the outside of animals: small, wet seeds may temporarily stick to skin, hair, and feathers. Mud, containing seeds, adheres to legs and coats as they walk. Some animals deliberately wallow in mud: the mud and the seeds that it contains may then be brushed off some distance away. In aquatic plants, small seeds or small floating ramets (such as duckweed, *Lemna* spp.: Lemnaceae) may stick to waterfowl: rather than simply contributing to dispersal distance, this also ensures that they reach the appropriate habitat in a landscape where water bodies may be discrete, far apart, or perhaps ephemeral.

The diversity of animals makes it more difficult to quantify the effectiveness of hooks and spines than it is to quantify aerodynamic or hydrologic properties. Animal coats can be either thick or thin and made of hair, wool, feathers, or they may be smooth; bodies may be close to the ground or raised as much as two metres above it; feet may be soft or hard. Thus the effectiveness of a morphological trait of a propagule will depend on the animal species and its size. However, we can use live animals of appropriate species or dummies covered with their coats to estimate experimentally (a) the ability to become attached to an animal, and (b) the length of time before an attached propagule then becomes detached.

Based on morphology, we would expect propagules with hooked appendages or barbs to attach better to passing animals than those without these structures; we should not forget, however, that hooked propagules may require strong forces to become detached from the plant, thus ensuring that they do not disperse when only knocked by neighbouring plant stems. We would also expect such propagules to be removed better when brushed against by curly animal coats than by straight hair, and by short animals rather than those with long legs (see Chapter 4). Few studies, however, have measured the proportion of propagules transferring from plants to animals. Mouissie et al. (2005) examined the propagules that became attached to sheep and cow dummies that were wheeled through either a heathland or a grassland. The greatest probabilities of attachment were for grasses with long awns, although some smooth propagules were found to have attached to the dummy sheep. The researchers also noted that parts of inflorescences or plant stems incorporating several seeds often became attached. Using a live sheep, they also reported a preponderance of propagules with bristles and hooks in the wool as well as smaller numbers of smooth propagules.

Retention of seeds from animal coats has been the focus of a number of studies. Although it is straightforward to develop laboratory tests for the duration of attachment, it is not easy to determine which traits have the greatest influence over retention. This is because traits may themselves be correlated: for example, few very small propagules have hooked appendages while few heavy propagules are balloon-like (Tackenberg et al. 2006). Over a large sample of species, the strongest predictor of retention time tends to be small propagule size (Tackenberg et al. 2006; Römermann et al. 2005). For a given morphology, smaller propagules tend to remain attached for longer. However, it should not be concluded that propagule morphology has no effect: heavy seeds remain well attached only if they have hooks or elongated appendages (Tackenberg et al. 2006). Graae (2002) compared the retention on a dog (a German wire-haired pointer) of propagules that were bristly, with one hook, or smooth, placed either on the back or sides. Seeds stayed on the dog longer if placed on its back and if they were bristly.

3.5 Miscellaneous traits

The aim of this chapter has not been to provide a complete compendium of dispersal traits. One mechanism not covered so far is the presence of traits that cause movement as a result of changes in humidity. Propagules of some species, such as wild oats (*Avena* spp.: Poaceae) and storksbills (*Erodium* spp.), have a twisted 'hygroscopic' awn. As humidity changes, the uncoiling of the awn has a corkscrew effect, causing the propagule to move across the ground. The distance moved may be only a few centimetres (Stamp 1989), but this may be sufficient for the seed to push itself into litter or to fall down a crack and ensure its burial. The length of the awn, the tightness of its coils, and the angle at which the awn is bent will all contribute to the distance moved. Movement along the ground has also been reported in wheat (*Triticum turgidum*: Poaceae), in which there are two awns without twists (Elbaum et al. 2007). Cycles of drying and wetting cause the awns to alternately bend apart, then back together: stiff hairs on the awns provide friction with the soil, and as a result the propagule moves seed-first along or into the ground.

Limited lateral movement can be achieved simply by seeds falling directly from the parent to the ground under gravity, although its importance is seldom recognised in discussions on dispersal. Depending on their mass, diameter, and height of release, propagules may hit the ground at considerable speed. Some fruits falling from tropical trees may weigh several hundred grams: since their kinetic energy on impact with the ground is $0.5 \times$ mass $\times$ velocity^2, the energy potentially available for lateral movement may be significant. The angle of the ground at the point of impact and its hardness (or the presence of objects such as other propagules) will determine the horizontal velocity and direction after impact. The ability of a propagule to roll, and hence the distance dispersed, will also be determined by its shape. If a seed is angular, its sides will impact repeatedly with the surface, reducing its kinetic energy.

In an unpublished study, diverse dispersal units of nine species were dropped from 50 cm on to a hard, rough, horizontal surface in the laboratory. The propagules varied in mass from 1.1 mg (*Brassica tournefortii*: Brassicaceae) to 650 mg (*Lupinus albus*: Fabaceae) and had surfaces that were hard and round, angular, bristled, or hairy. Mean distance achieved laterally away from the from the release position ranged from 3 cm for *Bidens pilosa*

(Asteraceae) to 27 cm for *Lupinus albus*; the greatest distance rolled by any seed was 54 cm.

Mean seed size is one of the least plastic of plant attributes (Harper 1977; Weiner et al. 1997). However, propagule mass *can* vary considerably within a plant: for example, coefficients of variation (CV) of 12% to 98% were reported for 39 species in Illinois by Michaels et al. (1988). Longden (quoted by Austin 1972) found that the CV of seed mass for indeterminate annual species ranged from 50–60%, whereas for determinate species the CV was only 10–25%. Presumably this difference in variability was because indeterminate species produce poor quality seeds as conditions deteriorate at the end of the growing season. This variation in mass will cause variation in the kinetic energy of the seeds and may contribute to variation in the distances that they roll. Variation in seed shape within a species does not appear to have been given a great deal of attention (other than for dimorphic seeds of the Asteraceae adapted for wind dispersal).

It should not be inferred from our discussion that attributes which enhance rolling have in any way evolved as adaptations to dispersal. For most plants, rolling will have only minor effects on dispersal distance (except perhaps where a plant is at the top of a steep, hard slope!). However, rolling may take a propagule far enough to reduce competition from other seedlings derived from the same maternal plant. A propagule may roll into a crevice or into a dip in which scarce water soaks. The short distance rolled may also be the pre-cursor to much longer movement by another vector. For example, the rounded shape of a coconut may help it to reach the water a few metres away from its parent tree; it will also roll more easily into the sea with retreating waves.

Although humans are animals, and many of the traits described elsewhere in this chapter apply equally to them, we have created additional modes of dispersal that do not apply to other species. In particular, the development of vehicles and farm implements has provided considerable opportunities to plants for dispersal. Propagules can become stuck to the outside of vehicles, collected and distributed by harvesting machinery, or dragged along by cultivation implements. Lonsdale and Lane (1994) found seeds of 88 species in samples taken from the radiators, wheel arches, and tyres of cars in Kakadu National Park, Australia. Two thirds of the species were grasses. As in the case of dispersal by animals, small-seeded species are likely to stick to machinery in mud. There is little evidence that any particular types of seed are moved preferentially in soil as it is cultivated or as it is moved by road-making machines, although this would be an interesting subject for future research.

The harvester is probably the most important machine dispersing seeds in crop fields (Cousens and Mortimer 1995). Species with wiry or unripe stems may become tangled in the front end mechanisms of harvesting machinery and require that the driver removes the material some time (and some distance) later. Plants that have seeds still on their stems at harvest time, such as those that share a similar pattern of growth and development with the crop, will have their seeds taken up into the harvester. If the seeds are small, they will be sieved or blown out from the grain and returned to the field, perhaps some distance away from the maternal parent. If they are of similar size to the crop seeds, they will be retained with grain in the hopper and may then be planted in another field, perhaps hundreds of kilometres away if the grain is sold for sowing. Thus, dispersal ability of weeds in cropping systems could in theory have evolved through selection for maturity date (see p. 15), retention of seeds on plants, and seed size.

3.6 Conclusions

It is clear from an examination of morphological and chemical traits that many species have evolved enhanced ability to be dispersed by particular vectors. A description of these traits, and their categorization into dispersal syndromes, has long occupied the attention of natural historians and ecologists. There is variation in these traits between species, between and within populations, and among propagules on the same plant. As we will see in Chapter 4, the movement of propagules by the vectors to which they are adapted has been measured on numerous occasions. However, *all* propagules can be moved, to a lesser or greater extent, by air and water; we can predict their potential for movement from the equations of fluid dynamics and measurement of physical attributes such as mass, volume, and area. We can also make empirical measurements, such as terminal velocity and the duration of flotation, which can be incorporated into equations or simulation models. Propagule mass and shape may be correlated with adhesive ability and duration of attachment to animals and vehicles; size, relative to the mouth of the animal, and chemical constituents can indicate which species might ingest the seeds and consequently how long they could be carried. As we will show in Chapter 4, our ability to predict the distances dispersed by animals is increasing rapidly. However, the vast diversity of animal vectors, with their associated diversity in physiology, metabolism, and behaviour, and the vast diversity of fruit chemical constituents, digestibility, and texture, means that our ability to measure propagule dispersability will remain crude in comparison with other vectors.

CHAPTER 4

Post-release movement of propagules

4.1 Introduction

Having left the parent plant, where will a given propagule come to rest: how far away and in what direction? As discussed in Chapter 3, we can assess the *propensity* of a propagule to be moved by a given dispersal vector on the basis of its physical, and perhaps chemical, traits. But the possession of traits suited to dispersal does not guarantee that the propagule will go anywhere: the influence of the dispersal vector is clearly crucial. Hence, the key to understanding the distances that propagules are *actually* dispersed is a detailed understanding of the behaviour of the dispersal vectors and the influence of their behaviour on a given propagule.

Like its mathematical counterpart, a dispersal 'vector' can be considered to have *magnitude* and *direction*. If these were to remain constant over time, then the prediction of dispersal distances would be straightforward. However, both magnitude and direction can vary considerably over time, both in the short and long term, with considerable implications for dispersal. Consider two hypothetical examples. If the abscission layer on a plumed seed is fully mature and the force required for release is slight, a very light wind on one day may move the propagule only a few metres in one direction, whereas on another day a gale force, gusty wind may move it hundreds of metres in a completely different direction, even if it is not fully mature. A bird may eat a fleshy fruit from a tree and then fly off a great distance where it defaecates the seed; alternatively, the same fruit may be overlooked by the bird and taken later by a monkey which removes the flesh and drops the seed straight to the ground beneath the parent tree.

The variation in the dispersal vector is seldom completely random. Vector magnitude and direction will often be autocorrelated at particular scales of time and space. If a wind is strong enough to remove a propagule from a plant, it is likely to still be blowing in the next few seconds, minutes, or even hours, while its direction may be similar over several days and particular directions will recur more often than others throughout the year. Water direction in non-tidal rivers, and in many parts of the oceans, is always in the same direction. If water flow is fast, it is likely to be fast and in the same general direction several metres or perhaps kilometres away. In tidal rivers, magnitude and direction of flow will cycle in a predictable manner.

A further aspect of variation in a dispersal vector is the modifying effect of the environment. A vector may take a propagule through local environments that differ considerably from that of the parent plant, modifying the behaviour of the vector and hence altering its rate and direction of motion. A flat plain will reduce the speed of water, trees will impede wind, and the availability of such things as food, water, display sites, and safety will determine the movement of an animal. In considering the post-release movement of a propagule, we must therefore consider the environment that it is travelling through.

The magnitudes and directions of the vector while a propagule is in motion thus together determine the propagule's *trajectory*. If magnitude and direction change frequently, the result is a trajectory that can be highly complex. In the period between ingestion and defaecation of a seed, for example, a bird may visit over twenty trees, moving in an assortment of directions and staying in each place for varying periods of time. Even propagules moved by wind may travel along erratic pathways (Zazada and Lovig 1983). The actual pathway followed, of course, is of little relevance to the population biology of a plant. More important is how far and in what direction away from its point of origin the propagule ends up. Has the trajectory taken the propagule away from the influence of the parent, into another vegetation type, into another part of the landscape, or across inhospitable regions? To predict the point where a phase of movement ceases, in relation to the point where the movement started, requires that we integrate the effects of the vector over that period.

In this chapter we discuss the main types of dispersal vector, the characteristics of their behaviour relevant to dispersal, their effects on propagule trajectory, and the final distances dispersed under their influence. For simplicity, we will consider vectors one by one, as if they act in isolation: wind, water, animals, and humans. The implications of sequences of dispersal vector acting on the same propagule will be considered in Chapter 5.

4.2 Movement through the air

There is seldom a day in any habitat on which there is absolutely no air movement. Many propagules, whether or not they are adapted for dispersal by wind, will have their release and/or dispersal influenced by the wind to some extent. The strength of the wind varies seasonally, daily, diurnally, by the minute, and by the second. It may have vertical as well as horizontal components, caused by turbulence and convection. Hurricanes and tornadoes have the potential to move propagules, and even larger objects, considerable distances. Although these are usually extremely rare events for a given location, they may be of great importance to the distribution of the species at a geographic scale (see p. 131). To understand the effects of wind, we therefore need to understand how its strength and direction vary, before considering propagule movement by a given wind strength.

There is an overall tendency for winds at a given location to blow in a particular direction—the *prevailing* wind direction. This varies predictably with latitude. The actual direction on a given day will deviate from this as weather systems pass by. Over a period of weeks or months, however, the wind direction will have a distinct bias. The strength of the wind is determined by gradients of pressure across the weather system (isobars close together on a weather map indicate strong winds). Other modifiers of wind strength and direction are associated with land masses and topographic features. Monsoon winds develop as the result of heating/cooling of continental land masses. Diurnal heating of the land in summer draws in winds from the sea, resulting in on-shore sea breezes. Depending on the topography and weather conditions, strong cold or warm winds can blow up or down from mountain ranges. Winds are, in general, stronger at higher altitudes and where there are gaps in the landscape for the air to funnel through. Of particular importance are weather patterns, such as storms, that create particularly strong winds, since in addition to high wind speeds these are associated with turbulence (Horn et al. 2001).

Wind speeds are usually reported as averages over some period (almost invariably only in a direction parallel to the ground), but often in meteorological records only the maximum gust speed may be reported. Since propagules may only be in the air for a matter of seconds or minutes, more frequent weather data are clearly required for mechanistic models. Although winds have a predominant direction parallel to the ground, turbulence involves air movement that has an additional vertical component. By lifting propagules upwards, vertical air flow will mean that they are in the air for longer and they may encounter even stronger air currents. For example, Nathan et al. (2002) and others have noted the presence of propagules, often quite heavy ones, on the tops of buildings at elevations greater than the tops of the trees from which they were released. These propagules could only have reached such heights through turbulence. On the basis of such observations, Horn et al. (2001) argue that the probability of a propagule being transported a long way by a strong wind depends more on the characteristics of the wind than on intrinsic differences between propagules.

Wind blowing across a surface, whether it is bare ground, grass, or forest trees, will encounter friction; the wind thus becomes slower close to the

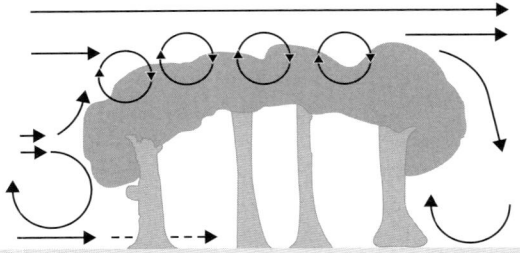

Figure 4.1 Simplified representation of the effect of a forest stand (cross-section) on wind flow.

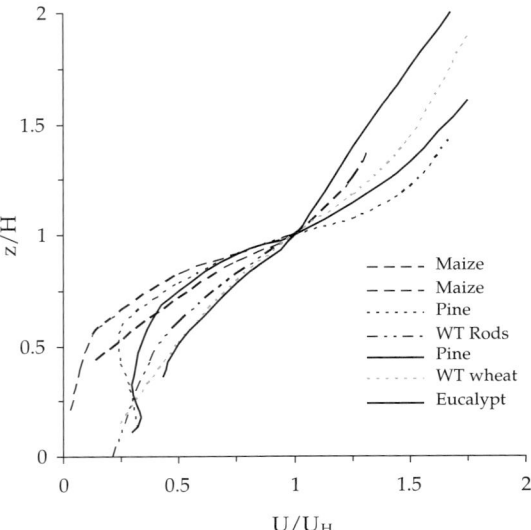

Figure 4.2 Relationships between wind speed (U) and height in the canopy (z) (redrawn from Finnigan and Brunet 1995, reproduced courtesy of Cambridge University Press). Both have been standardized, by dividing height by the maximum canopy height (H) and wind speed by the wind speed at the top of the canopy (U_H). WT indicates wind tunnel data.

surface. In this 'boundary layer', wind speed in relation to height can be approximated by a logarithmic curve whose coefficients reflect the effects of surface roughness (Monteith and Unsworth 1990). When the wind comes up against the edge of a plant stand, such as a forest, a proportion of its energy will be dissipated, being absorbed by stems and leaves, while some will be diverted over the top of the canopy (Fig. 4.1). Barry and Chorley (1998) record that a $2.2\,\mathrm{m\,s^{-1}}$ wind speed outside a Brazilian rainforest became $0.5\,\mathrm{m\,s^{-1}}$ at 100 m into the forest and was negligible 1000 m from the edge. As air passes over the top-most branches, local eddies will be produced (mechanical turbulence). A cross-section of wind speeds through canopy profiles (Fig. 4.2) shows the way in which wind speed decreases rapidly with height inside the stand. Nathan and Katul (2005) argue that the 'Kelvin-Helmholtz' eddies at the surface of a leafy canopy will be much smaller than the 'attached' eddies in a deciduous forest after the leaves have fallen. In attached eddies in a leafless canopy, the variance of the vertical component of wind speed is greater and propagules will move upwards for longer. Inside a leafy canopy, mean horizontal wind speed is less than for a leafless canopy, while immediately above a leafy canopy the wind speed is greater. Overall, the result is that more propagules will be uplifted from leafless canopies and these will travel longer distances than those uplifted from a leafy canopy. The authors note that many temperate species release their propagules in late autumn or early spring, when canopy leaf area is low.

For wind-blown propagules the consequences of wind currents in a plant stand can be deduced from Fig. 4.1. Propagules at the windward edge may be blown into the stand, but the distance will decline rapidly as the wind speed decreases. Propagules dropping down through the canopy will travel only a short horizontal distance, whereas those which get lifted upwards may get blown a very long way (Horn et al. 2001 refer to this as biphasic behaviour). At the leeward edge of the stand, large eddies will cause propagules to drop rapidly to the ground and perhaps get blown back in towards the stand. Thus, we expect preferential deposition of propagules in this location (Nathan 2005).

The effect of wind on the dispersal of simple particles, as well as propagules with more complex morphology, has been the attention of considerable theoretical research. The most simple model (first proposed for seeds by Dingler 1889—see Nathan et al. 2001) considers only horizontal air movement. It assumes simply that if an object is falling from a height H towards the ground at a constant vertical velocity W, and the wind moves it horizontally at a velocity U, then it will reach the ground at a distance

$$r = HU/W \qquad (4.1)$$

If the propagule falls quickly or the wind is light, then it will land much closer to its release point than if its terminal velocity is small and the wind is strong. To make the model more realistic, we can replace U by a function related to height (Monteith and Unsworth 1990), to account explicitly for the boundary layer effect, and then integrate over the duration of the flight. We can also allow for the fact that the mean wind strength will vary during a long flight, and we can replace W by $(W - W^*)$ to allow for increased turbulence (measured as a velocity acting upwards against the falling propagule) at increasing wind speed. So-called Gaussian plume and tilted plume models (Okubo and Levin 1989), based on particle diffusion, include variance parameters that allow for uncertainty in spatial coordinates over time due to turbulence: these result in a peaked probability distribution of settling distances rather than a single distance (Fig. 4.3a). If turbulence increases, without changes in other parameters, the mode of this distribution will move towards the point of origin while the variance, and hence the number of propagules deposited at longer distances, increases (Okubo and Levin 1989). Similar results were obtained by Greene and Johnson (1989) through the introduction of variance in propagule terminal velocity and in wind speed.

More advanced models (such as Coupled Eulerian-Lagrangian Closure (CELC) models) have recently been developed to allow for instantaneous wind speeds and more realistic simulation of vertical air movement within tree canopies and due to storms and thermal updrafts (Nathan et al. 2002; Tackenberg et al. 2003). These predict greater numbers of long-distance dispersing propagules than the cruder models. Nathan et al.'s (2002) model predicts a bimodal distribution of dispersal distances, with the peaks corresponding to the proportions of propagules uplifted or not. The largest peak, for non-uplifted seeds, was at a distance roughly equal to the height of the tree. For *Liriodendron tulipifera* (Magnoliaceae), greatest distances dispersed were predicted to exceed 10 km. Bullock and Clarke (2000) calculated that in the absence of turbulence it would have taken a horizontal wind velocity of 633 m s^{-1} to carry *Calluna vulgaris* (Ericaceae) propagules the observed maximum distance of 80 m, whereas Katul et al.'s (2005) CELC predicts that this distance could

be achieved with a horizontal wind velocity of only 10 m s^{-1} if combined with a vertical velocity standard deviation of 3 m s^{-1}.

Wind displays a frequency distribution of velocities over the course of the propagule maturation season, which may be described by a Weibull or a log-normal distribution (Soons et al. 2004b). Nathan et al. (2001) fitted a lognormal distribution to the frequencies of observed wind speeds and then selected random values from this for predicting the trajectory of each *Pinus halepensis* seed. The most significant aspect of the more complex wind dispersal models, however, and the cause of their more frequent predictions of long-distance dispersal, is the way that they handle variation in wind velocities. Both horizontal and vertical air movements do not display simple random fluctuations: variation in air flows are auto-correlated in both time and space (Soons et al. 2004a). If a propagule has just experienced an upwards air flow, it is likely to do so again along its flight path. For such individuals, sustained upward air movement takes them away from the boundary layer and into sustained, stronger lateral air currents and they are deposited further away from the parent.

Using their CELC simulation model as a basis and then making a series of simplifications, Katul et al. (2005) derived a simple analytical model for the frequency distribution of dispersal distances from a point source. Their equation is the same as the inverse Gaussian (or Wald) distribution, but with parameters derived from wind data and plant attributes rather than (as in Chapter 5) estimating the parameters by fitting the equation to observed dispersal distances:

$$w(r) = \sqrt{\frac{\lambda}{2\pi r^3}} \exp\left[-\frac{\lambda(r - \mu)^2}{2\mu^2 r}\right] \qquad (4.2)$$

where $w(r)$ is the probability density of a propagule landing at distance r; $\mu = H\overline{U}/V_t$; $\lambda = H^2/\sigma^2$; and $\sigma^2 = \frac{0.3h}{u_*}\frac{2\sigma_w^2}{\overline{U}}$. H is the height of release; $\overline{U}$ is the depth-averaged mean velocity within the canopy; V_t is the terminal velocity of a propagule; h is the height of the top of the canopy; u_* is the friction velocity above the canopy; and σ_w is the standard deviation of the vertical wind velocity. Like the tilted Gaussian plume and advection-diffusion models (Okubo and Levin 1989), the predictions from this model agree well with observed dispersal data (Katul et al., 2005).

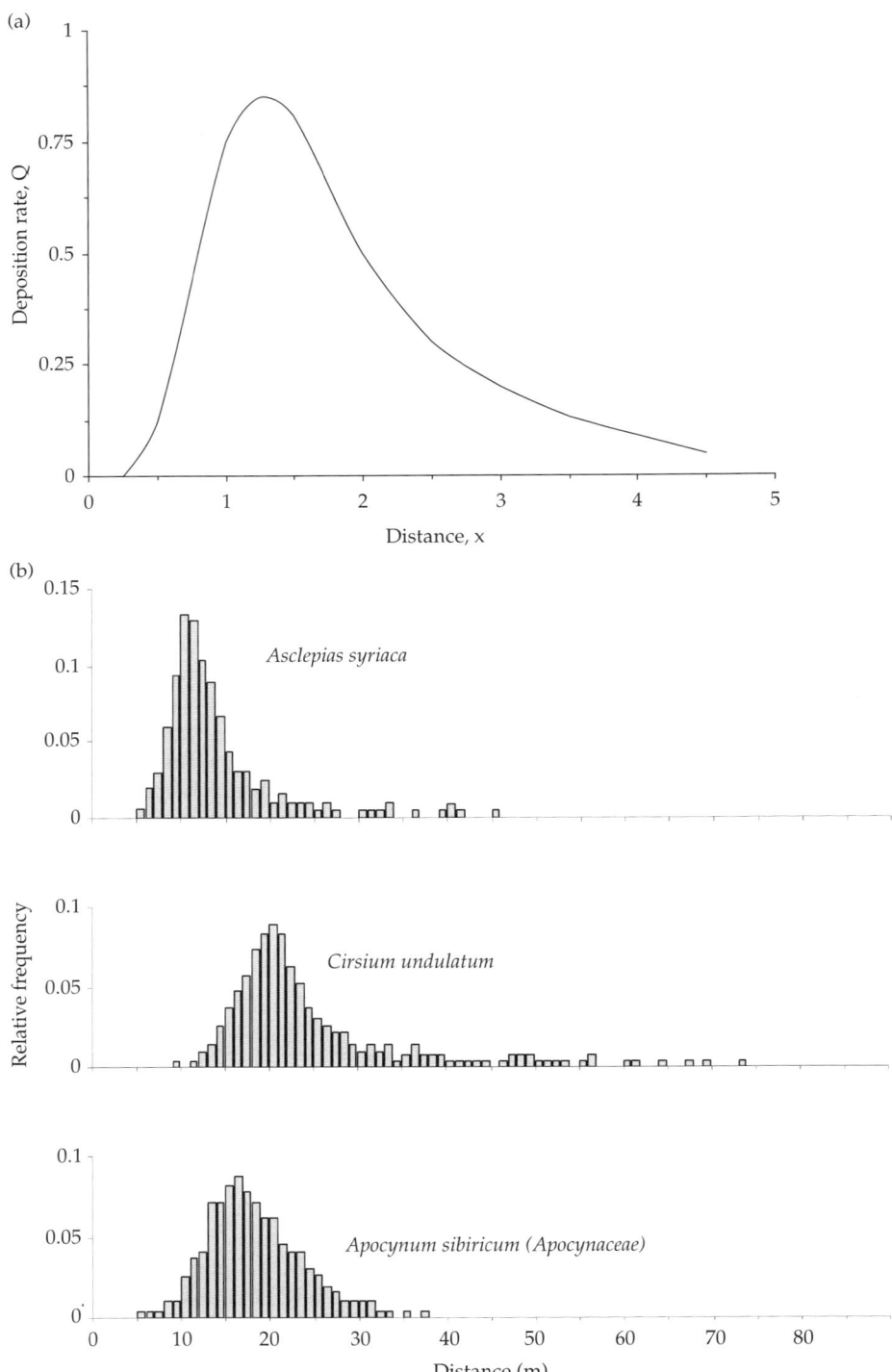

Figure 4.3 (a) Predicted distribution of dispersal distances from a point source in a constant wind velocity according to a tilted Gaussian plume model (redrawn from Okubo and Levin 1989, reproduced courtesy of Ecological Society of America); (b) observed distances from propagules released at plant height into winds of 2.8–4.2 ms^{-1} (redrawn from Platt and Weis 1977).

Considerable attention in the dispersal literature is given to the shapes of frequency distributions at long distances, that is, their 'tails'. An exponential tail implies a constant proportional deposition with distance: a 'thin' tail is no steeper than an exponential, while a 'fat' tail is flatter than an exponential. Although the tail of the frequency distribution of distances predicted by equation 4.2 is fat, it is less so than the other two models. The tilted Gaussian model predicts a tail proportional to $r^{-0.5}$, while the advection-diffusion model (Okubo and Levin 1989) has a tail proportional to $r^{-1-\beta}$ (Katul et al. 2005 quote a value of 0.15 for β). The tail of the inverse Gaussian distribution is proportional to $r^{-1.5}$ (Katul et al. 2005).

There have been many experiments in which seeds have been released from a fixed height (i.e. a point source) and the dispersal distances, along with the mean wind speed, have been measured (see examples in Fig. 4.3b). Usually the seeds are a sample from the population and thus vary in their aerodynamic properties (albeit usually to a limited extent see—Nathan et al. 2001). In all experiments, the frequency distribution of dispersal distances has a distinct peak and a long 'tail' (e.g. Greene and Johnson 1989; Katul et al. 2005) as predicted by most models of particle movement in air. The equations given in Table 5.1 may be useful for describing such distributions empirically (e.g. Higgins and Richardson 1999). Okubo and Levin (1989) summarized the results from several experimental release studies, finding that the distance of the mode of the dispersal frequency distribution was well estimated by HU/W (Greene and Johnson 1989, found the same for the mean distance). The greatest outlying studies from this relationship were for very light seeds that would have been more affected by turbulent air flow.

The environment through which a propagule is blown can influence dispersal in a number of ways. As we have seen, the nature of the habitat surface will affect wind speed and turbulence in the boundary layer. The height of vegetation around the maternal plant can act as a physical barrier, intercepting horizontal movement of propagules through the air (Cousens and Mortimer 1995). Achenes blowing from tall thistle heads (*Cirsium vulgare*: Asteraceae), for example, will travel further over short surrounding vegetation than when they are liberated from the centre of a thistle stand (Michaux 1989), since their angular descent through the air will be longer if there is nothing to stop their progress. By manipulation of habitat, McEvoy and Cox (1987) found that *Senecio jacobaea* achenes dispersed further over mown plots than over unmown plots. Both natural and artificial obstacles will also impede movement: tumbleweeds may roll for kilometres until they come up against a fence. Schurr et al. (2005) developed a model

Table 4.1 Examples of the times taken for intact seeds to pass through the guts of animals. A number of statistics are given, depending on what was presented in the source papers. Where ranges are given, this indicates the values for different plant or animal species or, in the case of the mode, the limits of the histogram bin with the greatest frequency.

	Minimum	Mean	Median	Mode	Maximum	Source
Various birds (9 sp.)	9–43 min			13–54 min	28–348 min	Levey (1986)
Various birds (3 sp.)	4–15 min		13–33 min	5–35 min	28–81 min	Murray (1988)
Hornbills (2 sp.)		2.6–5.8 h				Holbrook and Smith (2000)
Cassowary	0.9–7.7 h	3.3–27 h	2.9–30 h		10–40 h	Westcott et al. (2005)
Primates (4 sp.)		21–25 h				Poulsen et al. (2001)
Primates (14 sp.)*	2.75–38 h				84 h (one sp.)	Milton (1984)
Pig				1 day	9 day	Setter et al. (2002)
Sheep				1–4 day	8–31 day	Piggin (1978) and J. W. Heap (pers. comm.)
Horse				3–4 day	23 day	St. John-Sweeting and Morris (1990)

*Study using plastic beads as seed surrogates.

for movement along the ground surface, based on the equations of fluid dynamics (see p. 27) and a model for wind speed, in which obstacles were encountered by dispersing propagules.

The surface texture of the habitat will also affect the ability of a propagule to be blown along after it has reached the ground (Witztum et al. 1996). On some surfaces, such as sandy beaches, ice, or asphalt, this wind-facilitated movement may be more important than the initial flight through the air. The velocity along the ground will depend on the strength of the wind, the shape and mass of the propagule (see p. 27), the roughness of the surface (at a scale appropriate to the size of the propagule), and the slope of the terrain. Friction causes the velocity of the propagule to be less than the velocity of the wind. The effect of surface texture on seed movement is seen in the distribution of seeds and seedlings on the ground: even if the distribution of propagules landing per unit area is homogeneous, we usually observe aggregations of seeds in more retentive micro-sites, such as holes, vegetation, and areas with rough surface texture (Matlack 1989; Aguiar and Sala 1997; Chambers 2000; Bullock and Moy 2004). In order to include the effect of topography in a model, Higgins et al. (2003c) derived a frequency distribution of stopping distances from an elevation map. The map was divided into areas with slope either greater than or equal to one degree. The frequencies of lengths of flat ground in the direction of the prevailing wind were obtained by assuming that a seed blowing along the ground would stop when it reached a slope of greater than one degree. About 70% of seeds were predicted to have stopped by 50 m, and with an average distance of 59 m.

The movement of explosively dehisced propagules through the air can be predicted from equations of motion, based on the forces discussed in Box 3.1. Many applied mathematics texts present such simple equations of motion for theoretical projectiles in the absence of friction (i.e. no drag or lift). When friction is included, even without the inclusion of air movement and asymmetric propagule shape, the equations have no analytical solution and predictions must be made using numerical simulations (Swaine et al. 1979). From sensitivity analysis of such a model, Beer and Swaine (1977) concluded that initial velocity and coefficient of drag have major effects on dispersal distance. Optimum angle of release, achieving the greatest distance for a given propagule, was predicted to be considerably less than 45° (which would be the optimum in the absence of drag) and, at 14° for a height similar to mature trees of *Hura crepitans*, was well below the observed release angles of real fruits measured in the laboratory (35°). Although release height affects the optimum angle of release, it does not greatly affect dispersal distance: once drag has overcome the horizontal component of velocity, the direction of fall is vertical, so increased height results in no additional forward movement.

The distances moved by ballistic propagules are far less than those caused by wind. Many species which release their propagules explosively are herbs of less than a metre in height: maximum dispersal distances are 7 m or less (Swaine and Beer 1977). The greatest distance recorded is for *Hura crepitans*: a single propagule was found 45 m away from the base of an 11.2 m tall tree (Swaine and Beer 1977): depending on the point of origin in the canopy, the distance moved by the propagule would have been somewhere between 36 m and 54 m. Most published estimates of ballistic dispersal are from whole plants and do not differentiate between source positions on the parent.

Here are three examples where dispersal distances were recorded from individual fruits held above the ground at a fixed height considered relevant to the species. Seeds of *Vicia sativa* pods held 0.5–0.8 m above the ground were found by Garrison et al. (2000) to disperse their seeds up to 9 m, with an average of 3.4 m. Stamp (1989) found no difference between dispersal distances for fruits of *Erodium* spp. held 5, 10, and 20 cm above the ground. For six herbs, Stamp and Lucas (1983) recorded dispersal from individual fruits held at a height of 6 cm. Mean distances ranged from 0.24 m (*Impatiens capensis*: Balsaminaceae) to 3.29 m (*Geranium carolinianum*). All except *I. capensis* had frequency distributions with modes away from the source and with very short tails.

4.3 Dispersal by water

For buoyant propagules the great distances travelled by water in ocean currents and rivers provide the potential for extreme long-distance dispersal. Ocean currents, for example, can potentially transport seeds of coastal species over thousands of kilometres. Species such as *Caesalpinia bonduc* (Fabaceae) and *Entada gigas* (Fabaceae) have achieved pantropical distributions by this means of dispersal; their seeds are sometimes washed up on beaches in temperate regions (Murray 1986). (There are even Sea Bean societies, whose members collect propagules washed up along beaches, have web sites, and organize conferences!) Although coconuts are often cited as effective oceanic dispersers, there has been much debate over this: humans have certainly played some part in their dissemination (Ward and Brookfield 1992). While our attention is drawn to these extreme observations of dispersal, the energy provided by moving water is also important for transporting propagules at a very local scale, down slopes and into cracks and dips on the soil surface. Dispersal in water is also not limited to propagules that float: fast-moving water can transport non-buoyant seeds

Water can move quickly: river water in rapids may flow at $10–15\,\mathrm{m\,s^{-1}}$, ocean currents may be as fast as $1.3\,\mathrm{m\,s^{-1}}$, while in lakes there may be almost no movement. In steep areas, water velocity will be high, while the water will lose energy and slow down where the landscape becomes flatter. Propagules will thus travel faster in mountain streams than in rivers meandering across flood plains, though this does not necessarily mean that their final dispersal distance will be any greater.

Water speed will vary over time, either unpredictably, according to volume produced by run-off from storm events, or predictably, as when there is seasonal flow of melt-water from snowfall, or where seasonal changes in surface temperatures affect the strength of ocean currents. At some times of year, or over periods of several years in desert conditions, water flow may stop completely. Significantly, the direction of flow is often highly predictable. The paths of rivers may change little over hundreds of thousands of years. On land, water takes the path of steepest descent and its velocity will be related to slope. Rivers only flow in a single direction until they come under tidal influence, where their direction reverses on a regular cycle.

Within a water flow, the motion is qualitatively similar to that of air (Okubo et al. 2001; Ackerman 2002). There will be boundary zones on the bottom and sides of a river, where flow is reduced by friction, leading to speed profiles similar in shape to those in Fig. 4.2. Velocity is greatest towards the outside of a river bend and least on the inside (the process that leads to the development of meanders). Obstructions, such as fallen trees, will cause slack water behind them and may result in eddies in the horizontal and vertical planes. Eddies may also form where a stream widens and where flows from different rivers or currents meet. Turbulence will be produced in a river by obstructions, such as rocks, in the river bed and will even keep waterlogged propagules in suspension (Nilsson et al. 2002). Submerged plants will reduce current flow and dampen wave action. Plant propagules will tend to be deposited in those locations where water velocity decreases.

Engineers and oceanographers often study water movement in complex aquatic landscapes experimentally, by constructing scale models called 'flumes'. Different substrate profiles can be created, islands can be positioned to impede flow, artificial rivers can merge, and so on, all within a laboratory. A flume study of *Betula fontinalis* seed dispersal down a river found that greatest deposition along the river bank occurred in places where there were eddies, flow expansions, point bars, pool margins, and slack water (Fig. 4.4). Deposition along the watercourse varied with hydrological regime, with greater deposition occurring under descending water volumes and where flows were pulsed (such as when periodic dam releases occur). These experimental results are supported by field studies of seed dispersal, such as those by Schneider and Sharitz (1988) for seeds of *Taxodium disticum* (Taxodiaceae) and *Nyssa aquatica* (Nyssaceae) in a forested floodplain. Seeds became concentrated around emergent substrates, such as logs and trees. At a finer scale, *Zostera marina* seeds may become concentrated in dips in the substrate where water currents are reduced (Orth et al. 1994). Even in terrestrial environments, propagules may be concentrated in micro-topographic depressions as a result of rain-splash or surface water flow.

Distances dispersed from a point source in water have been studied only rarely. In one example, 22 mm painted pine cubes (having been shown to

(a)

(b)

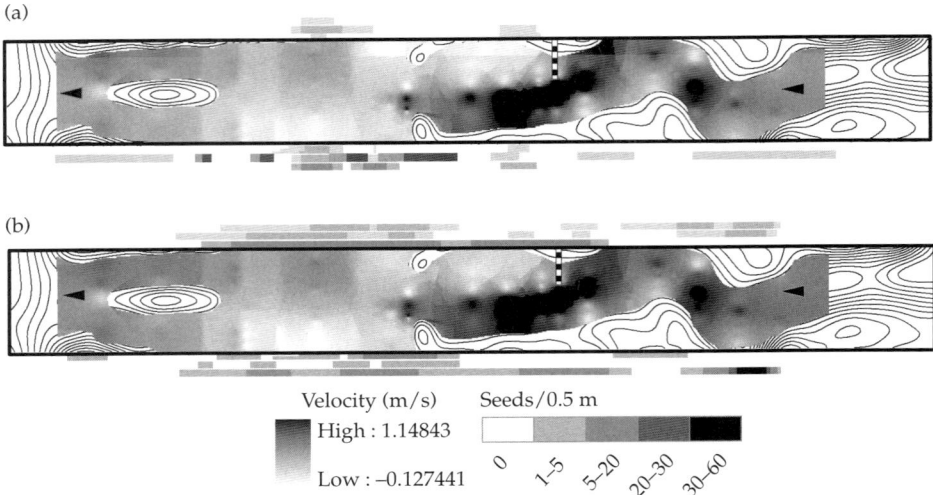

Velocity (m/s)   Seeds/0.5 m

High : 1.14843

Low : –0.127441

0   1–5   5–20   20–30   30–60

Figure 4.4 The numbers of *Betula fontinalis* seeds deposited along shorelines in flume experiments under (a) an increasing flow regime and (b) a pulsed flow regime. Three sets of sidebars above and below the flume indicate the mean number of seeds deposited: early dispersed seeds are shown adjacent to the flume; late dispersed seeds are furthest from the flume. The water velocity measured in the experimental section of the flume at a steady discharge of 30 L s^{-1} is shown as a greyscale within the flume. The bathymetric contour interval is 10 cm. Arrows indicate direction of flow; the barred vertical line upstream of the eddy indicates the location of a structure used to cause flow separation (modified from Merritt and Wohl 2002, reproduced courtesy of Ecological Society of America).

disperse similarly in water to sunflower seeds) were released at five points along a river in spring flood (Andersson et al. 2000). The river banks were later searched to recover the cubes. Stranding of cubes increased where flow rates decreased. The maximum recorded distance dispersed during the study period was 152.5 km. Frequency distributions of distance were peaked and skewed to longer distances (Fig. 4.5). Mean distances ranged from 1.1 km from the most upstream release site to 40.3 km from the most downstream site. It should be noted that experimenters often release seeds or seed mimics into the middle of a river. Clearly, if they fall from the parent into the river margin, they may remain in the boundary layer and move only very short distances. In a study of dispersal down a mountain stream in Spain, plastic beads (as seed mimics) were placed on the stream banks, away from the running water, from where they were dispersed by periodically increased water levels following storms (Hampe 2004). The peak of bead density moved downstream by over 100 m over 8 months.

Many propagules move at the air–water interface, with either buoyancy or surface tension keeping them raised somewhat above the water. The velocity of these propagules will depend on the viscosities and velocities of the two media. Thus, the velocity of the propagule will be somewhere between that of air and water, with the cross-sectional area that is situated within each medium determining the extent to which velocity is closer to one than to the other.

4.4 Dispersal by animals

Animal dispersal vectors range from the largest of land vertebrates, elephants and rhinoceros, to small invertebrates, such as beetles and ants. Even among one order, the birds, dispersers range from large, flightless species, such as the ostrich and cassowary, to some of the smallest passerines. The behaviour of each species is unique and complex.

Direct observation of the trajectories of individual propagules is usually impossible, since they are hidden from view or too far away from the observer, the routes travelled may be impossible for observers to follow (through the air, through high forest canopies, or under the water) and seeds from several feeding events may be indistinguishable. Trapping or

(a)

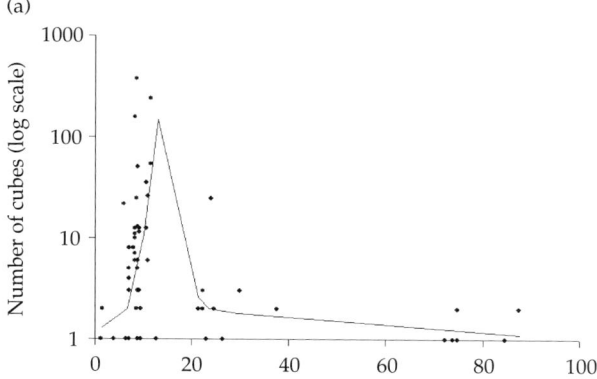

(b)

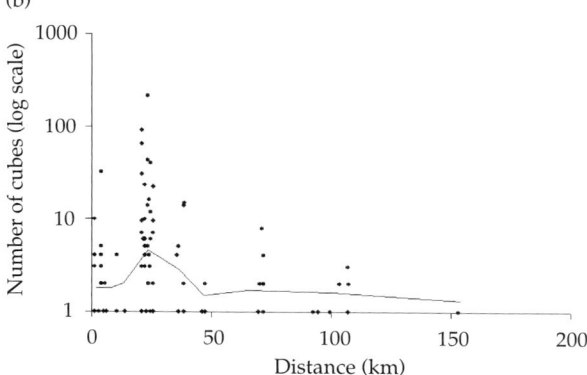

Figure 4.5 The dispersal patterns of 22 mm pine cubes released in 1992 at two sites on the Vindel River, Sweden. Sandsele (a) is further upstream than Vindelgransele (b), such that few cubes from site (a) reached site (b). Each dot represents a 50-m long section of the riverbank (redrawn from Andersson et al. 2000, reproduced courtesy of Blackwell Publishing). Note different scales.

mapping of seeds after they have reached the ground is also not normally an option: seeds are spread over a very large area, the numbers of seeds deposited per unit area is therefore low, and there may be multiple sources of propagules within the study area. Thus we must use indirect methods: studying the behaviour of the dispersal vectors rather than the seeds themselves (the same principle that has considerably increased our understanding of dispersal by air) and then predicting dispersal distances using simple models.

In Chapter 3, we introduced three main ways in which seeds are dispersed by animals: ingestion and subsequent defaecation or regurgitation; deliberate removal for later consumption; and inadvertent external attachment. We will follow the same classification here. The distances resulting from each of the three dispersal methods is dependent on the detailed behaviour of the animal species, the rate at which they move, the landscape within which

they are moving, and the plant and animal community with which they interact. Some overlap with the previous chapter is thus unavoidable, since animals may exhibit different behaviours, resulting in very different dispersal distances, depending on the traits of the propagules.

We must also bear in mind that there are many ways that animals *fail* to disperse seeds that are adapted for dispersal by them (or rather, cause them to disperse only very short distances). Animals may leave fruits unpicked on the plant, or they may overlook them on the ground beneath the parent; they may accidentally knock fruits directly to the ground, they may pick and then immediately discard some fruits; they may eat only part of a fruit (such as the flesh) and discard the remainder (including seeds); or they may fail to brush past a plant, leaving its adhesive propagules to fall to the ground as the supporting structures degrade. These events may often be an important component of the overall frequency distribution of dispersal distances and they must not be ignored if our aim is to quantify movement of the propagule population as a whole. In reviewing the literature on dispersal by animals, however, we should note that a great many studies only quantify the distances that seeds are successfully transported away from the influence of the parent.

4.4.1 Dispersal by ingestion

The distance that a seed moves from its source as a result of ingestion by an animal depends on (1) the time taken for it to be defaecated or regurgitated, and (2) the rate and pattern of movement by the animal. There is variation, both random and systematic, in each of these quantities: not all seeds pass through the gut at exactly the same speed, while animal speed and direction can change frequently. Dispersal thus results in a frequency distribution of dispersal distances which, as we will see later, is usually highly skewed. We can derive this distribution from the two component processes using a statistical model. Let $w_R(r)$ be the probability density function (pdf) of the dispersal distance (R) of a seed. If $w_T(t)$ is the pdf of the time from ingestion to ejection (T) and $w_{R|T}(r|t)$ is the conditional pdf expressing the displacement distance of the animal (the horizontal distance from the point of ingestion) at a given time, then

$$w_R(r) = \int w_{R|T}(r|t) w_T(t) dt \qquad (4.3)$$

Although we might fit an appropriate equation to the observed frequency distribution of ejection times to obtain $w_T(t)$, it is by no means straightforward to obtain an equation for $w_{R|T}(r|t)$. We therefore usually convert both time and distance into discrete variables and use a tabular approach to calculate the probability of dispersing to a given distance interval. The corresponding discrete version of equation 4.3 is

$$\Pr(r) = \sum_t \Pr(r|t) \Pr(t) \qquad (4.4)$$

where $\Pr(r)$ is the probability that dispersal is into distance category r, and so on (Murray 1988; Holbrook and Smith 2000; Higgins et al. 2003b). Our estimates of these probabilities are simply the

observed relative frequencies of seeds ejected in each time interval, and of the animal being in each distance interval at a given time interval. For example, $\Pr(r|t)$ might be given by the following table, where time intervals are each four hours and distance intervals are 100 m:

	$t=1$	$t=2$	$t=3$	$t=4$	$t=5$
$r=1$	0.9	0.7	0.5	0.3	0.1
$r=2$	0.1	0.2	0.2	0.2	0.2
$r=3$	0	0.1	0.1	0.2	0.3
$r=4$	0	0	0.1	0.1	0.2
$r=5$	0	0	0.1	0.1	0.1
$r=6$	0	0	0	0.1	0.1

(note that each column adds to 1.0). Thus, animals are close to the origin of a seed for the first eight hours and then slowly spread out over greater distances. Similarly, $\Pr(t)$ might be given by

$t=1$	$t=2$	$t=3$	$t=4$	$t=5$
0.0	0.5	0.3	0.1	0.1

where the majority of seeds are ejected between four and twelve hours. The probability of a seed dispersing to $r=2$ (i.e. 100–200 m) will then be

$$(0.1 \times 0) + (0.5 \times 0.2) + (0.3 \times 0.2) + (0.1 \times 0.2)$$
$$+ (0.1 \times 0.2) = 0.20$$

and so on for all distance categories.

To predict dispersal distances, we thus need to quantify these distributions and to understand the factors that cause them to vary.

4.4.1.1 Passage rates

The rate of passage of seeds through animals has received considerable attention, from casual observations of animals in zoos (e.g. Ridley 1930) to rigorous experimental feeding experiments. Table 4.1 presents results obtained for a range of animals. The time taken from ingestion to defaecation depends to a major extent on the length of the gut. In general, the larger the animal, the longer and larger the gut, and the longer the time taken between ingestion and defaecation (Calder 1984). In primates, for example, there is a positive correlation between body mass and the time taken for the first seeds to be passed (Milton 1984). Retention times for birds are mostly very short, since they typically have very simple intestinal passages (hence, the likelihood of long-distance seed dispersal via migrating birds is very low). However, there is considerable variation in gut morphology among birds of similar sizes (Jordano 1992). Several studies have found that passage rates for birds specializing in feeding on lipid-rich fruits are slower than those feeding on carbohydrate-rich fruits (Levey and del Rio 2001). Passage through ruminants, with their complex intestinal systems, may take a very long time: observations on sheep ingesting *Solanum elaeagnifolium* (Solanaceae) have found small numbers of viable seeds defaecated up to one month after ingestion (although peak throughput was two to three days: J Heap pers. comm.).

In a study of nine bird species, Levey (1986) found that the range of defaecation times within a species was consistently greater than the modal distance. In the most extreme case, an emberized finch, the earliest seed was defaecated after 31 minutes, the mode was after 54 minutes, while the longest passage time was 316 minutes. Frequency distributions

of passage time are peaked and skewed towards longer times (Fig. 4.6), especially for ruminants, and may approximate to the log-normal distribution (Vellend et al. 2003). In considering food passage through ruminants, Pond et al. (1988) used a two-compartment model, with one compartment releasing material according to a gamma distribution and the other following an exponential distribution.

A number of morphological and chemical attributes of seeds can potentially affect gut passage rate. Seed movement through a bird species may differ according to the species of plant whose propagules are eaten (Sun et al. 1997). For example, among ten legumes Gardener et al. (1993) found that the species with larger, denser seeds and a lower proportion of hard seeds were defaecated sooner by cattle. Differences in passage rate between different sizes of seed, however, are inconsistent among studies (Holbrook and Smith 2000). Some seeds contain chemicals that may have either a laxative or constipatory effect (Wahaj et al. 1998), while fruits with flesh attached firmly to the seeds may have longer passage rates.

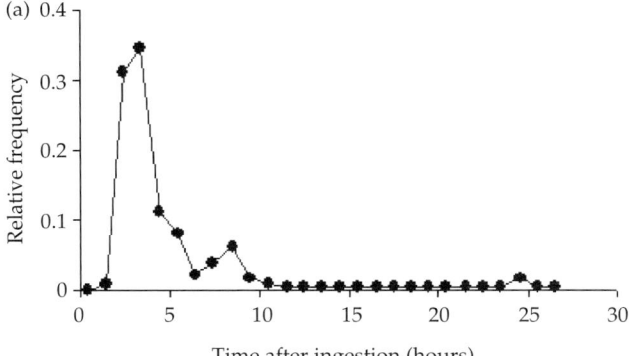

Figure 4.6 Rates of *Annona glabra* seed passage through (a) cassowary and (b) pig (redrawn from Setter et al. 2002, reproduced courtesy of Plant Protection Society of Western Australia). Note different scales.

Behavioural characteristics of the animals may also have an influence over passage rate. For example, birds may defaecate immediately before they take off or when they alight on a perch. They may fast and change gut functioning prior to and during migration (Figuerola and Green 2002). The amount of food ingested may have a significant effect on passage rate, with food passing more rapidly through an empty stomach (French 1996). It is important to recognize that most studies on seed passage rates are determined in confined enclosures and this may bias our estimates of dispersal: actively foraging animals may have different passage rates from caged animals (Sorensen 1984).

In many birds, seeds are often regurgitated rather than defaecated. Flesh is stripped from the seeds in the upper parts of the gut. By regurgitating the seeds, the gut is less full of indigestible material and thus more can be consumed in a given time. The time between ingestion and regurgitation is less than that between ingestion and defaecation. For example, Levey (1986) found that the median time to regurgitation of *Psychotria marginata* (Rubiaceae) by the manakin was around half the time (8.3 minutes) taken for seeds to pass the full length of the gut (17.7 minutes). Larger seeds (in relation to the gape width of the animal) are more often regurgitated (Stiles 1992). Thus, for a given animal, dispersal distance for large seeds is likely to be less than for smaller seeds.

A particularly important outcome of dispersal via defaecation is that deposition will often occur as clumps of seeds. For example, in a study of gorillas and chimpanzees faecal clumps contained means of 41 and 18 seeds (Poulsen et al. 2001). There is considerable variation, however, depending on what else the animal has been eating: numbers of *Cola lizae* (Sterculiaceae) seeds in gorilla faeces ranged from 1 to 240 in a study by Tutin et al. (1991), with a mean of 38. Seeds passed in faeces may later be spread out somewhat through secondary dispersal by trampling, or they may become more aggregated if hoarded by rodents: regardless, the result will be a relatively aggregated distribution. This will have consequences for competition in the next generation and may negate the often-stated advantages of dispersal in terms of reduction of competition (Potthoff et al. 2006). Generalist feeders will also mix seeds from different plant species in their faeces, resulting in inter-specific competition among the resulting seedlings (Stiles and White 1986). Regurgitated seeds will tend to be deposited singly, or in smaller clusters of single species than for defaecated seeds.

The number of seeds surviving ingestion (and therefore dispersed) very much depends on the particular species of animal, seed size, and the characteristics of the seed coat. Chewing is an important component of food processing in ruminants and many seeds are destroyed in this way. Large seeds are more likely to be masticated than small seeds, some of which may be swallowed directly along with fruit or leaf biomass (see p. 42). Gut passage tends, on average, to increase the speed of seed germination (Traveset and Verdu 2002) although there is considerable variation between studies. In birds, survival of ingested seeds will depend on gut morphology and, in particular, on the presence of a gizzard (Jordano 1992).

In various places we have commented on defaecation patterns in birds and mammals. It is important in volant species to keep body weight low to minimize the energy required for flight, and therefore swift passage through the gut and early defaecation is more likely in birds. Furthermore, birds maintain a higher body temperature than mammals and have a higher metabolic rate than mammals of equivalent size, which can also be associated with rapid passage of material through the digestive tract.

4.4.1.2 Displacement distance

The position of an animal within a landscape at the moment of seed deposition will be highly animal-specific and depends on a number of factors, including:

- Visitation rate. This will depend on the behaviour of the animals and on the availability of food sources. Some animals may be relatively nomadic, browsing as they travel, whereas others may home in on particular sources of food. The number of visits to a given food source will be greater if there are few alternative food sources in the vicinity, or if that source is particularly laden. Hence, the return time to the source plant will depend on coarseness of the spatial distribution of the food in relation to the motility of the animal.

• Feeding duration. Durations of visits to food sources may be very short, perhaps to remove a single propagule, or lengthy, during which time large numbers of propagules are ingested. If the duration is greater than the time taken for gut passage, then the first seeds ingested could be defaecated under the parent tree. The last seeds ingested, however, will tend to be dispersed away from the source. Visits may be curtailed by interactions with more aggressive individuals or species.

• Locations of other resources. Time will be required to travel to, and access, water and essentials such as salt. In rangelands, for example, food sources may be several kilometres from water holes and food close to water may become exhausted. Long distances may be traversed in minutes by birds, in hours by large terrestrial mammals, and not at all by small mammals or insects.

• Reproduction. Many animals display as a prelude to mating. This requires time and may take them to specific locations in the landscape, such as those used habitually for leks. During the breeding season, time will need to be spent obtaining nest material and then taking food to the nest.

• Dormitory location. Some animals, such as bats and sheep, have habitual sleeping sites, while even a single night spent in one location will increase the likelihood of a seed being deposited there. For example, in one study 50% of gorilla faeces were deposited at sleeping sites (Tutin et al. 1991).

• Territories. Many mammals and birds have areas which they contest and defend, and boundaries to be patrolled and marked. Thus, their movement within the landscape is restricted. Larger animals tend to have larger territories (Calder 1984). If a food plant is on the edge of a territory, it is likely that dispersal will be asymmetric, with the next movement being parallel to the boundary or inwards.

• Display sites. In addition to pre-reproductive displays, many species announce their ownership of territories, which they have obtained by contest. There are often particular locations that are favoured for display, such as tall trees, tall shrubs within grassland, or fence-lines on farmland. As with dormitories, time spent at a display site will mean

a greater probability of seeds dispersing there by defaecation.

• Weather. In harsh weather conditions, animals may shelter or reduce their travel (Raemaekers 1980).

Several of these activities involve movement towards particular places within the landscape. Dispersal towards a specific location from different sources is often referred to as 'directed dispersal'. For example, particular perches may be used regularly to display, to eat, or to sleep. Because more time is spent on them pro rata than in places in the landscape where the animals are merely in transit, the likelihood of defaecation, and therefore of seed deposition, is thus higher at perches. For example, male bellbirds feed in trees of Ocotea endresiana (Lauraceae), but then fly to song perches in gaps within the forest, where they also deposit seeds (Wenny and Levey 1998). Similarly, birds feeding on Prunus mahaleb in Spain tend to perch in trees and shrubs, thus tending to disperse seeds into dense vegetation. Seed deposition of bird-dispersed species into old fields has been increased by erecting artificial perches (McDonnell and Stiles 1983). It has also been suggested that control of invasive plants by birds could be improved by focusing management around roosts and other perches, or by introducing perches to concentrate defaecation in particular locations (Gosper et al. 2005).

In principle, the behaviour of animals can be expressed in purely quantitative statistical terms, by defining the duration of stops and the speeds and directions of movement between each type of activity. The spatial movement of animals within a landscape is recorded either by radio-telemetry or freehand mapping using visual observations: these can be used to calculate the statistics of moving animals, but usually behavioural information is not detailed enough to be able to assign times to specific activities. The distances covered within a day can be considerable. For example, arboreal monkeys have been found to move typically about 1 km in a day (Poulsen et al. 2001), while chimpanzees and gorillas may travel up to 7 km per day (Goodall 1986) and 2.6 km per day (Tutin et al. 1991) respectively. This distance may not, of course, be in a straight line, because of factors such as territorial behaviour and habitat preferences. Chimpanzees in Goodall's work had a maximum range in the course of a year of only $14 km^2$. Average home ranges of two hornbill species in West Africa were measured at $27 km^2$ and $29 km^2$, although individuals varied between $9 km^2$ and $44 km^2$ (Holbrook and Smith 2000). Small mammals will have much smaller home ranges, for example usually less than $0.01 km^2$ in the case of pine squirrels (Steele 1998). Fences will constrain farm animals, limiting the maximum distance that they can move. Hence, although a sheep may move up to 4 km per day in a paddock (Tribe 1949) and a cow may travel 2.8 km (Hancock 1953), the distance at any time from a source plant may be less than 100 metres. In open rangelands, cattle in one study were found to move an average of 5.3 km per day, compared with 6.1 km per day for sheep, and 9.6 km per day for goats (Cory 1927). From such tracking studies, displacement from either an arbitrary point or a specified location (e.g. a fruiting plant) can be plotted as a function of time, generating time-dependent frequency distributions of displacement distance (e.g. Westcott et al. 2005).

Animal behaviour changes during the day and is often crepuscular, with peak activity near dawn or dusk in many species and long daytime or night-time periods of inactivity. Such variation in behaviour may result in very different dispersal distances for seeds ingested at different times of day. Rather than combine data for all times of day to produce a single overall displacement vs time relationship, it is often more realistic to determine separate relationships for each of the various phases within a 24-hour period and then to model dispersal differently for each phase (e.g. Westcott et al. 2005; Russo et al. 2006).

Animal movement can be modelled in a number of ways, the simplest of which is a 'random walk' (often referred to as simple diffusion). In each (infinitely small) time interval the animal is assumed to move in a random direction by an infinitely small step. The result is that the probability of an animal being at a given location decreases with distance from its point of origin, according to a two-dimensional normal probability density distribution; the variance of the distribution increases with time, causing the normal distribution to become wider and flatter. As we will see on p. 66, this model for displacement has been used to predict dispersal of propagules on the outside and via the insides of animals.

The direction taken by a real animal in one time unit, however, is likely to be correlated with the direction taken in the next: real animals may move for considerable distances without markedly changing course. A 'correlated random walk' model is thus more appropriate. In such a model, angles are distributed according to a frequency distribution that reflects the tendency to continue in similar directions; the step size can also be chosen to be appropriate to the time scale at which field movement data are collected. Limits to movement, such as territorial boundaries, can also be incorporated. Kareiva and Shigesada (1983) derived equations for calculating the mean square displacement distance resulting from a correlated random walk with a fixed step length.

Movement patterns of wandering albatrosses and spider monkeys, measured irrespective of the particular activity involved, have been shown to be approximate to the assumptions of another model, the Lévy walk (Viswanathan et al. 1996; Ramos-Fernández et al. 2004). In this model there is spatial scale invariance in step length and temporal scale invariance in the duration of intervals between steps (i.e. in both cases the frequency distribution followed a negative power law). If they have wider applicability, Lévy walk models may be of use in simulating animal displacement from source plants and hence in calculating seed dispersal distances. Clearly, however, models of this type are very crude in comparison to the advanced meteorological models of dispersal in air, since they take no account of factors that cause the vector to change in magnitude and direction. A more mechanistic approach is needed for simulating animal movements within a landscape, based on quantifying the times spent in the various daily activities, the association between these activities and particular habitat types, and the development of behavioural 'rules' which appear to dictate switching between them. A useful basis for such future work might be 'agent-based' methods within a GIS, so that we can vary behavioural parameters and the spatial distribution of habitats, and thus predict the resulting animal movements (Topping et al. 2003; Pitt et al. 2003).

Detailed behavioural data are available for some species and these could be used to construct such a model. For example, silvereyes feeding on *Coprosma quadrifida* (Rubiaceae) spent an average of 0.68 minutes in a bush, swallowed 10 seeds per minute and returned an average of 2–6 hours later (French et al. 1992). Birds leaving *Prunus mahaleb* trees usually fly less than 30 m away (Jordano and Godoy 2002). The habitats to which they move next varies with bird species: for example, 92% of next stops by robins were recorded as being in shrubland, compared with 86% of next stops by mistle thrushes in pine trees. Importantly, these habitats differ in their survival probabilities for the plant species. However, few studies (with the exception, perhaps, of studies of primates) quantify all daily activities: watching a single tree and timing the animal visits is comparatively easy, whereas following animals throughout the landscape at a distance where details of their behaviour can be ascertained is often extremely difficult.

4.4.1.3 Dispersal distance

Frequency distributions for dispersal distance, predicted using equation 4.3 and empirical data, are usually peaked and skewed towards longer distances. Predictions for dispersal of four tropical fruits by a hornbill are shown in Fig. 4.7. Maximum dispersal distance is over 6.5 km, while the mode is either less than 500 m or between 500 and 1000 m depending on the plant species (Holbrook and Smith 2000). In comparison, for three smaller tropical birds with much shorter seed passage rates, the predicted mode is 50 m or less (Murray 1988). Vellend et al. (2003) predicted a mode at 200–300 m and a maximum at 3.7 km for white-tailed deer dispersing *Trillium* spp. (Liliaceae), using movement data from the literature and a two-compartment food passage model (Pond et al. 1998). The distribution showed a reasonable fit to the log-normal distribution, but with the tail of the log-normal over-estimating the number of long-distance dispersers. Russo et al. (2006) predicted a tri-modal distribution of dispersal

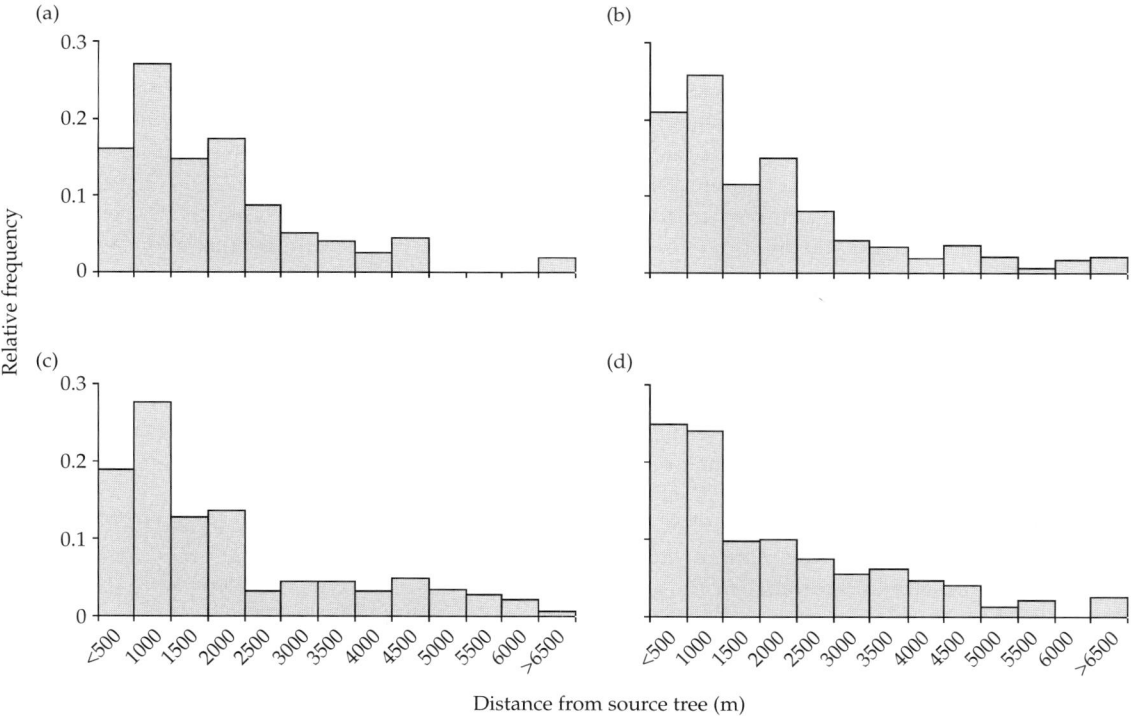

Figure 4.7 Dispersal distances resulting from passage through the hornbill *Ceratogymna atrata* by fruits of four plant species (redrawn from Holbrook and Smith 2000). (a) *Xylopia hypolampra*; (b) *Maesopsis eminii*; (c) *Rauwolfia macrophylla*; (d) *Cleistopholis patens*.

distances for *Virola calophila* (Myristicaceae), reflecting different types of behaviour exhibited at different times of the day by spider monkeys. Instead of using distance, Sun et al. (1997) estimated the probabilities of seeds being dropped beneath the canopies of successive trees visited by three species of turaco. The greatest probabilities were found to be for deposition in either the next tree after the source tree, or the one after that. All of these bird studies showed a long tail for the distance distribution, with deposition of a few seeds at long distances.

Instead of using empirical observations of displacement distance, Morales and Carlo (2006) used two approaches to predicting dispersal distances. Firstly, they assumed that animal movement followed a random walk and that gut passage rates were random selections from a gamma function. They obtained the frequency distribution of dispersal distances as

$$w(r) = -\frac{2}{r\Gamma(a)} \left(\frac{\lambda r}{2} \right)^a \left[2aK_{(\eta)}(\lambda r) - \lambda r K_{(\eta+1)}(\lambda r) \right]$$

(4.5)

where $\lambda = \sqrt{1/bD}$, a and b are respectively the shape and scale parameters of the gamma distribution, D is the diffusion constant (a measure of the rate of movement in the random walk), $K_{(\eta)}(\cdot)$ is a modified Bessel function of the second kind of order η, and Γ is the gamma function. Although the mathematics will be beyond most ecologists, the important outcome is that, as we have seen in dispersal from a point source by other vectors, this function has a shape that is zero at the origin, peaks relatively close to the source and then extends into a long tail. The position of the peak, the overall flatness of the distribution, and the mean dispersal distance are determined by the relative sizes of the rate of movement of seeds through the gut and the rate of movement of the animal over the ground. Even if the detail of animal movement deviates from a pure random walk and the seed passage distribution differed from a gamma distribution, these same general outcomes would be true.

Morales and Carlo's (2006) second model replaced the random walk with a very simple process-based model of bird feeding behaviour (p. 64). Food plants were placed at random, in discrete locations within the landscape. Bird activity was divided

into periods perching in fruiting trees and flights between perches. The length of time for each perching event was chosen from a gamma distribution. Birds then selected their next destination (always a food plant) according to an index based on the number of ripe fruits on the tree and the distance away (implicitly assuming that birds could both perceive their environment and make decisions in this way). Birds then flew in a straight line, at a set velocity, to the tree with the greatest index value. The number of (single-seeded) fruits eaten while perching at the next tree depended on the number of fruits on the tree and the number of fruits already in the gut (birds would no longer feed when they were full). Defaecation was assumed to occur, regardless of whether the bird was perched or flying, when a seed had been in the gut for a length of time taken from a gamma distribution. Food trees were allocated to positions in the landscape according to particular patterns and densities. Thus, from a random starting tree, the trajectories of birds (and hence seeds) through the landscape was simulated. Predicted dispersal distance frequency distributions were well-described by the Weibull distribution. The scale of dispersal was primarily determined by the degree of plant aggregation, whereas the shape of the distribution was affected mainly by the density of frugivores. Usually (though not in every case) the tails of the distributions were 'thin' (see p. 55), with slopes somewhere between those of an exponential and a normal distribution. Frequency distributions with tails fatter than exponential were found more often when food plants were aggregated and when there were higher numbers of dispersers present. Clearly there is considerable potential for making processed-based dispersal models that are ever more realistic and which characterize the behaviours of different animals.

The predictions from Morales and Carlo's second model illustrate an important point. Dispersal frequency distributions are likely to be *context-specific*, since animal behaviour will change in response to a wide range of factors, such as food plant abundance, diversity of food sources, abundance of animal competitors, presence of predators, complexity of the landscape, and so on. If one of these factors changes, then the resulting distribution of dispersal distances is likely to change. For example, at some times of the year birds will be looking after a nest, whereas at other times they may be nomadic. Some animals habitually travel along well-established paths and dispersal will be related to these: in different locations there will be different networks of paths. Carlo (2005) found that the addition of fruiting plants of *Cestrum diurnum* (Solanaceae) into a patch of *Solanum americanum* (Solanaceae) increased the number of visiting frugivores. The same effect did not occur when merely the density of *S. americanum* was increased. The mean dispersal distance for fruits fed upon by flying foxes in Tonga increases with the abundance of bats (McConkey and Drake 2006): in crowded trees squabbling forces some animals, especially juveniles, to leave after only short visits, thus leading to greater seed dispersal away from the source.

Thus, we cannot expect dispersal distances measured in one time or place to adequately describe dispersal of that plant by that animal under all circumstances. This context-specificity of dispersal distances is both a constraint on our interpretation of field data and a challenge to researchers. We need to know the magnitude of variation in distance distributions and to understand the most influential factors. We can do this to some extent by repeating field studies in multiple situations, although the effort required will often be prohibitive. The development of realistic, process-based models will therefore play a key role in addressing this issue and in raising hypotheses for field researchers to test.

In the previous discussion, we have assumed that there is only one phase of propagule movement. Once seeds are defaecated or regurgitated, that is not necessarily the end of their movement: secondary, tertiary, or further phases of dispersal may move some or all seeds. The recipient environment will determine the availability of further dispersal vectors. For example, the habitat may contain dung beetles, which cut faeces into balls (perhaps containing small seeds), which they roll away and then bury (Estrada and Coates-Estrada 1986); if faeces are ejected into an aquatic environment, secondary movement may take place via water. These later phases of movement should be brought into calculations of frequency distributions of dispersal distance, but published examples are

rare: we will return to the topic of combining the effects of different dispersal vectors in Chapter 5.

4.4.2 Dispersal by deliberate removal

Locomotion may precede food consumption for a number of reasons. Some animal species never remain long at a food source, preferring to take propagules to another place (such as a perch) where they process them. This is part of their usual routine: they habitually 'take-away' rather than 'dine-in'. Once away from the source, they may swallow the whole item, they may remove the most nutritious tissues and discard the remainder (perhaps containing seeds) on to the ground, or they may store them for later recovery. In other species, it is the interactions between animals that may dictate their dining behaviour. Food source plants can become crowded, fights may occur or established hierarchies force some animals to flee, taking fruits with them.

In the simplest of cases, removal of a propagule from the source plant represents just a temporary relocation of the feeding position: rather than consume the flesh of a fruit in the parent plant, the fruit is taken to a suitable perch, perhaps away from annoying competitors, where the flesh is removed and the seeds discarded. In Panama, the bat *Artibeus lituratus* flies with individual 18–23 g fruits of *Dipteryx panamensis* (Fabaceae) (equal to a third of the animal's body weight) up to hundreds of metres from the parent tree, only then gnawing off the flesh and discarding the seeds (Bonaccorso *et al.* 1980). Many old-world monkeys have cheek pouches in which they carry fruits. *Cercopithecus* monkeys, for example, carry arillate fruits of Myristicaceae in their cheek pouches, then later eat the flesh and spit out the seeds (Gautier-Hion et al. 1985). New World primates do not have cheek pouches.

Many colonial insects habitually take food back to their nest. Seeds with lipid-rich appendages (eliaosomes) are transported by worker ants to their colony, where the appendage is eaten and the remaining, still viable, seed discarded. This may leave a concentration of seeds at the nest site, above or below-ground, or seeds may be taken to the border of their territory and discarded (Gorb et al. 2000). The distance that the seed is moved thus depends on the distance of the source plant from the nest, and hence on the densities of food plants and nests within the habitat. In a review of the world literature, Gómez and Espadaler (1998) found that the mean ant-dispersal distance was 0.96 m, with a range from 1 cm to 77 m. This distance varies with both ant species and environment (Andersen and Morrison 1998); the frequency distribution of dispersal distance can be peaked (Andersen 1988) and skewed, either negatively or positively.

At times of high food availability, some animals, particularly birds and rodents, will take nuts and bury them for future consumption when food is scarce. This is known as larder-hoarding if there is a large cache and scatter-hoarding if there are many, smaller caches (Price and Jenkins 1986). The plants benefit from the inefficiency of the exercise: some nuts will fail to be recovered and they can germinate where they are buried. Often this will be well away from the influence of the parent tree, and may be in a different habitat type. For example, nuts may be preferentially moved from forests to early-successional habitats, thus facilitating the succession process (Vander Wall 2001).

If the season is good, more nuts may be stored than are later required. Some caches may be recovered, perhaps moved, and then re-buried. Prodigious quantities of nuts can be hoarded: for example, Chettleburgh (1952) estimated that 35 European jays stored 63,000 acorns (*Quercus* spp.) during a 10-day period. The success rate in retrieval can be extremely high (Vander Wall 2001) but a small proportion of a very large number results in at least some nuts producing seedlings. Animals may show preferences in the species that they retrieve: for example, Barnett (1977) found that squirrels more often recovered hickory nuts (*Carya glabra*) than acorns (*Quercus alba*), perhaps because they are easier to detect by smell (Price and Jenkins 1986).

Within a species, there may be size-selection by hoarders. The spiny pocket mouse preferentially consumes larger nuts of the palm *Astrocaryum mexicanum*, but caches similar proportions of large and small nuts (Brewer 2001). Small nuts tend to be buried further away; in this study, all buried large nuts were recovered, but only 70% of small nuts.

The proportion of nuts taken from a tree and cached depends on the season and the abundance of food. Nuts and seeds may be stored in hibernacula, for example by European dormice, for use during and at the end of their winter hibernation. Hoarding is usually in climates with distinct seasons or where rainfall is unpredictable, such as temperate regions and deserts (Smith and Reichman 1984). However, there are tropical exceptions: the agouti in forests of Costa Rica, for example, scatter-hoards pods of *Hymenaea coubaril* (Fabaceae) (Hallwachs 1986). 'Mast' seed production is likely to favour hoarding and results in reduced recovery rates due to storage in excess of requirements. However, there are few estimates of the proportions of nuts being cached (Price and Jenkins 1986). In one example, Darley-Hill and Johnson (1981) estimated that during a 27-day period, blue jays cached 54% of an acorn (*Quercus palustris*) crop, while 20% were eaten and the rest predated by weevils.

The distance that nuts are moved by hoarders depends on the animal. Distances of up to 100 m are typical for rodents, while corvids can disperse nuts several kilometres (Vander Wall 2001). Such dispersal by birds has been suggested as the reason for the rapid spread of some tree species after the last ice-age (see Clark et al. 1998a). Short dispersal distances can be measured by tagging nuts with pieces of metal, placing them in 'cafeterias' and then finding them with a metal detector after dispersal. Using this method, Xiao et al. (2005) found that frequency distributions of distances to seeds cached by rodents had modes within 5 m of cafeteria, and with greatest distances of over 35 m. The longest and flattest tail to the frequency distribution was for the largest seeded species, *Lithocarpus harlandii* (Fagaceae). A proportion of seeds were recovered by animals and then re-cached. However, Sork (1984) found that most (>99%) *Quercus rubra* acorns at a cafeteria were predated; the density of cached seeds declined monotonically with distance along transects starting from the edge of the 5 m diameter area where acorns were initially located. Other mark-and-recapture methods have included radioactive labelling, magnetic transponders, and a 'spool and line' technique (Steele and Smallwood 2001).

4.4.3 Discarding seeds at the source

Many animals are messy or wasteful feeders: during the peak maturation season there will often be large numbers of uneaten, or partly eaten, propagules found underneath a plant. Mammals feeding in forest trees may strip the pulp from fruits and discard the seeds, which fall directly to the ground: large primates such as chimpanzees remove the flesh of large-seeded fruits and spit out the seeds. Amongst birds, fruit handling behaviour is often categorized into species that are 'gulpers' (who swallow intact seeds) or 'mashers' (Levey 1987). Levey (1986) observed that mashers, who masticate food before swallowing, often worked seeds to the side of their bills and then dropped them. They may ingest small seeds, but discard large ones (Levey 1987). Jordano and Godoy (2002) found that an average of 70% of the seeds found beneath *Prunus mahaleb* trees originated from the parent immediately above them.

Although the dispersal distances of seeds discarded at the parent plant are very short, it is important to recognize the importance of this fraction of the population. They may constitute a significant component of the total crop: for example, Estrada and Coates-Estrada (1986) estimated that 76% of seeds were dropped under the source canopy by monkeys and 22% by birds. Forest redtail monkeys have been observed to spit out 83% of *Strychnos mitis* (Loganiaceae) seeds within 10 m of the source and 56% were spat out within less than a metre (Kaplin and Lambert 2002). Even if a frequency distribution of distances dispersed by ingestion has a mode some considerable distance away from the source plant, when this is combined with the numbers of seeds falling beneath the parent without ingestion (p. 101) the true frequency distribution for the plant may be found to have a single peak very close to the origin or it may be bimodal. Thus, discarded seeds have considerable influence over the mean dispersal distance and it is vital that they are taken into account in dispersal studies.

4.4.4 Dispersal on the outside of animals

The distance that a seed moves from its source after it has adhered to the outside of an animal will depend on (1) the retention time on the animal, and (2) the distance the animal has moved away from the parent plant during that period. Once again, there will be a frequency distribution of retention times and a time-dependent frequency distribution of displacement distances.

Retention times for either propagules or mud on animals have been measured under artificial conditions (e.g. Römermann et al. 2005) and for real animals in the field. For example, (Fischer et al. 1996) placed painted propagules of *Bromus erectus* and *Helianthemum nummularium* (Cistaceae) on two sheep ('Lotte' and 'Berta'). They were inspected daily to record the number remaining. Over half were lost in the first day. Loss thereafter was slower, with 8% of *B. erectus* remaining after 7 weeks and 17% of *H. nummularium* remaining after 5 weeks when the experiment was terminated. Sorensen (1986) describes experiments on the retention of burdock (*Arctium* spp.: Asteraceae) burrs by snowshoe hares in enclosures. All burrs placed on hares were removed or lost within one day and most within two hours. Burrs attached to feet remained for shorter periods than those placed on the backs (presumably more noticeable to the animals and more accessible to grooming). Burrs attached at high density on the backs had shorter mean retention times than those at low density; small burrs were retained longer than large burrs.

The surface of the animal will have an important influence, since objects will become more easily entwined in long, curled fibres. Seventy-five percent of artificially placed *Xanthium occidentale* burrs, for example, remained on sheep after 25 days compared with none remaining on horses and cattle (Liddle and Elgar 1984). Some propagules may become so firmly attached to animals that they will not be lost until moulting occurs, perhaps months later.

The behavioural components involved in determining the displacement distance will be similar to those contributing to dispersal by ingestion. However, the distances travelled during the retention times for externally-carried propagules may be very much greater than for ingested seeds due to the greater retention times. Nomadic movements by grassland herbivores in search of water or food, may result in movements over hundreds of kilometres. Some animals migrate long distances and can potentially move seeds great distances. Various species of goose, for example, migrate over 4,500 km between the Arctic and their winter feeding grounds (Owen 1980).

A simple model that has been used to better understand dispersal distances on the outside of animals involves two assumptions: (1) that animal movement is a random walk, resulting in a radially symmetrical probability distribution of locations, with a variance that increases with time; and (2) that a constant proportion of propagules dislodges per unit time, resulting in an exponential decline in the number attached over time. This is very similar to the model described in equation 4.4. The resulting probability density function for propagules landing at a point at a distance r from the source can be shown to be

$$v(r) = \frac{\rho}{2\pi D} K_0 \left(r\sqrt{\frac{\rho}{D}} \right) \qquad (4.6)$$

where $K_0()$ is the modified Bessel function of the second kind, order zero (Levin et al. 2003). $K_0()$ is proportional to $\log_e(1/r)$ at short distances and to $e^{-r}/\sqrt{r}$ at long distances. However, this equation cannot be compared directly to the frequency distributions of dispersal distance that we have described for other dispersal vectors, since it does not combine the results of dispersal in all directions. To do this, we need to integrate equation 4.6 through $360°$, that is,

$$w(r) = \int_0^{2\pi} v(r)d\theta = 2\pi r v(r) \qquad (4.7)$$

For the relationship given in equation 4.6, the distribution of dispersal distances $w(r)$ is peaked, with an intercept at zero distance and skewed to longer distances. Its scale is dictated by the ratio of the rate of seed detachment (ρ) to the rate of animal movement (given by the diffusion rate, D).

As we discussed on p. 64, animal movements are more likely to follow a correlated random walk. Using a correlated random walk model based on seed attachment times on cattle and sheep dummies and published data on wood mice and fallow deer, Mouissie et al. (2005) predicted probability distributions of dispersal distance that were peaked and right-skewed. Modal distances for seeds applied to the backs of animals were about 400 m for sheep, 100 m for cattle and deer, and less than 5 m for wood mice. The furthest 1% of seeds were predicted to travel at least 2.8 km for sheep, 800 m for cattle, 450 m for deer, and 12 m for mice. The assumption of a constant rate of deposition of propagules may also not be appropriate in many circumstances. Those propagules lost soonest may be those that are less firmly attached or that are attached to parts of the animal that are more likely to become dislodged, such as the legs and belly and areas that have shorter hair. Thus, it might be more appropriate to model with rates of deposition that decrease over time.

Although many propagules are clearly adapted to being transported on the outside of animals and some will be transported long distances, few of them may ever become attached to an animal. They will stay on the plant until eventually falling to the ground. Thus, the proportion of seeds dispersing in this way is often very low. As an extreme example, almost all *Galium aparine* (Rubiaceae) seeds produced in a cereal crop will fall straight to the ground, even though each is covered in small hooks. The proportion of propagules that successfully attach to animals can be estimated indirectly, by trapping those that fall directly to the ground under the parent plant and by counting those still on the plant. This has, perhaps surprisingly, seldom been measured. In one example, Liddle and Elgar (1984) found that up to 26% of *Xanthium occidentale* burrs were removed by sheep over a 14 week period, compared with 6–11% by horses. Many simply fell to the ground or were 'lost'.

4.5 Dispersal by humans

Propagules can be dispersed on the outside of humans, caught in clothing (Vibrans 1999), trouser turn-ups, and stuck in footware. They may be removed while in the field in a suitable habitat, or at places of residence in conditions less suitable for establishment. Retention on trousers and shirts has been shown to decline exponentially with time by Bullock and Primack (1977). Seeds can also be ingested and defaecated by humans, either directly to the ground or through the use of treated sewage as manure: tomato seedlings, for example, often emerge from human manure. These methods of dispersal are identical to those of other animals and we will not deal with them any further.

The unique influence of humans comes through their use of machinery, such as for farming operations, road-making, or transport (p. 48).

Most vehicles, however, transport propagules in ways very similar to animal vectors. Cars collect mud and seeds on the under-side of their chassis, under wheel arches, on tyres, and around the radiator. Seeds that have become attached to various parts of vehicles either fall off, are washed off by driving in the wet, or are removed when the vehicle is cleaned at the end of a journey. In a study of cars entering Kakadu National Park in northern Australia (see p. 49), individual cars were found to be carrying up to 789 seeds from 15 species, although many cars carried no seeds at all (Lonsdale and Lane 1994). The mean was about 6 seeds per car. Vehicles may travel many hundreds of kilometres along highway systems before being cleaned, while others have a home range of just a few square kilometres and are cleaned each Sunday. This variation may be a result of the type of vehicle (ranging from large haulage vehicles to children's' bicycles) or the behaviour of the owner (adventurous off-road drivers to in-city commuters). Eighty-eight percent of the visitors to Kakadu were from another state, mostly over 3000 km away, of which 20% had driven off-road on the way to the park.

There are also vehicles that ingest and then defaecate seeds! Combine harvesters take up plant material, process it to extract the crop seeds, and then eject the remainder, including weed seeds, to the ground. 'Ingestion' takes place along the wide cutter-bar at the front on the machine, and 'defaecation' is in a narrow swath that issues from the rear of the machine. As was the case for passage through an animal gut, the distances that seeds are moved by a harvester can be obtained by substituting (1) the probability distribution of the time taken to pass through the machine, and (2) the probability distribution of the vehicle being at a particular displacement at a particular time, into equation 4.3.

The rate of throughput of material in modern harvesters can be considerable. In a study by Howard et al. (1991) most seeds moved through the harvester faster than the vehicle moved forwards, resulting in net movement of seeds 'backwards' from the point of entry (Fig. 4.8b). The rate of throughput will vary with the crop species and the peculiarities of harvester design, so that net forward movement of weed seeds can also be observed in some studies (Ballaré et al. 1987). If it is assumed that the harvester travels at a constant speed in a straight line, variation in dispersal distances will be governed only by variation in the rate at which seeds pass through the harvester. The distribution of rates of passage of crop seeds through a harvester is typically peaked and skewed towards longer distances, showing good fit to the inverse Gaussian distribution (Whelan 1988), a function that can be derived from a linear Brownian motion. As with animal intestines, there will be many places inside a harvester where seeds can become trapped in corners or eddies, only to re-join the main stream later on. Hence, although many seeds pass rapidly through the machine, a few seeds may be returned to the field some considerable time (and hence distance) later. Ballaré et al. (1987), for example, found that some seeds of *Datura ferox* (Solanaceae) travelled at least 100 m from their source. Howard et al. (1991) suggested that the distribution shape in Fig. 4.8b arises because throughput is the combination of a normal distribution of 'fast' seeds and a slower, exponential distribution of 'slow' seeds.

Unlike animals, the movement of harvesters is very predictable. They travel mostly in straight lines up and down a field, or in a spiral, working towards the centre (Monjardino et al. 2004). The displacement of the harvester at the point of ejection of a seed will depend on both the speed of the vehicle and the seed's point of uptake in relation to turning points within a field. The speed of the harvester may decrease as the vehicle turns tight corners and may be adjusted by the operator in response to the quantity of material being cut. Hence, both the dispersal distance and direction will depend on location within the field.

Some weed seeds may never emerge from the harvester, being taken up into the hopper with the crop grain. These seeds will then either be sieved out and discarded at a mill, processed with grain and hence destroyed, or remain as impurities in grain for kept for sowing. The number of weeds taken up into the harvester can be considerable: in a survey of farmers' crop seed about to be sown in southern Australia (i.e. seeds that have not been successfully removed from the crop grain), Moerkerk (2002) found over 40 species as contaminants and up to 9 species in any single sample. If the crop seeds retained from the harvester are sown on the same farm, dispersal of the

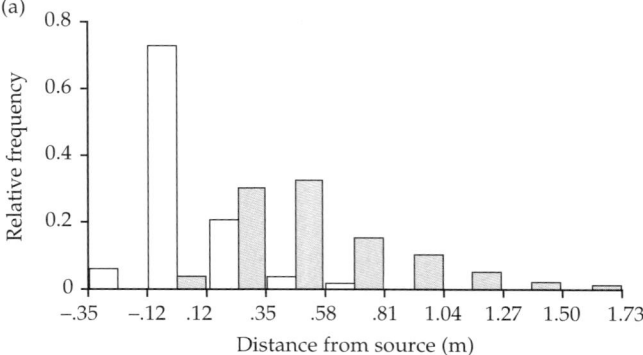

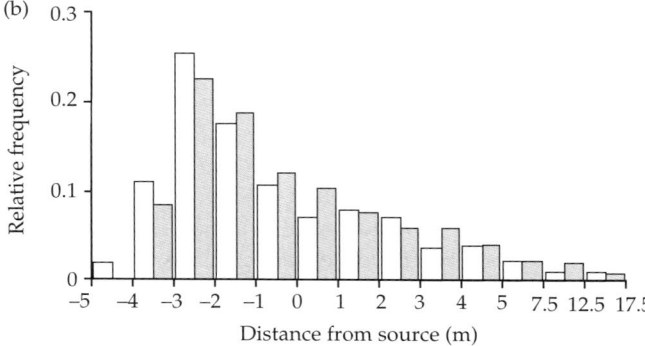

Figure 4.8 (a) Dispersal of seeds of canola, used as a weed seed surrogate, by cultivation, via spring tine (grey) and seed drill (white) (redrawn from Howard et al. 1991, reproduced courtesy of BCPC), (b) the spread of *Bromus* seeds by passage through a combine harvester (redrawn from Cussans and Marshall 1990, courtesy of IACR): *Bromus sterilis* (white); *Bromus interruptus* (grey). The direction of travel of the vehicle is left to right, with positive distances indicating forward dispersal. Note the different scales. For (a) dispersal was taken as the positions of seedlings (i.e. effective dispersal) and assuming that all seeds germinated or died.

weeds will be restricted to within the farm boundaries. However, if sold to a merchant, they may be dispersed tens or hundreds of kilometres (merchants do not guarantee 100% purity of the seed, only that impurities are below a given level). Harvesters may also move plant material without ingestion: stems with attached seeds may become clogged in the tines of the header, allowing seeds to be dragged a considerable distance along the soil surface.

Tillage implements or earth-moving machinery may drag propagules along with soil. The distance that propagules are moved will depend on the residence time on the moving 'wavefront' of soil material at the blade-soil interface. The fluidity of particles (including seeds) in this front will vary with soil type, compaction, and moisture content. Many seeds probably move at the same rate as soil particles, since the seeds may be stuck within soil aggregates. However, segregation of different sizes of material may occur and very large seeds may thus move different distances than very small seeds. The width of the soil wavefront will depend on the size of the tillage blade: retention on a bull-dozer will

clearly be greater than on a narrow tine and seeds will thus be moved further. Many studies of seed dispersal by tillage machinery have used plastic beads as seed surrogates, rather than real seeds, because it is easier to see where they go. Results for tined cultivators and seed drills show that mean dispersal distance is short, with just a few seeds travelling over a metre (Fig. 4.8a); movement perpendicular to the direction of the implement is slight (Howard et al. 1991). By sowing plastic beads of different colours at different depths, it has also been possible to describe the vertical components of movement by tillage implements (Cousens and Moss 1990).

Data on the movement of seeds by single tillage events have been used to predict the outcomes of multiple tillage events. A simple matrix model was used by Cousens and Moss (1990) to predict vertical distributions of seeds within the soil profile arising from repeated cultivations and with annual inputs of new seeds from reproduction. It was shown that repeated use of a mouldboard plough would mix seeds throughout the cultivated soil horizon, whereas repeated use of tined implements would

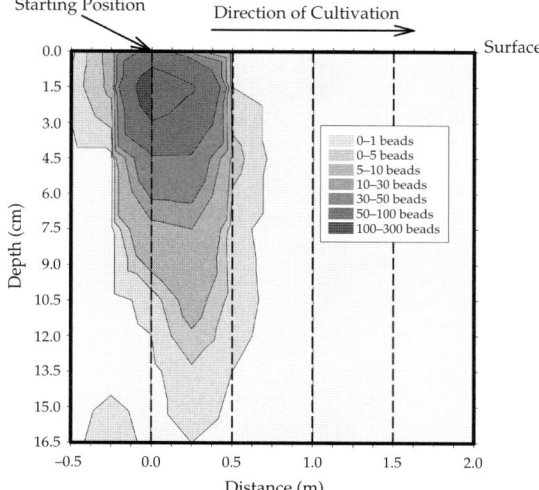

Figure 4.9 Distribution of seeds within a soil profile as a result of tillage with spring tines, predicted from a gamma frequency distribution of dispersal distances and vertical movement in the soil profile according to a beta distribution (Mead et al. 2003, reproduced courtesy of Association of Applied Biologists).

result in a distribution that declines rapidly with depth. Horizontal dispersal of seeds in the direction of cultivation equipment has been modelled by Brain and Marshall (1999), by assuming a gamma frequency distribution of dispersal distances from a point source; Mead et al. (2003) then added vertical movement in the soil profile to this model by assuming a beta distribution in that direction (Fig. 4.9). For the parameter values used, both the gamma and beta frequency distributions were peaked and showed a reasonable fit to the data.

For some machinery, the rate of throughput and velocity of travel can be ignored: dispersal is simply a matter of where a vehicle starts and stops. For stock carried on board a ship, faeces and straw will be cleaned out at discrete locations, such as ports. Trucks transporting animals on land may disperse some material in transit, but most seeds will again be disposed of at the destination or at border de-contamination stations.

4.6 Conclusions

As we have shown in this chapter, we now have a considerable ability to predict the trajectories of individual propagules in air and water under specific conditions (though considerably less attention is given to water). This has been achieved through the development of a mechanistic understanding of the dispersal vector and the consequent increase in realism of the models that have been developed. Thus, we now understand the particular conditions under which propagules will be moved very long distances. However, most studies are still concerned with species in which there are clear, often extreme adaptations to the vector in question. Although our models *can* be used to predict the movement of poorly adapted propagules (and there are many such species), this is seldom done. In contrast to our elegant models of air and water, our models of movement by animals are very empirical, based largely on descriptive data from tracked animals. As a result, our calculations are highly context-specific: we cannot, with any confidence, make predictions for a different habitat even for a single species. One of the biggest challenges for ecologists is to achieve a much more mechanistic understanding of the movement of the animal vectors, so that we can build more realistic models and thus make advances in understanding comparable to those recently achieved for wind dispersal. We accept that this will be very much more difficult and that the models will be less precise than for wind and water, but the rewards will be considerable.

We conclude with two notes of caution. Although the frequency distributions of dispersal distances from artificial releases under average conditions (such as the mean wind speed) may be similar in shape to data obtained from propagules caught in traps around plants, we would not expect the distances to be the same as those resulting from a real plant over the entire course of its dispersal season. For example, artificial releases have no threshold force for release to occur; for real plants, no dispersal will occur in low wind speeds insufficient to detach the propagules, while many propagules may become detached from the parent under extreme conditions before the threshold detachment force has reached its minimum. As we will show in Chapter 5, the prediction of real frequency distributions of dispersal distance may be very complicated and has seldom been attempted.

Patterns of dispersal from entire plants

In Part A, we considered the various factors that together determine the trajectory of an individual propagule. From a particular location and under defined conditions, how would it be moved by a particular vector or set of vectors? We described experiments in which a number of similar propagules were released from a point location and considered various models that predict the distance dispersed from a point source by a given magnitude and direction of a dispersal vector. The resulting dispersal distances will always have a frequency distribution, since the dispersal vectors, such as wind and water flow and animal behaviour, are, to some extent, unpredictable. In any dispersal experiment, no matter how hard we try to avoid it, there is always going to be some level of variation in the experimental conditions and the propagules used. Together these experiments and models allow us some level of understanding of the processes involved in dispersal. However, they do not tell us what will happen to the population of propagules dispersing from a real plant: this is the subject of Part B.

As we saw in Part A, different propagules on the same plant may be liberated from very different points in space, which can result in very different dispersal distances. The physical and chemical properties that determine propagule trajectories under the influence of different vectors also vary. Perhaps most importantly, the release times, and therefore the conditions under which propagules disperse, can vary considerably. Propagules do not all mature at the same time: they have a season, which may stretch over a number of days, months or even years, during which different propagules develop, mature, and then become detached. During the course of the

season, the relevant dispersal vectors may vary by orders of magnitude in strength and through any direction; some vectors may be entirely absent during some parts of the season, while several vectors may coincide at other times. Complicating matters still further, different propagules within the population may be moved by different vectors, or by combinations of vectors acting in sequence. In one season an extreme event may occur, causing very long dispersal distances, while there may be many successive seasons during which such extreme events do not occur.

There are two approaches that we can take to determine the scale and pattern of dispersal around entire plants:

- We can locate seeds on the ground after propagules have dispersed, thus describing the 'seed shadows' around plants. As we will see, there are a number of reasons why this is often difficult to achieve. In principle, however, we can learn about the ecological causes of variation in seed shadows by comparing case studies of post-dispersal distributions from different species and in different habitats; we can also vary the habitat conditions experimentally (though this is seldom done).

- We can model dispersal from entire plants, by combining the processes explored in Part A. Sensitivity analysis can then be used to understand the factors that result in different distributions of propagules. Again, modelling is often not easy and inevitably involves assumptions that may not, in hindsight, prove to be acceptable. To obtain an acceptable level of realism may require complex models and mathematics that is beyond the scope of many ecologists.

As with all branches of science, the most effective approach is a combination of theoretical and empirical studies. Ideas for modelling arise from field observations; support for model predictions can be obtained by comparisons with existing data, new field studies of appropriate species and habitats, and through experimental manipulations.

This section contains just one chapter, but it is the most crucial part of the book. It links the chapters in Part A with those in Part C. Dispersal from entire plants is the outcome of the various components described in Part A and is the basis for the predictions made in the chapters within Part C. As we will show, however, it is perhaps the weakest aspect of our knowledge of dispersal and its ecological implications.

CHAPTER 5

Patterns of dispersal from entire plants

5.1 Introduction

The population of propagules emanating from a single plant may consist of just one or two individuals, or perhaps hundreds of thousands. Even for highly fecund individuals, it is unlikely that any two propagules will follow exactly the same trajectory as they disperse (an exception to this might be seeds ingested in a single mouthful by an animal and then voided in the same faecal deposit). Their unique trajectories spread the population of propagules out around their parent plant. The pattern that they form when they come to rest, often called the maternal plant's 'seed shadow', determines where in the landscape the individuals in the next generation must establish, and hence indirectly the abiotic environment and intra- and inter-specific interactions that the progeny experience. As a result, the pattern of dispersal of seeds ultimately plays an important part in the fate of the population over multiple generations (as we will demonstrate in Part C). The aim of this chapter is to describe the information that we have obtained from empirical studies of dispersal and the theoretical framework that is being developed, to help us to understand how particular shapes and scales of propagule distributions arise. Considerable attention will be given to research methodologies, since our ability to make observations is severely constrained, particularly with respect to the large areas over which propagules of many species fall and the overlap between the shadows of neighbouring plants.

Before we begin, however, it is essential to first distinguish between two ways of depicting dispersal data. They are related mathematically, both give rise to dispersal 'curves', but these curves have quite different meanings and shapes. An awareness of their differences is essential for ecologists to communicate their results unambiguously and to accurately interpret the published literature.

5.2 Ways of summarizing dispersal patterns

Dispersal data commonly take one of two forms: a point pattern of the positions of individual propagules; or propagule densities (numbers per unit area) estimated at discrete locations. In the following discussion, unless otherwise stated, we assume that all seeds come from a single maternal parent.

5.2.1 Point patterns

Let us begin by assuming that we know the position of every individual propagule around the parent plant, which we can plot on a map. The location of each propagule can be represented either in Cartesian (x, y) coordinates, or as radial (angle θ, radius r) coordinates from the centre of the parent. The two coordinate systems are inter-convertible, since $r = \sqrt{x^2 + y^2}$ and $\theta = \tan^{-1}(x/y)$.

Figure 5.1a shows a point pattern for the seeds of *Erythronium grandiflorum* (Liliaceae) in relation to their maternal parent. There are a number of ways in which we might extract summary information from such a pattern. The mean distance dispersed, $\bar{r}$, is simply the average of all values of r (for Fig. 5.1a $\bar{r} = 34.6$ cm). Another useful statistic is the centroid (centre of gravity), the mean of the Cartesian coordinates in each direction, $\bar{x}, \bar{y}$. The net movement of the population of propagules away from the parent can be measured by the position of the centroid in relation to the centre of the parent plant (Osada et al. 2001; for Fig. 5.1a the centroid is 11 cm from the parent). The maximum dispersal distance, r_{max}, may often be singled out for comment (for Fig. 5.1a $r_{max} = 100$ cm), especially given the importance of extreme events for the rate of spread of a population over generations (see p. 131). However, the maximum distance is often a statistical outlier, a single, often very extreme, event. Purely by chance, it may differ considerably from one plant to the next. Maximum distance is also likely to be under-estimated: the probability of missing a propagule in a field study increases with distance from the parent, since the area to be searched increases with the square of the radius.

The area over which propagules are dispersed can be calculated in a number of ways. The method used in Fig. 5.1b is to fit a convex hull to the point pattern. This is defined as the smallest convex polygon (one in which all internal angles are less than 180°) that contains all points: an analogy is that it would be the shape taken on by an elastic band stretched around the set of points (for Fig. 5.1a area = 1.65 m^2). As with r_{max}, individuals at the periphery may be extreme outliers and thus have a strong influence on the calculated area.

Probably the most commonly used method for summarizing point patterns is to construct separate histograms of θ and r (thus assuming that the distributions of the two variables are independent). For the angles of dispersal, it is common to illustrate the frequencies of different angles by a 'rose' diagram (a directional histogram: Fig. 5.1b); this will indicate any tendency to disperse in particular directions. 'Circular' statistical methods are available for calculating the 'mean vector' (= mean direction) and for testing for directional uniformity (Batschelet 1981): for Fig. 5.1b the mean direction is 251°; Rayleigh's test shows significant departure from a uniform distribution of directions, p < 0.001.

A histogram constructed from the values of r gives the frequency distribution of dispersal distance (Fig. 5.1c). The height of each histogram column represents the number of propagules n_{ij} landing within an annulus around the parent defined by an inner and an outer radius, r_i and r_j respectively (i.e.

n_{ij} is found by counting the number of propagules landing between two adjacent concentric circles in Fig. 5.1a). The relative frequency in each histogram bin, n_{ij}/N where N is the total number of propagules, is the observed probability of a dispersal distance R occurring between those limits, $P_o(r_i \leq R < r_j)$.

The shapes of histograms can be affected considerably by the choice of starting point (though for dispersal this is invariably zero) and bin width. An alternative to a histogram is a smooth curve produced by a statistical technique known as 'kernel density estimation' (Silverman 1986). Identical distributions (e.g. the normal distribution), each with area $1/N$, are centred on each data point; the kernel density estimate, or 'empirical pdf', is obtained by summing these distributions. Like histograms, the width of each distribution (e.g. the standard deviation) needs to be chosen to produce a satisfactory curve, but there are algorithms for doing this automatically. Although kernel density estimates have been advocated for use in dispersal data analysis and subsequent spatial modelling (Lewis et al. 2006), a common problem is that they give positive estimates at negative distances.

We can also fit an equation of an appropriate shape to the relative frequency data (such as the Weibull function fitted in Fig. 5.1c), both to smooth the data and to provide input into some individual-based models of population dynamics (see Chapter 6). This equation, which we will refer to here as $f(r)$, is the *'probability density function'*, or simply 'pdf', of a distance R occurring. The estimated probability of a dispersal event occurring between distances r_i and r_j is given by integrating the pdf between these distances

$$P(r_i \leq R < r_j) = \int_{r_i}^{r_j} f(r)dr \tag{5.1}$$

Table 5.1 gives a number of equations that have been used for $f(r)$, several of which can be fitted to a set of observed distances using standard statistical packages. Most have infinite tails. Although in reality infinitely long dispersal distances are clearly impossible (cf. Kot et al. 1996), the probabilities in an infinite tail decline so rapidly that they become effectively zero within a very short distance (Sokal and Rohlf 1995, p. 100).

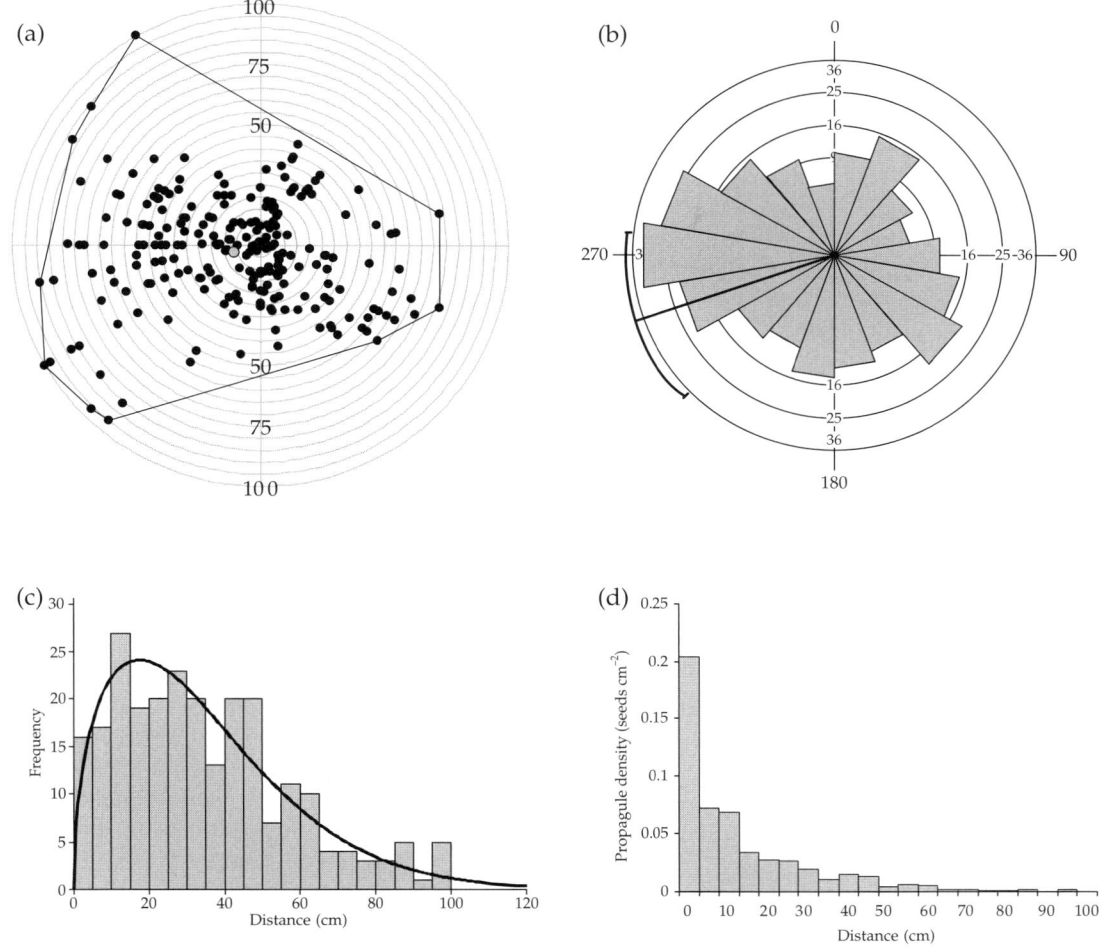

Figure 5.1 Analysis of a point pattern of *Erythronium grandiflorum* seeds (original data, Trail 403, from G. D. Weiblen, published previously in Weiblen and Thomson 1995). The base of the parent plant is at the origin; data are combined from ten individual plants. (a) shows the positions of seeds, the centroid of the pattern (grey circle) and the convex hull (black lines) drawn around all points; concentric circles are in increments of 5 cm, with the annuli between them corresponding to the histogram bins in (c) and (d); (b) is a directional histogram showing the frequencies of different angles relative to the parent—the mean vector is shown at 251° along with its 95% confidence interval; (c) is a histogram of the frequencies of dispersal distances with a fitted Weibull function; (d) frequencies converted to average propagule densities.

Caution is warranted at this point, since confusion could arise from the multiple meanings of the term 'density'. 'Probability density' in the context of a pdf is a statistical concept and does not, in this instance, refer to ecological use of 'density' as the number of propagules per unit area. As we have discussed, $f(r)$ allows us to calculate the probability of a *distance* occurring (in any unspecified direction). We will hereon refer to $f(r)$ as the 'distance pdf'. The area under any pdf is, by definition, unity:

$$\int_{0}^{\infty} f(r)dr = 1 \tag{5.2}$$

just as the sum of the relative frequencies in a histogram is 1.

As we will see in Part C, however, many spatial population models are based on a different pdf, which relates directly to the ecological concept of

Table 5.1 Examples of pdfs used in published studies of the frequency distribution of dispersal distance, i.e. $f(r)$. Note that each of these can be converted to $g(r)$ by dividing the right hand side of the equation by $2\pi r$ (see equation 5.9). References to example uses of the equations are given. Note that these equations may also be used to fit data from controlled releases from point sources (the type of data described in Chapter 4), but the shapes and scales of distance pdfs from point releases may be very different from natural dispersal from entire plants.

	Equation	Mean dispersal distance
[1]Lognormal	$f(r) = \frac{1}{ar\sqrt{2\pi}} \exp\left(-[(\ln r - b)/a]^2/2\right)$	$\exp\left(b + a^2/2\right)$
[2]Inverse Gaussian (or Wald)	$f(r) = \sqrt{\frac{a}{2\pi r^3}} \exp\left(-a[r-\mu]^2/[2r\mu^2]\right)$	μ
[3]Weibull	$f(r) = ab^{-a}r^{a-1}\exp\left(-[r/b]^a\right)$	$b\Gamma\left(1+a^{-1}\right)$
[4]Gamma	$f(r) = [r^{(a-1)}\exp(-r/b)]/[\Gamma(a)b^a]$	ab
[5]Beta (with maximum r)	$f(r) = \frac{1}{r_{max}}\left(\frac{r}{r_{max}}\right)^{a-1}\left(1-\frac{r}{r_{max}}\right)^{b-1}\frac{\Gamma(a+b)}{\Gamma(a)\Gamma(b)}$	
[6]Exponential $g(r)$ converted to $f(r)$	$f(r) = r\exp\left(a - br\right)$	
[3]Mixed Weibull	$f(r) = \sum\left[a_ib_ic_i(b_ir)^{c_i-1}\exp\left(-(b_ir)^{c_i}\right)\right]$	

[1]Greene et al. (2004, used in its $g(r)$ form); [2]Whelan (1988); [3]Higgins and Richardson (1999) – note that the models were fitted to point source releases, but the model is useful in fitting data from entire plants; [4]Watkinson and Powell (1997); [5]Cousens and Rawlinson (2001), used for point source but ay be appropriate for $f(r)$; [6]Weiblen & Thomson (1995).

density. By dividing the frequency within each distance class n_{ij} by the area of the annulus $\pi(r_j^2 - r_i^2)$, we can convert the histogram of distances in Fig. 5.1c to a histogram showing the average density of propagules landing in each annulus (Fig. 5.1d). The 'probability density' of a propagule landing in an (infinitely small) area at a point located a distance r from the parent can be obtained by dividing the densities in each of the columns in Fig. 5.1d by the total number of propagules, N. Note the very different shapes of frequency vs distance and density vs distance: density declines very rapidly at short distances. We will return to this issue later.

If distance and direction are clearly dependent on one another (i.e. the distribution of propagules is 'anisotropic'), we may choose to divide the data into different sectors and to calculate frequency distributions and mean distance separately for each sector. For example, in Fig. 5.1a 18% of the propagules dispersed in the northeast quadrant by a mean distance of 23.6 cm, whereas 31% dispersed in the southwest quadrant by a mean of 41.0 cm. Indeed, to reduce effort in the field or to allow sufficient time to search for the most distant propagules, we may choose only to map some sectors and to leave others unrecorded (Howe et al. 1985).

Point pattern data are not common: small seeds are hard to locate amongst litter or dense vegetation and it is difficult to locate the furthest individuals. Examples are therefore mostly limited to small plants whose propagules are poorly adapted for dispersal (e.g. *Erythronium grandiflorum*: Weiblen and Thomson 1995), large plants with heavy propagules (e.g. *Cycas armstrongii* (Cycadaceae): Watkinson and Powell 1997), or under laboratory conditions with an artificial substrate (e.g. *Cardamine hirsuta* (Brassicaceae): Salisbury 1961).

5.2.2 Density at sampled locations

A complete census of *every* propagule emanating from a given parent is rarely possible. For large plants producing copious quantities of seeds, and for dispersal vectors that move propagules over distances greater than just a few metres, the task is so great that we can only *sample* the pattern. By analogy with meteorology, rather than trying to map where every raindrop lands, we measure the amount of rain falling in a small area (a rain gauge) at known locations. Sampling of the 'seed rain' is done in a similar way, although the number of propagules arriving in the small area (usually some form of trap) is assessed rather than their combined depth. The number of propagules $n(r,\theta)$ in a trap divided by trap area $a(r,\theta)$ provides an estimate of the local density $\rho(r,\theta)$ of dispersed propagules at coordinates r,θ:

$$\rho(r,\theta) = \frac{n(r,\theta)}{a(r,\theta)} \tag{5.3}$$

where r and θ are measured relative to the centres of the parent plant and the centres of the traps (note that Cartesian coordinates x,y can be interchanged with r,θ in this equation). Depending on the sampling design, a may or may not be constant (see p. 109). We can thus visualize the overall pattern of dispersed propagules on the ground as a three-dimensional surface of propagule density which we have sampled at discrete locations. If the sample traps are spaced out around the parent, perhaps in a grid, we can use a contouring program to map this surface. Such maps are rarely produced (though see Augspurger and Hogan 1983, Augspurger and Kitajima 1992 for exceptions).

Rather than sampling on a grid, it is more usual to estimate propagule density at a series of points along one or more radial transects away from the parent. Effectively, each transect is assumed to describe an infinitely thin slice for a particular θ (often related to compass bearings) through the three-dimensional surface of propagule density. Usually each transect is treated separately and an equation is fitted, such that for a specified θ

$$\rho(r,\theta) = \omega_\theta(r) \tag{5.4}$$

where $\omega_\theta(r)$ is a one-dimensional function of the distance from the parent along that particular transect (statistical issues of fitting models to data are considered on p. 109). Example equations that have been used for $\omega_\theta(r)$ are given in Table 5.2. The fitted curves from transects at different angles can be compared graphically (Fig. 5.2) and their parameters can be used to show how shape and scale of the density vs distance relationship vary with direction.

The probability density function (pdf) for a propagule landing in an infinitely small area at a distance r from the parent is

$$g(r,\theta) = \omega_\theta(r)/N \tag{5.5}$$

where N is the total number of propagules dispersed. If we are prepared to assume that there is radial symmetry (isotropy, i.e. exactly the same density vs distance relationship in all directions) it is straightforward to estimate N by integrating propagule density across all combinations of r and θ:

$$N = \int_0^{2\pi} \int_0^\infty \omega(r)\,dr\,d\theta$$
$$= 2\pi \int_0^\infty r\omega(r)\,dr \tag{5.6}$$

(note that the conditioning variable θ is usually omitted from g and ω because under the assumption of radial symmetry their values at a point r,θ are determined only by r). Division of $\omega(r)$ by N also ensures that $g(r)$ sums to unity, a requirement for a pdf. Henceforth, we will refer to $g(r)$ as the 'density pdf' (see also Peart 1985) since it is directly related to propagule density and to distinguish it from $f(r)$, the distance pdf. Some equations that have been used for $\omega(r)$, for example power functions, do not have a finite solution to equation 5.6 and therefore $g(r)$ cannot be determined. Although such equations may give reasonable fit over the range of the data, they should be extrapolated with great care.

Rather than beginning with a one-dimensional density vs distance equation for $\omega(r)$ and converting it to a two-dimensional symmetrical pdf (as above), we can choose $g(r)$ directly from among a library of potential symmetrical pdfs whose properties are well known (Table 5.2). There is a wide range of alternatives which may be considered as arbitrary equations that fit data well, or they can be chosen according to their derivation from diffusion theory or other plausible assumptions (Clark et al. 1999; Katul et al. 2005; Snäll et al. 2007). Functions differ in their shapes at very short distances; they may be flat or steeply declining at the source, or they may incline to a peak away from the source. At longer distances, a pdf may have a tail that declines more slowly than an exponential ('fat-tailed') or its tail will be as steep as an exponential or steeper ('thin-tailed') (There are a range of definitions for fat-tails that are used in other fields, such as economics, but following Kot et al. (1996) this definition has become established in ecology). Two functions may have apparently similar shapes, but they may differ considerably in their ability to fit observed data. Some functions are particularly flexible and can take on very different

Table 5.2 Examples of equations proposed or previously fitted to density vs distance data [$\omega(r)$] and for density pdfs [$g(r)$]. $\omega(r)$ can be converted to $g(r)$ by dividing by the normalization constant (which must first be derived—see equation 5.7). $g(r)$ can be converted to $f(r)$ by multiplying by $2\pi r$.

	Equation	Comments
ω (r)		
[1]Power	$\omega(r) = ar^{-b}$	Has infinite density as $r \to 0$.
[2]Power	$\omega(r) = a(1 + br)^{-c}$	This form allows the power function be converted to $f(r)$, since it has a finite density at $r = 0$.
[3]Negative exponential	$\omega(r) = a\exp(-br)$	Density at origin $= a$.
[4]Weibull	$\omega(r) = a\exp(-br^{c})$	Shape depends on parameter values
[5]Half normal	$\omega(r) = \dfrac{2}{ka\sqrt{2\pi}}\exp(-r^2/[2a^2])$	Plateau at $r = 0$
[5]Half Student t	$\omega(r) = \dfrac{2}{ka}\dfrac{\Gamma([b+1]/2)}{\sqrt{b\pi}\,\Gamma(b/2)(1+(r^2/a^2)/b)^{(b+1)/2}}$	
[6]Cauchy	$\omega(r) = \dfrac{1}{k\pi a}\left(1 + \dfrac{r^2}{a^2}\right)^{-1}$	Steep at $r = 0$. Cannot be converted to $f(r)$.
[7]Mixed model	$\omega(r) = ar^{b} + c\exp(-mr)$	Cannot be converted to $f(r)$. Mixed models can incorporate additive combinations of any of the other equations, or multiples of the same equation.
g (r)		
[8]Exponential (Laplacian)	$g(r) = \dfrac{1}{2\pi a^2}\exp(-r/a)$	Thin tail. Concave at origin.
[8]Normal (Gaussian)	$g(r) = \dfrac{1}{\pi a^2}\exp(-r^2/a^2)$	Thin tail. Plateau at $r = 0$
[8]Generalised exponential	$g(r) = \dfrac{c}{2\pi a^2 \Gamma(2/c)}\exp(-r^c/a^c)$	Fat tail and steep at $r = 0$ for $c \leq 1$ or thin tail and plateau at $r = 0$ for $c > 1$.
[8]2Dt	$g(r) = \dfrac{a}{\pi b}\left(1 + \dfrac{r^2}{b}\right)^{-a-1}$	Shape depends on parameter values; Gaussian at large a, Cauchy as $a \to 0$
[4]Log-normal $f(r)$ converted to $g(r)$	$g(r) = \dfrac{1}{a(2\pi)^{1.5}r^2}\exp(-[\ln(r/b)]^2/[2a^2])$	Peaked very close to source,

[1]Clark *et al.* (2005) – note that these authors constrained the model to be constant above and below certain distances, to allow it to be converted to a *pdf*; [2]Wallinga *et al.* (2003); [3]Nadeau and King (1991); [4]Greene *et al.* (2004); [5]Clark *et al.* (2005) – note that an additional constant has been added to allow for fitting to density data; [6]Blanco–Moreno *et al.* (2004) – additional constant added; [7]Bullock and Clarke (2000); [8]Clark *et al.* (1999).

shapes according to the data to which they are fitted. Greater flexibility to fit data can also be obtained by using an additive (weighted) combination of two or more *pdfs*: these so-called 'mixed models' are considered to be more realistic of the real world, in which there is often more than one dispersal vector, each moving a proportion of propagules according to a different pdf. For example, we might combine a thin-tailed with a fat-tailed function to produce a pdf with hybrid characteristics. Adding together two thin-tailed pdfs creates a model that displays fat tailed characteristics over moderate distances (although adding two thin tails never makes the extreme tail fat). We will return to the choice of equations on p. 90.

Whether we begin with $\omega(r)$ or with $g(r)$, the aims of fitting models to dispersal data are the same: (a) to produce the best descriptor of the empirical data, so that (b) when we use the data to model processes that involve dispersal there will be a high probability that we will make sound predictions and discover biological rather than purely mathematical phenomena. Even after filtering out the 'noise' in our data, the many interacting biological processes and environmental variables make it unlikely that any two-, three-, or even four-parameter equation will fit

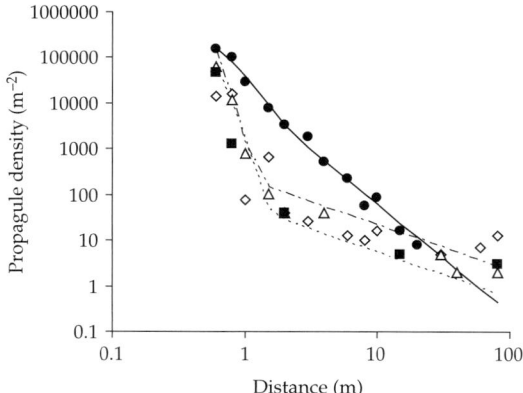

Figure 5.2 Dispersal of *Calluna vulgaris* in four compass directions (redrawn from Bullock and Clarke 2000): SE △; SW ▪; NE ●; NW ◇. Note both axes are logged: zeros therefore cannot be displayed. Lines are the model $\omega(r) = a\exp(-br) + cr^{-d}$ (see Table 5.1) fitted separately for each transect (the model was unable to be fitted to the SW transect).

the underlying relationship perfectly (indeed, since trap data are discrete observations, there can never be a perfect fit to a continuous function, even if the data have been sampled directly from the pdf that they are being fitted to). There is unlikely to be one 'correct' model in any single circumstance and certainly no single, most appropriate equation for all circumstances. Thus, we need a range of *pdfs* that, in most circumstances, allow us to select one that fits to an acceptable level of tolerance '*both near and far*' (Clark et al. 1999), as well as in the middle. We also need to design our sampling regimes so that, for an acceptable level of effort, we can achieve a satisfactory description of the scale and the key aspects of the shape of the underlying relationship. We will return to the issue of sampling design at the end of the chapter.

5.2.3 Comparison

We have introduced and contrasted two distinct ways of looking at dispersal patterns: point patterns, from which we can estimate dispersal distances and angles directly, as well as characterizing many other aspects of spatial pattern; and density surfaces, which we sample and then describe by equations. This dichotomy also reflects the way in which many population models

are structured: 'individual-based' models predict the positions of each plant or propagule, while 'continuous' models predict a density surface. It is crucial when examining dispersal data throughout this chapter, as well as the models in the rest of the book, to recognize the difference between data produced by the two approaches and to treat them unambiguously.

While the two probability density functions that we can derive from dispersal data, the distance pdf $f(r)$ and the density pdf $g(r)$, along with the density vs distance relationship $\omega(r)$ can all be referred to as dispersal 'curves', they are not the same thing and they are usually very different shapes. However, they are clearly related and we can convert one to the other if we are willing to assume radial symmetry, by integrating $g(r)$ in 360° around the parent:

$$f(r) = \int_0^{2\pi} g(r)$$
$$= 2\pi r g(r) \qquad (5.7)$$

(Peart 1985). The multiplier r in the right hand side of the equation is critical, since it completely changes the shape of the function. Fig. 5.3 shows the shapes of selected mathematical functions for $f(r)$ and their corresponding shapes of $\omega(r)$ and $g(r)$. This distinction between the shapes of distance and density pdfs is more than just academic: if we are to compare field data with the predictions of models, we must ensure that we are comparing like with like (see Box 5.1). If a simulation model is based on a density pdf, for example, then we must make sure that the function we use in the model is a shape appropriate for $g(r)$ (or make the conversion from $f(r)$ to $g(r)$ first). If a theoretical model predicts a particular outcome (e.g. an accelerating population edge) for a particular shape of pdf, field data need to be compared with the appropriate type of pdf if we are to decide whether or not that behaviour is likely in our species of interest.

We note that some empirical studies convert trap count data to counts per annulus around the parent, using simple geometry rather than curve-fitting (as indeed we did in the other direction—see Fig. 5.1d), and state that this transformation 'corrects' for the reduction in the proportion of a circular area sampled by a fixed trap size. Thus it is perhaps implied

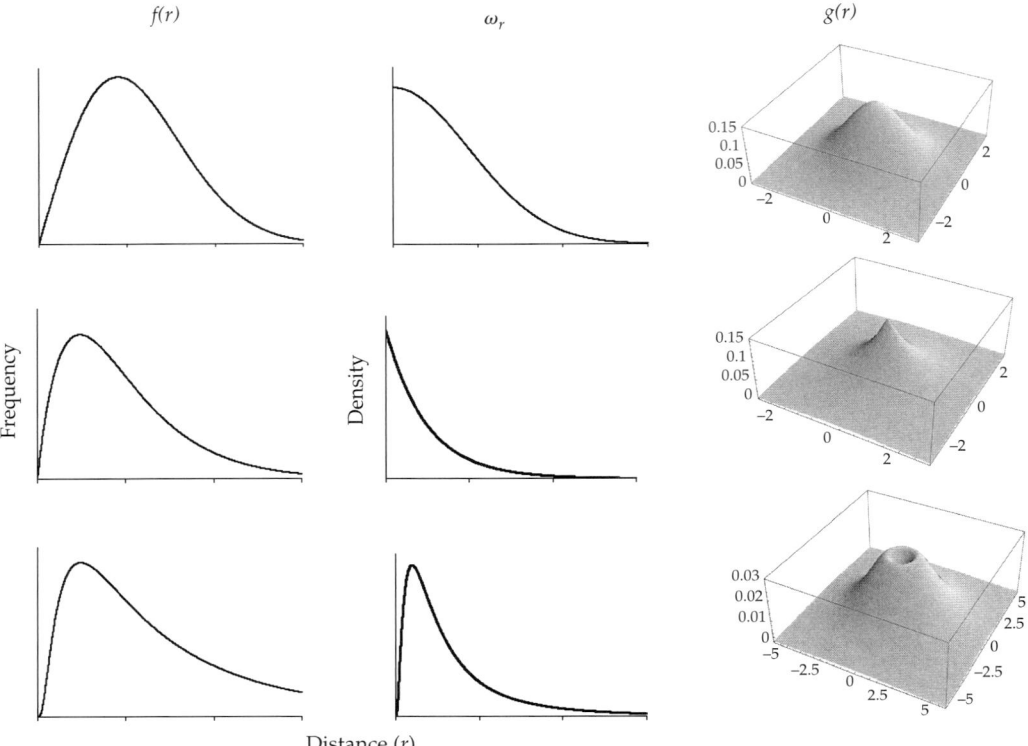

Figure 5.3 Possible shapes of curves for probability of dispersing a given distance [$f(r)$], and the corresponding density at a given distance [$\omega_\theta(r)$] and density pdf [$g(r, \theta)$] (after Peart, 1985). For example, when a Gaussian $g(r)$, which declines sigmoidally from a density 'plateau' at the origin, is converted to $f(r)$ it becomes the Rayleigh distribution, a special case of the Weibull function (Tufto et al. 1997; Clark et al. 1998b; Snäll et al. 2007) which is peaked and skewed towards longer distances. A Laplacian (negative exponential) $g(r)$ converts to an $f(r)$ that is a special case of the gamma distribution, again with a peak. In general, if $g(r)$ has a finite intercept or is zero at $r = 0$, its equivalent $f(r)$ then must approach zero at $r = 0$ and hence $f(r)$ will have a peak: mathematically, it must do. Since the possible dispersal distances are infinite at one extreme and bounded by zero at the parent stem, we expect $f(r)$ to be right-skewed. A lognormal $f(r)$ (as well as certain cases of the Weibull and gamma), which has its origin at zero and is concave at very short distances, retains its zero origin when converted to $g(r)$, but is then convex at short distances densities; when shown as a three-dimensional surface, this $g(r)$ thus resembles a volcano. At the other end of the distributions, their 'tails' also differ in shape, since $f(r)$ is proportional to $rg(r)$. If the tail of $g(r)$ declines exponentially with distance, for example, its equivalent $f(r)$ will necessarily have a fatter tail than an exponential (though the difference may not be great). However, for particular parameter values both $f(r)$ and $g(r)$ may be thin or they may both be fat.

that trap counts are somehow wrong. We would argue that neither way of presenting the data is wrong: they are simply two alternative ways of looking at data. Neither is incorrect, they are just different.

Another issue in which the distinction between $f(r)$ and $g(r)$ is important concerns the calculation of the mean dispersal distance. If we measure individual dispersal distances in the field, then the calculation of the mean is trivial: it can be calculated directly from the data as $\sum r/N$. The calculation of

the mean distance from density vs distance transect data is less easy. An equation (ω_r) needs to be fitted to the data, then (assuming dispersal is in two dimensions and isotropic) converted to $f(r)$ using equations 5.5 to 5.7. The mean can then be found by two methods: simulation, taking a large number of random distance values from $f(r)$ and then calculating $\sum r/N$; or, for those with mathematical ability, using the moment generating function

$$M(s) = \int_0^\infty \exp(sr)f(r)dr \qquad (5.8)$$

where s is a dummy variable. The mean can then be calculated as

$$\mu = \left[\frac{d}{ds} M(s) \right]_{s=0} \tag{5.9}$$

Consider the simple case of an exponential function for density vs distance along a transect, $\omega_r = b \exp(-br)$. Although the mean for an exponential distribution in one dimension is $1/b$, since the distribution of seeds is in *two* dimensions, not one, use of the moment generating function shows that the mean distance will be double (i.e. $2/b$). Clark et al. (1998b) give an equation for calculating the mean distance for a two-dimensional function that

includes the normal and exponential as special cases. For some functions, however, the moment generating function cannot be calculated, in which case simulation can still be used to give an estimate of the mean.

5.3 Empirical entire-plant dispersal data

There are many published case studies of dispersal, covering a wide range of plant species, in many environments, and dispersed by very different vectors. Rather than attempting a comprehensive review and meta-analysis, our aim here is to give

---

Box 5.1 Curves and kernels: avoiding inconsistency and confusion

Unambiguous terminology is important in any branch of science, otherwise there is the potential for miscommunication. We have deliberately avoided the use of both 'dispersal curve' and 'dispersal kernel' in this chapter, even though both are widely used terms for empirical dispersal data. This is because they have been used inconsistently, in some cases to mean $f(r)$ or in other cases to mean $\omega(r)$ or $g(r)$. Yet, as we have seen, the alternative pdfs take on very different shapes. Thus, two researchers debating the shape of a dispersal curve/kernel (for example whether or not it is leptokurtic or whether its tail is fat or thin) can become completely confused, since one may be referring to the shape of $f(r)$ while the other may be referring to $\omega(r)$. As a result, it is possible that the wrong shaped function could be incorporated into a model as a result of loose terminology.

Modellers now appear to use 'kernel' to refer to any pdf representing the movement of propagules, in some cases equivalent to our $f(r)$ and in other cases equivalent to our $g(r)$, which they then define unambiguously by assigning a particular equation to the kernel. However, 'redistribution kernel' or 'dispersal kernel' was first introduced in papers on models that treated space as one-dimensional (e.g. Kot and Schaffer 1986); in that particular case only, the pdf of a propagule dispersing a given distance (i.e. $f(r)$) is the same shape as the probability density of propagules landing in an infinitely small area at that distance, i.e. $g(r)$. Those papers, as well as more recent papers on two-dimensional models, often state that the kernel is the 'pdf of moving a given distance' (or similar wording), which in turn may have resulted in the current use by many field ecologists of $f(r)$ as the dispersal kernel.

To try to achieve greater consistency and therefore less confusion, Nathan and Muller-Landau (2000) defined dispersal kernel as 'the (pdf) of the location of seed deposition with respect to source, yielding the probability of a seed landing per unit area as a function of the distance from its source' (i.e. our $g(r)$); they gave the name 'distance distribution' to $f(r)$. Few researchers appear to have adopted their nomenclature consistently. Higgins et al. (2003a) stated that 'dispersal kernel' should be used for offspring density and not for propagule density (perhaps because modellers often do not account explicitly for propagule mortality), but again this re-definition has not been widely adopted. Cousens and Rawlinson (2001) referred to $g(r)$ as the two-dimensional pdf (since density involves area, a two-dimensional quantity) and $f(r)$ as the one-dimensional pdf. However, this could be confused with models that treat space as being either one- or two-dimensional. Willson (1993) referred to the frequency distribution of dispersal distances as the 'circular distribution', recognizing that it represents the combined dispersal in all directions. The relationship between disease symptoms, which are dependent on propagule density, and distance along a transect is known by plant pathologists as a 'disease gradient'.

Rather than try again to redefine terms that are already in widespread use, we will consistently use 'distance pdf' to refer to $f(r)$ and 'density pdf' to refer to $g(r)$. We will use 'kernel' only when discussing models (in Chapters 6–8) and not when discussing empirical data and we will define it unambiguously with reference to $f(r)$ or $g(r)$.

examples, using the approaches which we have just introduced, and summarize what they tell us about dispersal. Methodologies used in field studies differ considerably: in particular, the sources of propagules may be individual plants, lines or blocks of plants, the edges of an extensive forest, or a number of plants distributed throughout a mixed community. Extreme care needs to be taken to ensure that the data from different types of situation are viewed separately. We will deal firstly with the straightforward situation in which the propagule shadows from the nearest conspecific individuals do not overlap, then turn our attention to the more complicated situation in which shadows from neighbouring plants overlap.

5.3.1 Non-overlapping shadows

Non-overlapping propagule shadows provide the easiest context in which to examine dispersal. All seeds found on the ground or in traps are considered to have originated from the same maternal parent, because there are no close neighbours of the same species. Many of these situations will be highly artificial, however, such as isolated plants growing in parks or gardens or plants growing in pots that are either taken to a location completely lacking that species or are placed in a laboratory. It can, however, be argued that isolated plants may be encountered at the margin of a species that is spreading into new territory. A subjective decision has to be made as to whether or not in a plant's true habitat the closest neighbours are far enough away for their shadows to be separate; an incorrect decision will inflate the 'tail' of the pdf of the focal plant. Note also that we consider only the primary phase of movement away from the parent (as in most published studies).

5.3.1.1 Individually-mapped propagules

First we consider studies that have mapped all propagules around a parent, since these lead naturally and directly to an assessment of the distance pdf and statistical parameters of distance. For logistical reasons, however, the plant species are inevitably those with limited dispersal, with primary dispersal not usually enhanced by wind or animals, but including species where seeds are forcibly ejected

from the parent. As a result, the mean distance dispersed and the area of the plant's seed shadow are small.

Raphanus raphanistrum is an annual weed, growing to around a metre in height in cereal crops; in the absence of competition they can reach 1.5 m in diameter, sagging under the weight of their seed pods. Pods can reach around 7 cm in length. When growing in the open, pods are produced throughout the plant's volume, whereas in a crop most pods are located close to the top of the canopy. In the absence of a mechanical harvester, most pods fall directly to the ground. Thus, the area over which pods disperse is not much greater than the vertically projected area of the plant. The frequency distributions of dispersal distance are strongly peaked and with only a short 'tail' (Fig. 5.4). The mode, mean, and the maximum distance become closer to the base of the parent in the presence of a crop. Of three equations fitted to the data, the Weibull most often gave the best fit to the distance frequency data, followed closely by the gamma distribution (Taghizadeh 2007) while the log-normal consistently fitted poorly. When converted to density vs distance, the distributions take on a more sigmoidal shape, with a plateau at short distances.

Cycas armstrongii usually has one stout stem (though sometimes they may branch), typically up to 3–4 metres in height. Propagules are 2–4 cm in diameter and up to 80 of these are held in a ring around the top of the trunk (Watkinson and Powell 1997). Distance pdfs from plants of 1.85 and 3.23 m in height have been found to have a mean of 58 cm and 38 cm respectively (Fig. 5.5b, c). A shorter plant (1.28 m) had a distribution with the mode at the shortest distance category and a mean dispersal distance of only 21 cm (Fig. 5.5a). The gamma distribution showed an excellent fit to all three distance histograms; all three histograms in Fig. 5.5 convert to steep, monotonically declining density vs distance relationships.

Most data sets of mapped propagules follow a similar pattern to the two case studies discussed above: a unimodal distance pdf, peaked either under the parental canopy or not far from it and with a short tail (e.g. Figs. 5.1; Carey and Watkinson 1993; Augspurger and Hogan 1983; Theide and Augspurger 1996). They also convert to a steep,

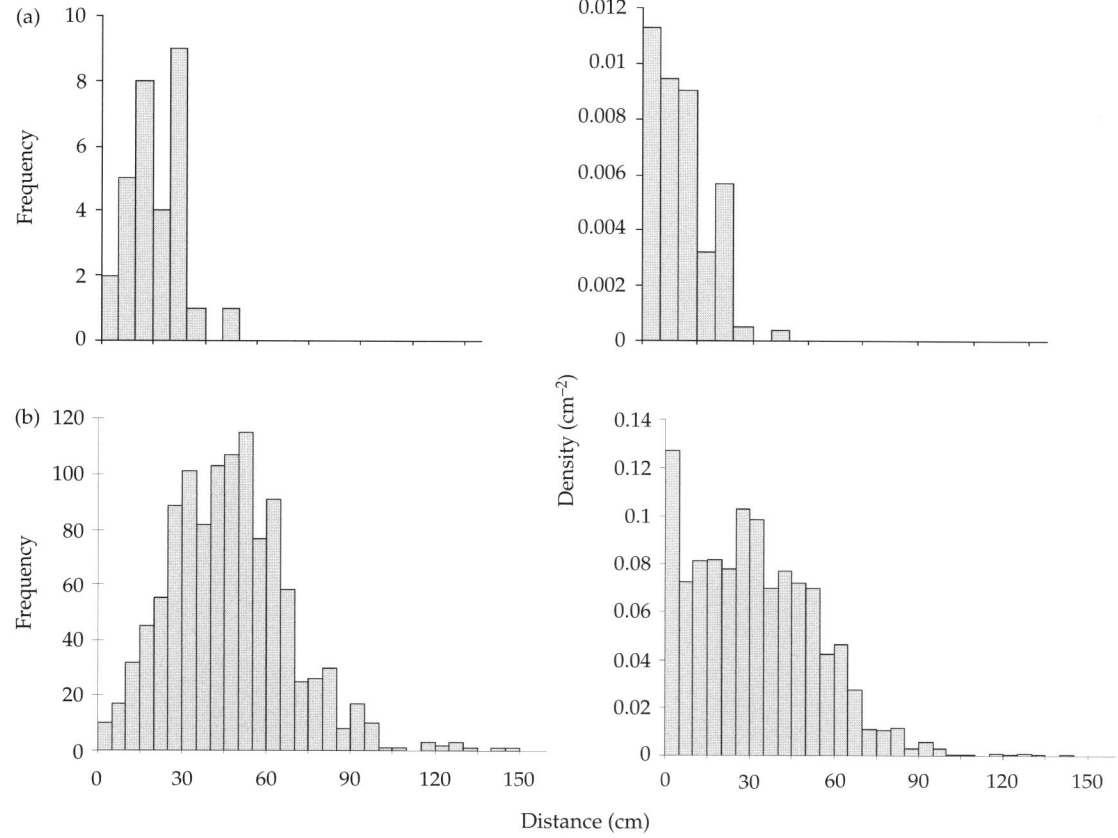

Figure 5.4 Density-dependence of dispersal in *Raphanus raphanistrum* (Taghizadeh 2007, reproduced courtesy of the author). (a) Grown in the presence of 180 wheat plants m^{-2}; (b) grown in the absence of competition. Left hand side: frequency distributions of dispersal distances; right-hand side: densities in annuli at increasing distances.

monotonically declining density pdf, often found to be leptokurtic, that is, steeper at the source than a two-dimensional normal distribution. In a meta-analysis of published data, Willson (1993) found that for herbaceous species the mode was invariably less than 1 m; for wind-dispersed trees the average distance of the mode from the parent was 27 m, compared to 9 m for species dispersed by vertebrates (though the data included some studies of artificially released propagules).

There are some studies, however, that appear to have flat (Fig. 5.5a) or steeply descending (Fig. 5.6) distance pdfs at the lowest densities. As a consequence of equation 5.7, this should not be possible:

for a finite density at the source, the distance pdf should approach zero at the source. One plausible explanation is that the distance pdf for these species does, in fact, have a zero intercept, but this is obscured in the process of constructing a histogram. If the bin width chosen for the histogram is sufficiently large, the first histogram column may straddle the (real) peak in the $f(r)$ distribution, making the first column the one with the greatest frequency. The steeper the peak in $f(r)$ (i.e. species with very limited dispersal) or the longer the tail, the more likely this is to occur.

Some data sets have an apparently multimodal $f(r)$ (e.g. Hoppes 1988) and they are thus poorly

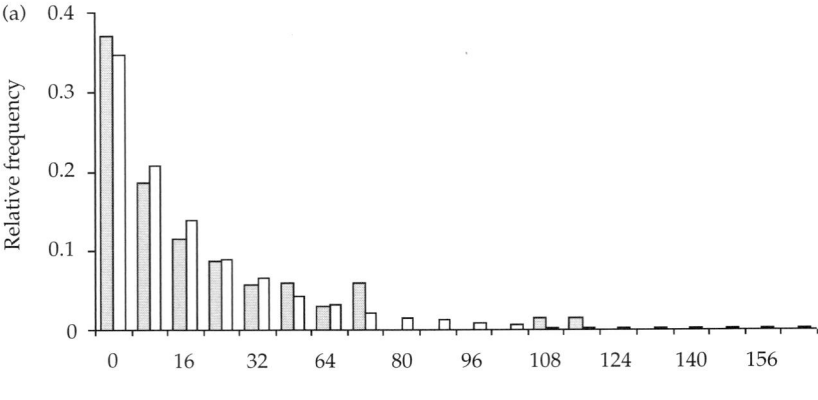

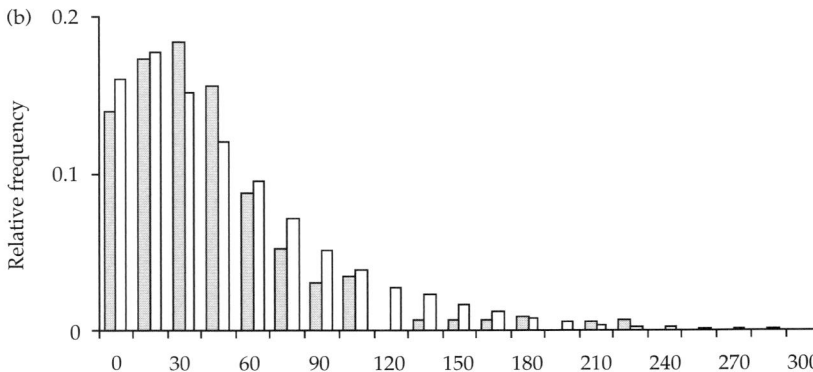

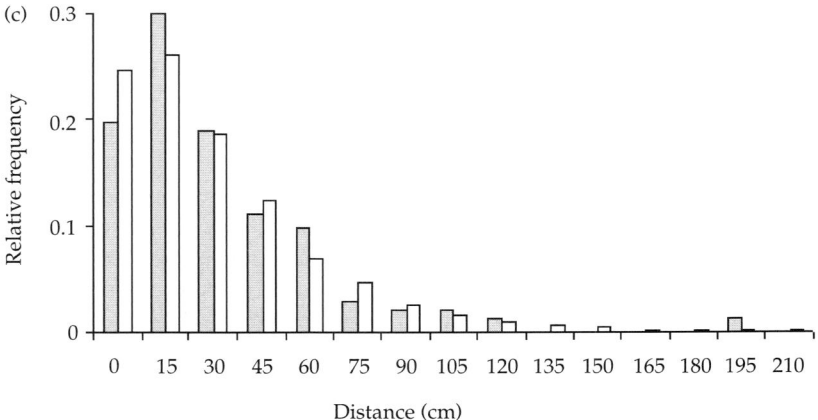

Figure 5.5 Observed (filled) and fitted gamma (open, $f(r)$) relative frequency distributions for the probability of *Cycas armstrongii* dispersing between two given distances from three plant heights (redrawn from Watkinson and Powell, 1997): (a) 1.28m, (b) 1.85 m, (c) 3.23 m. Note the different axis scales.

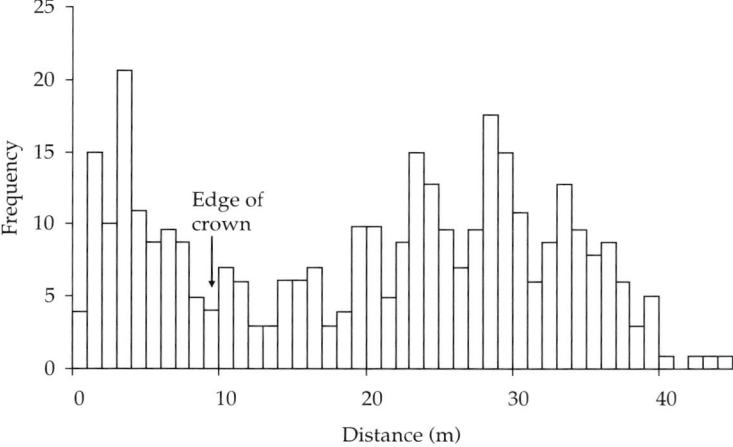

Figure 5.6 Frequency distribution of distances dispersed by seeds ($n = 345$) from a *Hura crepitans* tree (redrawn from Swaine and Beer 1977, reproduced courtesy of Blackwell Publishing).

described by any simple mathematical function. There are a number of reasons for such distributions, including:

• Lack of replication, especially in plants producing only a few propagules from a small number of discrete fruits on a plant, will result in uneven distance distributions reflecting the idiosyncrasies of single plants (e.g. *Cardamine hirsuta*: Salisbury 1961). If the data from several replicate plants are combined (as in Fig. 5.1), the peaks may then disappear. However, since real plants in real habitats may produce only a few propagules, the distributions should not necessarily be considered as erroneous.
• Multimodal distance distributions may result when the effects of multiple dispersal vectors, each with their own peak, are superimposed. We will explore this further on p. 103. The bimodal distance distribution of *Hura crepitans* seeds (Fig. 5.6) was hypothesised by Swaine and Beer (1977) as being due to two dispersal routes for their explosively ejected seeds: some seeds travel unimpeded away from the canopy, while others rebound internally amongst the tree branches. The edge of the tree's crown corresponded with the trough between the two peaks of the distribution.
• Animal behaviour can also generate irregular, multimodal dispersal distributions where food sources or suitable perches are located at discrete,

perhaps widely-spaced, sites within the landscape. Birds and bats may have habitual roosting sites and in the breeding season many animals travel to and from their nest. Occasional trees within an open community, such as a grassland or a desert, offer a very patchy environment that restricts the movements of tree-dwelling birds (Wenny and Levey 1998). Waterfowl may move between discrete water bodies. In a study of the mistletoe *Phoradendron californicum* (Viscaceae), the great majority of seeds were wiped on branches in the tree of origin (causing a peak in the distance distribution at a very short distance) and in a small number of nearby trees (Fig. 5.7).

5.3.1.2 *Propagules caught in traps*
It is not feasible to map individual propagules of species dispersed by wind and animals. Our knowledge of dispersal from entire plants in these species is dependent on data collected from traps and quadrats, usually along linear transects, which we then assume are point estimates of the density of propagules. The results from a study of dispersal in an African tropical forest (Fig. 5.8) are typical. All curves, showing $\omega(r)$ on a log-scale, indicate that the highest densities of seeds in traps are found closest to the maternal parent; density then declines steeply, with few seeds being found in traps 60 m from the source (if density was to be presented on an untransformed scale, the decline would be far steeper).

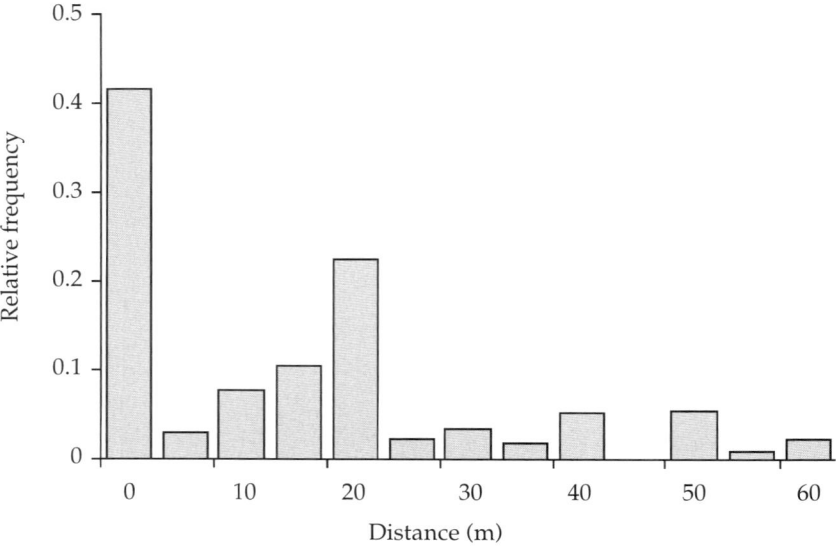

Figure 5.7 Relative frequency distribution of dispersal distance from the parent plant in *Phoradendron californicum* seeds, obtained from visual observations of birds (redrawn from Overton 1996).

Trees that were categorized as being mainly dispersed by wind, birds, or monkeys showed broadly similar trends (although detailed analysis suggested that $\omega(r)$ for wind-dispersed species was more convex at the source). The estimated mean dispersal distance was slightly greater for bird- or monkey-dispersed species than for those dispersed by wind; however, there was considerable variation among species within each category. In our experience (though not based on a formal analysis) most published density vs distance graphs are steep at the origin, decline rapidly with distance and within the range of the data are well-fitted by either an exponential or power function. Several modellers make similar observations about the shape of $\omega(r)$: that its shape is usually leptokurtic, '*with more seeds landing close to the origin than in a normal distribution*' (e.g. Kot et al. 1996; Lewis and Pacala 2000) and they quote examples of studies of propagule density.

There are some important statistical pitfalls to be aware of when examining the overall shape of $\omega(r)$. Like any data, densities in traps include sampling error as well as true variation in $\omega(r)$; fitting curves (equations) to data is a way of trying to summarize the shape and scale of the underlying relationship between density and distance. Thus, a number of studies fit several equations to densities along a transect and compare their goodness of fit (as in Fig. 5.8; see Snäll et al. 2007 for an analysis of pollen and fungal spore dispersal). Poor sampling intensity or high variance, however, will result in equations with very different shapes fitting equally well, especially if extrapolated away from a limited range of trap distances. Even in cases where sampling intensity is considerable and an equation may explain a very high proportion of the variance in the data, it may still give a misleading interpretation of shape: can we expect a simple mathematical function to fit an entire seed shadow perfectly? Greene and Calogeropoulos (2002) note that although both the 2Dt and the log-normal distribution (in their case a log-normal $f(r)$ transformed to density using equation 5.7) are both more flexible than some alternative functions, if either model is to fit well in the tail, it may well over-estimate densities close to the source (this phenomenon is known by some statisticians as 'the tail wagging the dog': cf. Skarpaas et al. 2004). If we are just comparing the tails of the distributions (see p. 81), two functions having very different tail shapes will often fit data almost equally

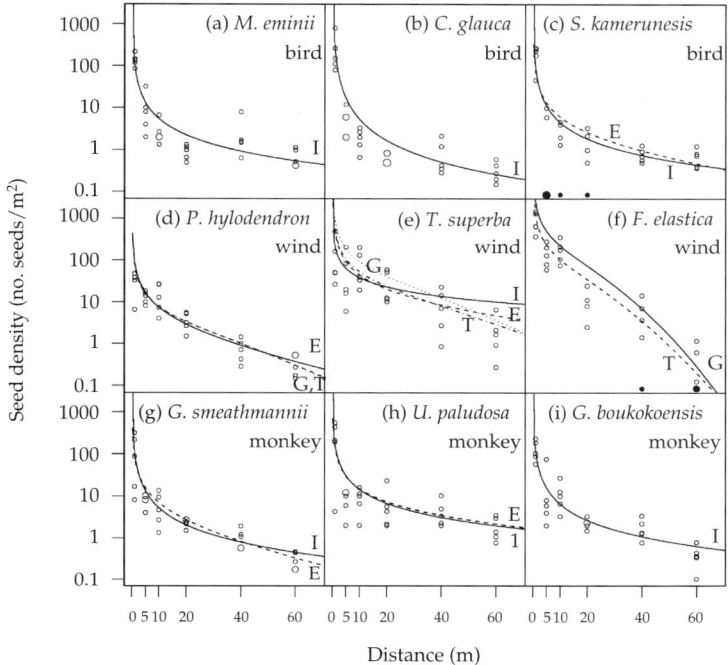

Figure 5.8 Seed densities in traps at increasing distances from trees in a Cameroon forest (modified from Clark et al. 2005, reproduced courtesy of Ecological Society of America). Letters are for best fitted equations: G (Gaussian), T (Student t), E (exponential). Species are: *Cleistopholis glauca, Maesopsis eminii, Staudtia kamerunensis* (Myristicaceae), *Gambeya boukokoensis* (Sapotaceae), *Garcinia smeathmannii* (Clusiaceae), *Uapaca paludosa* (Euphorbiaceae), *Terminalia superba* (Combretaceae), *Pteleopsis hylodendron* (Combretaceae) and *Funtumia elastica* (Apocynaceae).

well (Wallinga et al. 2002). This is because of the very low probabilities in the tail and the consequently low numbers of propagules observed in each trap.

Clark et al. (1999) suggested that a density pdf that is convex close to the source (e.g. a two-dimensional normal distribution) could be expected to result from a distributed source of propagules, such as the broad crown of a large tree. They also considered that a 'skip distance' (a gap between the source plant and the first propagules) is unlikely in real stands. In contrast, Greene et al. (2004) argued that a very short skip distance would be realistic, since the presence of the stem of the parent precludes seeds landing at exactly zero distance. Harper (1977) comments that 'poverty of seedfall' close to the parent is characteristic of isolated plants, but it is unclear whether he is referring to density pdfs or distance pdfs. Unfortunately, there are few data sets that begin at short enough distances to be able to detect the true shape close to the parent (many researchers would perhaps

argue that propagules landing beneath the parental canopy have *failed* to disperse and are therefore of little relevance). Certainly there are some studies in which the density pdf has a clear mode away from the source (Fig. 5.9; e.g. Cremer 1966). We can speculate on the reasons for these: some animals may take fruits away from the source before eating the flesh and discard the seeds in a non-fruiting tree nearby (p. 67); wind-dispersed seeds may only be liberated from the parent on days when there are strong winds that carry all seeds well away from the parent (p. 51). We would expect a peak in $\omega(r)$ where a dispersal vector is predominantly unidirectional (such as strong winds or water in a river): all individuals will be funnelled in a single direction and the shape of $g(r)$ will approach that of $f(r)$ (and as noted previously, $f(r)$ is expected to be peaked). Although not for single plants, Bullock et al. (2003) for *Rhinanthus minor* (Scrophulariaceae) (patch source), Howard et al. (1991) for *Bromus*

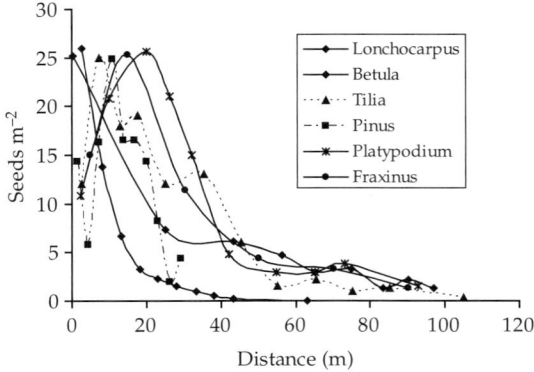

Figure 5.9 Densities of seeds in traps at increasing distances away from single trees within forests (redrawn from Greene and Calogeropoulos 2002, reproduced courtesy of Blackwell Science). Species are: *Fraxinus excelsior* (Wagner 1997); *Lonchocarpus pentaphyllus* (Fabaceae: Augspurger and Hogan 1983); *Platypodium elegans* (Fabaceae: Augspurger 1983); *Pinus strobus* (Rudis et al. 1978); *Betula uber* (Ford et al. 1983); *Tilia americana* (Tiliaceae: Greene and Calogeropoulos 2002).

spp. (line source), and Shirtliffe and Entz for *Avena fatua* (patch source) all found peaked density distributions in one direction where machines, which are predominantly uni-directional vectors (p. 71), were involved in dispersal.

There has been considerable ecological interest in the shape of the tail of $\omega(r)$. This has been prompted by the results of model predictions of rate of population spread (which we will examine in Chapter 6). One of the best studies designed to examine tail shape is shown in Fig. 5.2. There were traps at a large number of distances and sampling intensity (trap number) was increased to maintain the precision of density estimates at the greatest distances. The fit to both power and exponential functions was poor; a composite of the two functions, with the power function's fat tail (declining more slowly than an exponential), gave a much better description of the data, both near to the source and in the tail.

In a meta-analysis of published data, Portnoy and Willson (1993) found that a greater proportion of data sets had a $\omega(r)$ tail that fitted the power function better than an exponential The data sets had fewer densities than in Fig. 5.2 and the inclusion of points close to the mode may have biased the estimates of the asymptotic tail shape in some cases. The analysis also included both density and distance

frequency data, as well as different types of source (single plants, stands of plants, or artificial releases from point sources). It would be interesting to repeat the analysis with a more extensive collection of data.

So far, most of our discussion has been on distance and not direction. Many researchers routinely place traps along transects in different compass directions away from a source plant. Where wind is a major dispersal vector, we usually find that more propagules are dispersed, and often to greater distances, in the direction of the prevailing winds. Examples of such anisotropic dispersal include *Calluna vulgaris* and *Erica cinerea* (Ericaceae) (Fig. 5.2), *Vulpia ciliata* (Poaceae) (Carey and Watkinson 1993), and *Senecio jacobaea* (McEvoy and Cox 1987). Marshall and Butler (1991) reported asymmetrical density distributions for *Rumex obtusifolius* (Polygonaceae) when growing in the open, but almost isometric dispersal when growing behind a windbreak. Maps showing contours of density around a source tree also show distinct asymmetry (e.g. Augspurger and Hogan 1983; Augspurger and Kitajima 1992 for tropical wind-dispersed trees). If there is 'directed dispersal' in animal-dispersed species we would again expect to see anisotropic dispersal. However, there are few examples where this has been demonstrated formally.

5.3.1.3 Seedling densities

It is logical to argue that the distribution of successful offspring is more relevant to population dynamics than the positions of seeds (though it is the seed bank rather than seedlings that may ensure recovery after some major environmental perturbations). It is the plants that successfully establish which constitute the next generation, that form the spatial patterns that we observe in vegetation, and eventually produce their own offspring. Thus, ecologists often refer to 'effective' or 'realized' dispersal. Hence we can map seedlings rather than seeds and estimate their local population density; indeed, it is far easier to locate and identify seedlings than to locate small seeds amongst litter or vegetation. Distributions of seedling density follow the same sorts of patterns as seed density and the same equations are often fitted to them. For a hardwood forest in Quebec, Greene et al. (2004) found that in most cases the log-normal [for $f(r)$, converted to $g(r)$], Weibull and *2Dt* fitted seedling density data equally well. This was despite their very different shapes, the log-normal having zero density at zero distance and the 2*Dt* approaching a constant density at short distances. The similar fit may have been because of a lack of data near the source tree. Where the models gave very different fits, the log-normal was best more often than the other two functions.

Until this point in the book, we have deliberately avoided the concept of effective dispersal. This is because realized dispersal confounds dispersal *sensu stricto*, with all other processes that lead to successful seedling establishment, including diseases, competition with previous generations, seed predation, and grazing. Thus, for example, intense competition from a maternal parent and high levels of predation where post-dispersal seed densities are greatest, could lead to seedling density vs distance relationships that are very low at the origin and thus have a distinct mode. If we are to *understand* realized dispersal, rather than to simply map seedlings, we need to understand all the component processes and to model them, but that would be beyond the scope of a book of this size and would divert the reader away from the core topic. However, realized dispersal does emerge in later chapters when considering the consequences of dispersal such as Janzen–Connell patterns (p. 154), as well as later within the present chapter.

5.3.2 Multiple source plants

We now consider dispersal in situations where there are multiple parental plants whose propagule shadows overlap. Such situations may already exist in the field, such as within a habitat in which the species is common, at the edges of dense stands of a species (e.g. forest edges), or they may be created for the sole purpose of examining dispersal experimentally. One common reason for creating multiple plant sources is in species with low fecundity, to ensure that sufficient propagules are trapped to be able to describe their distribution. It is assumed that a small patch of plants may, in a species that disperses over large distances, give a pdf equivalent to a single plant. In the case of stands of trees, foresters often estimate the deposition of seeds in traps along lines perpendicular to the forest edge: they are interested in natural

re-population of clearings or expansion into non-forest habitats and not in the dispersal distribution from an individual plant. The pattern of propagule density vs distance is the net effect of dispersal from large trees at the forest edge and the many (often smaller) trees behind them, but there is no desire by foresters to disaggregate the propagule shadows of individual plants.

5.3.2.1 Patterns of seed rain from multiple plants

Dispersal from line sources has been examined in plants grown in the field and under more artificial laboratory conditions. Usually dispersal is measured along transects perpendicular to the line. In a study of the rare *Bromus interruptus* and its very common relative *B. sterilis*, Howard et al. (1991) found that dispersal from a line of plants growing in a cultivated field was described reasonably well by a normal distribution with its mode at the centre of the line. Although there were strong prevailing winds, this distribution reflected the combined wind strengths and directions during the entire period of dispersal. In contrast, for plants of *Liatris cylindracea* (Asteraceae) placed in a unidirectional wind in a laboratory, Levin and Kerster (1969) found that the greatest propagule density was away from the line of plants. There are relatively few studies of this type to make any further generalizations.

For patches of plants, O'Toole and Cavers (1983) showed that the shape and scale of the density distribution can be strongly affected by the vegetation within which the dispersing species is growing. When Proso Millet (*Panicum miliaceum*: Poaceae) was growing as a weed in a bean crop, seed density on the ground peaked outside the weed patch; however, density declined steadily from the block centre and extended to a shorter maximum distance when the weed was growing in a much taller maize crop. Although the *P. miliaceum* may have grown taller in the maize, the stout, tall maize stems would stop the weed from falling over and hence restrict its dispersal. In the case of a heavy infestation of the explosively dispersing hemlock dwarf mistletoe (*Arceuthobium tsugense*) on a single western hemlock tree (*Tsuga heterophylla*: Pinaceae), where the patch of plants varied in elevation, density increased to a mode away from the source, then declined almost linearly with distance (Smith 1973). In fact, we would expect a group of parasitic plants spread throughout a tree to show a dispersal pattern similar to the tree itself, provided that they had similar dispersal adaptations. We will return to the issue of canopy shape on p. 101.

A forest is a patch that is so large that it can be considered to have a single edge. Dispersal may not, however, simply be scaled up from smaller patches. Large forests may affect the local climate, and thus change dispersal trajectories through the air; larger forests may contain a different suite of animal dispersers (and predators) and different population sizes. Typically, density of seeds perpendicular to a forest edge exhibit a smooth, concave decline with increasing distance and are often well-described by an exponential function (Fig. 5.10; Greene and Johnson 1996). Thus, density along a transect through the stand edge may appear like a normal curve (e.g. Cremer 1966), just as Clark et al. (1999) argued might be the case for a large tree with a broad canopy.

Simple models have been used to predict what to expect from a line or patch source (for a mathematical derivation of these, see Snäll et al. 2007). Treating each plant as an identical point source on a grid, each dispersing seeds according to the *2Dt* density pdf, Greene and Calogeropoulos (2002) predicted that the decline in seed density with distance would be flatter from both line and patch sources than from an individual plant (cf. Willson 1993). The model did not predict a peak density away from the patch, as sometimes occurs. Greene and Johnson (1996) simulated dispersal from forest stands using a mechanistic model of wind dispersal, again assuming that trees were point sources arranged in a grid. They predicted that for a leafless canopy, seed density would decline approximately exponentially with distance from the forest edge. For a full canopy, the decline was predicted to be steeper than an exponential.

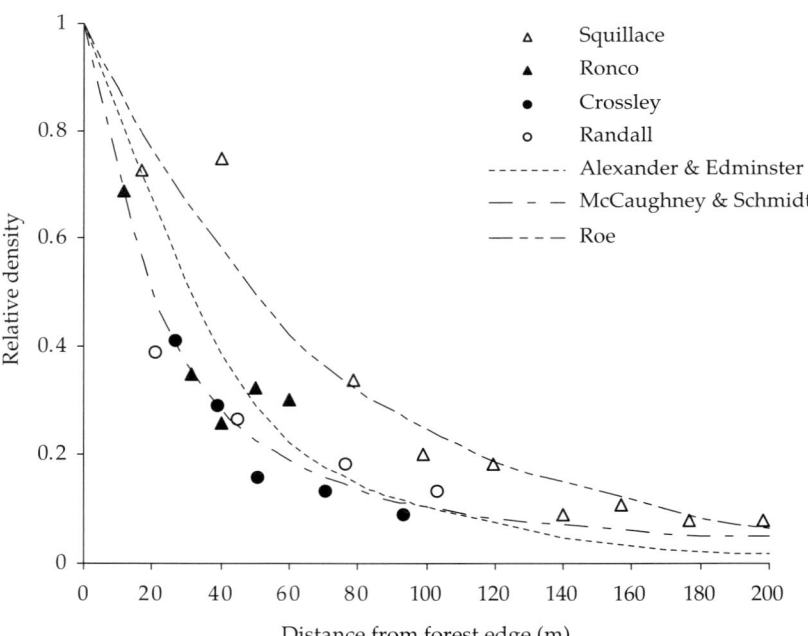

Figure 5.10 Observed and predicted seed dispersal of *Picea* (Pinaceae) with distance from forest edge in seven studies. Species are: *P. rubens* (Randall 1974) and *P. engelmannii* for the other six studies. Figure re-drawn from Greene and Johnson (1996), reproduced courtesy of Ecological Society of America.

5.3.2.2 *Estimating dispersal where plant propagule shadows overlap*

Most of our knowledge about dispersal from individual plants comes from the often unrealistic situation in which they are a long way from others of the same species. We therefore need methods that allow us to measure dispersal distributions where neighbouring propagule shadows overlap. The simplest method is to mark propagules prior to dispersal with paint, a dye, or (in the case of nuts) metal tags, provided this does not impede their separation from the parent. When collecting propagules in traps, or searching for them on the ground, we can then tell which ones came from that parent. This has been used successfully for seeds of herbs with relatively short dispersal distances (e.g. Keddy 1980; Watkinson 1978; Rew et al. 1996; Donohue 1998). However, it can be difficult to apply paint where plants are closely intermingled. The use of biologically marked individuals, such as different seed colour morphs (e.g. *Panicum miliaceum*), overcomes this problem under experimental conditions.

Molecular methods and isozymes are another way of identifying the maternal parents of propagules caught in traps (Godoy and Jordano 2001; Richardson et al. 2002; He et al. 2004), although they can be very expensive and impractical for very large numbers of propagules. Molecular techniques include RAPD, AFLP, RFLP (both nuclear and cytoplasmic), and microsatellites (or SSRs). Co-dominant markers such as microsatellites are more informative than dominant markers such as AFLP (Ouborg et al. 1999), while the use of isozymes is often hampered by a lack of variation. In an excellent example of how these methods can be used to quantify dispersal, Jones et al. (2005) determined the sources of wind-dispersed seeds of *Jacaranda copaia* (Bignoniaceae) in a Costa Rican forest. Using microsatellites, they characterized all 305 adult trees in an 84 ha plot, as well as a sample of 606 propagules collected in traps over two years within a central 50 ha area. Assuming that the sampled distances were a random sample from the true dispersal distribution, they compiled histograms of dispersal distances. Dispersal distances of up to 711 m were recorded, with a mode at 9.7 m. If seeds are found in such an exercise that have markers not present in the adult population, this can indicate longer-distance dispersal from outside the study area.

To date, isozyme and molecular methods have been used more often to link seedlings, rather than seeds, to their parents—realized dispersal. For example, Meagher and Thompson (1987) mapped all adults and seedlings in a population of the dioecious *Chamaelirium luteum* (Liliaceae). Maximum likelihood analysis of isozyme patterns was used to determine the most likely male and female parents of seedlings and hence dispersal distances. The resulting frequency distribution of distances is shown in Fig. 5.11. Distances from female parents

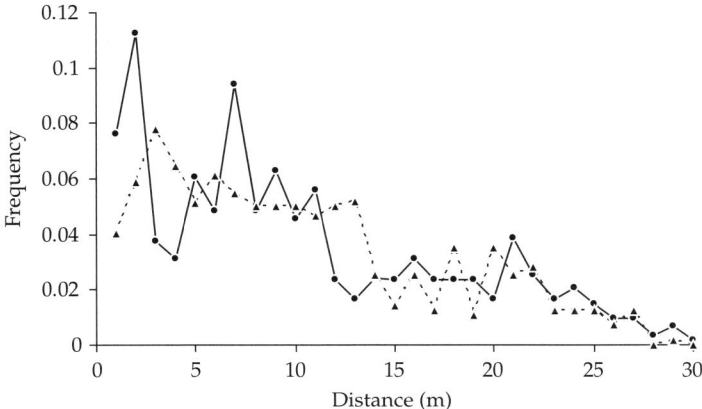

Figure 5.11 Frequency distribution of dispersal distances for pollen (——) and seeds (- - -) of *Chamaelirium luteum* from isozyme analysis of mapped plants (redrawn from Meagher and Thompson 1987, reproduced courtesy of Ecological Society of America).

(seed dispersal) and male parents (pollen dispersal) in this instance are similar. In a study of *Quercus macrocarpa*, however, 14% of saplings had no parent in the vicinity and must have arrived as acorns from outside the population; 71% had only one nearby parent, indicating extensive gene flow via pollen (Dow and Ashley 1996). Even though long-distance dispersal was demonstrated, these methods do not help us to determine the shape or scale of the tail of the dispersal distribution. If molecular techniques are applied to seedlings rather than seeds, errors of interpretation may arise because seedlings produced locally, by parents that are now dead, have emerged from a long-lived seed bank. Thus, although parental matching is very informative for perennials, it will be less use for annuals (Ouborg et al. 1999).

We do not have to be able to identify the parents of individual seeds within traps in order to estimate dispersal distributions. Each plant effectively has its own density pdf and the number of propagules landing in a given trap is determined by the combination of distances along the pdfs from all sources. We can write this as a series of equations and then use maximum likelihood methods to obtain the values of parameters in the model that best fit the data (Ribbens et al. 1994). The suitability of alternative density *pdfs* can be determined by comparing the residual variances after fitting each of them to the data. The method is commonly referred to as 'inverse modelling' (Clark et al. 1999) because it determines the best model and its parameter values from the data, rather than predicting the data from a model using independently obtained parameters. However, it is really just an extension of regression or 'curve-fitting': instead of one x-variable we have many. A number of assumptions are required to use this method. Typically, it is assumed that: the density pdf from each source plant is identical; the number of propagules released by each plant is proportional to some measure of plant size; and propagules disperse equally in all directions. The equation for the density pdf must also be specified (see Table 5.2). In principle, all of the assumptions can be modified, but the resulting increase in model complexity may lead to poor parameter estimation. Provided that the data are sufficient for the exercise, more complex models can be fitted than in the previous example.

For example, a model of spatial population dynamics over generations can be fitted to time-series data on spatial pattern, to derive $g(r)$ parameter values that would best predict the observed dynamics (Humston et al. 2005).

One of the most widely cited uses of inverse modelling is Clark et al.'s (1999) study of seed data from traps in temperate broad-leaf, mixed conifer, and tropical floodplain forests. They found that the 2*Dt* model was consistently better than the two-dimensional normal or exponential, although the residual variances from the three models were often very similar. More seeds were estimated by the 2*Dt* model to have landed very close to the parent and a greater proportion travelled longer distances than for the alternative models. Where the normal was the best model, this tended to be for animal-dispersed species rather than those dispersed by wind. In their study of *Jacaranda copaia*, Jones et al. (2005) found that mixed models consisting of the weighted sum of two pdfs gave better descriptions of their multi-source data than the 2*Dt* or other single functions for $g(r)$, even when allowance was made for the increased number of parameters (more parameters generally leads to greater flexibility to fit data; models with different numbers of parameters can be compared using the Akaike's Information Criterion: Akaike 1973). Greene et al. (2004) reported that the ability to distinguish between alternative functions was poorer for inverse modelling from multiple seed or seedling sources than when fitting to single-source transect data. They found that differences in the fit between the log-normal $f(r)$, Weibull, and 2*Dt* were usually minor.

Inverse modelling is also used to estimate realized dispersal distances from allele frequency data in the adult population. However, a large number of assumptions must be made. We will begin by discussing F_{ST}, the standardized among-population variance in allele frequency, introduced by Wright (1951). F_{ST} can be estimated in a number of ways (see, for example, Raybould et al. 2002). Populations vary in allele frequencies as a result of gene flow, genetic drift, selection, and mutation. If markers can be considered to be selectively neutral and mutation rates are low relative to gene flow, the genetic divergence among populations will be largely determined by the amount of dispersal (e.g. Neigel 2002).

Assuming that an infinite number of equal-sized island populations exchange migrants at a constant rate, regardless of their positions, and that the system is in equilibrium, Wright obtained the well-known equation

$$F_{ST} = \frac{1}{(4N_e m + 1)} \qquad (5.10)$$

where N_e is the effective population size and m is the migration rate. Thus $N_e m$ is the number of migrants per generation which, given the infinite island assumption, is of little help in calculating dispersal distances. However, if we calculate F_{ST} for all possible pairs of populations and plot $\log(N_e m)$ against $\log(\text{distance})$ between them, we obtain an 'isolation by distance' (IBD) regression line (note that various other measures related to F_{ST} can also be plotted). The slope of this line gives us a quantitative measure of the importance of gene flow: the less the slope (the more similar the populations), the greater the gene flow. The next step is to develop a model that predicts the IBD relationship from a density pdf (Tufto *et al.* 1996, 1998; Palumbi 2003). The usual assumptions for this are that populations occur as a circular series of equally-spaced stepping-stones linked to adjacent populations by dispersal. A population size has to be assumed, which is constant, along with an equation for the density pdf (usually a Laplacian, an exponential in both directions). Using maximum likelihood, inverse modelling then estimates the mean dispersal distance that would most closely result in the observed IBD regression. Fig. 5.12 shows estimates obtained in this way from published IBD data for marine plants (both algae and angiosperms) and terrestrial plants. It should be noted that the values are very high in comparison with directly measured values from point source experimental propagule releases (Kinlan and Gaines 2003) as well as the examples of entire plant dispersal earlier in this chapter. Long dispersal distances may be possible in strong marine currents (though they will not usually be bi-directional), but the values for terrestrial plants are surprising: there is a wealth of evidence to show that although maximum dispersal distances may be several hundred metres in some cases, the mean distance is much closer to the parent plant. Clearly, any number of the many assumptions of the IBD inverse modelling may be

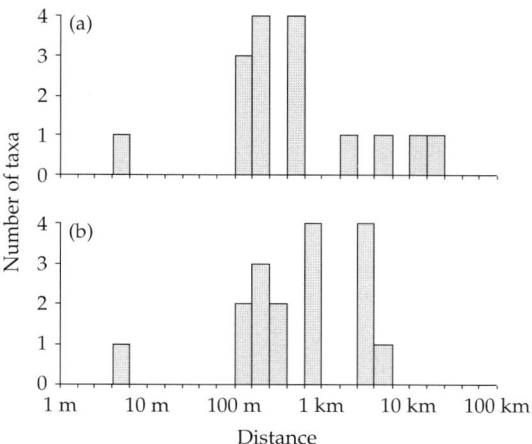

Figure 5.12 Indirect genetic estimates of mean dispersal distances (log scale), derived from published Isolation By Distance data (redrawn from Kinlan and Gaines 2003, reproduced courtesy of Ecological Society of America): (a) Terrestrial plants; (b) marine algae and angiosperms.

incorrect. Although these indirect genetic methods of estimating dispersal distance show promise, considerable caution should be shown until they can be properly validated.

5.4 Predictions of entire-plant dispersal from models

What determines the shape and the scale of the distance pdf and the density pdf? Empirical case studies are, by definition, context-specific. The examples in the first part of this chapter described what happened to the propagules dispersing from a particular plant, or stand of plants, growing in a particular place over the course of the species' dispersal season. This does not mean that the distributions of propagules will always be that way. If we go to another site or another year, we would expect dispersal distributions to vary. But by how much? What factors have the most influence? Can we predict the changes to the scales and shapes of dispersal distributions that will occur if we perturb the environment or we move to a different situation?

In this section we will discuss models that attempt to answer such questions. They take the component processes that are likely to have the greatest impact on individual propagule trajectories (i.e. using the information in Chapters 2–4) and predict the resulting distributions of the 'population' of propagules from an entire plant achieved over the duration of its dispersal season. Validation of the results (comparison with a real case study) is often difficult, especially if our interest is in the tail of the dispersal distribution. The results should be regarded as best guesses and the conclusions as hypotheses that can be tested in the future. Our confidence in the predictions depends largely on our level of understanding of the processes that we include and the confidence with which we make particular assumptions and simplifications in formulating the model. It is clearly unrealistic to develop a single model that incorporates every process for every dispersal vector. Hence, all of the models focus on particular propagule-vector combinations and particular issues. The issues that we will raise are: temporal variation in dispersal vectors; the combined influences of multiple vectors; and the influence of plant architecture.

5.4.1 Effects of environmental variation

Throughout the dispersal season of any plant, the strength and direction of dispersal vectors will change repeatedly. The various positions of propagules on the plant may determine their exposure to vectors, while the number of propagules ready to be released and the force required for their removal will change as individuals mature. In the case of wind, its speed varies considerably, from virtually zero to gale force or greater, on timescales of days, minutes, or seconds; direction also varies, but not as frequently as speed. The duration of flight of a wind-adapted propagule may typically be several seconds, a few minutes, or in extreme cases hours; the duration over which propagules mature may be several days or weeks; the range of release heights may be several metres for a tree, or just a few centimetres for an herb such as *Taraxacum officinale*. We now have a well-developed ability to model the flight of propagules through the air from a point source under a given wind speed or a frequency distribution of speeds (Greene and Johnson 1989; Nathan et al. 2000, 2001, 2002; Tackenberg 2003; Tackenberg et al. 2003; Soons et al. 2004a, b; Soons and Ozinga 2005; Nathan and Katul 2005; Katul et al. 2005).

To predict wind dispersal for the propagules of an entire plant, we need to bring the biological and environmental variation together with a flight model. By simulating the trajectories of each of a large number of propagules, choosing parameters from appropriate frequency distributions, we can produce a frequency distribution as a composite of their distances and directions. One of the best examples of this approach is the study by Nathan et al. (2001) of seed dispersal by *Pinus halepensis*. Their model assumed that:

- wind is the only dispersal vector;
- seed release sites are located at the centres of flat $1\,m^2$ cells and are normally distributed as a proportion of tree height;
- terminal velocity of propagules is normally distributed;
- timing of seed release varies with relative humidity and temperature; a regression equation was calculated from total numbers of seeds collected in traps, mean relative humidity, and maximum temperature in the interval between seed collections; seed releases were then distributed among source cells at random from a normal distribution;
- wind speed is taken at random from log-normal (horizontal component) and normal (vertical component) distributions, determined by fitting the distributions to observed data over the whole dispersal season; vertical speed was assumed constant during the trajectory of a seed; horizontal wind speed declined according to a logarithmic profile;
- dispersal distance was calculated from equation 4.1 (but making $W =$ terminal velocity minus vertical wind speed);
- direction was taken at random within a 45° sector and was constant during flight; the sector was determined from weather station data for each interval;
- no further movement took place once the seeds reached the ground (i.e. there is only primary dispersal).

The predicted distance pdf, $f(r)$, was peaked at 5–10 m, with an extended tail to at least 100 m (Fig. 5.13). When converted to average density vs distance (ω_r) this distribution approximates to an exponential decline, at least over the first 50 m. Sensitivity analyses of parameters indicated that the

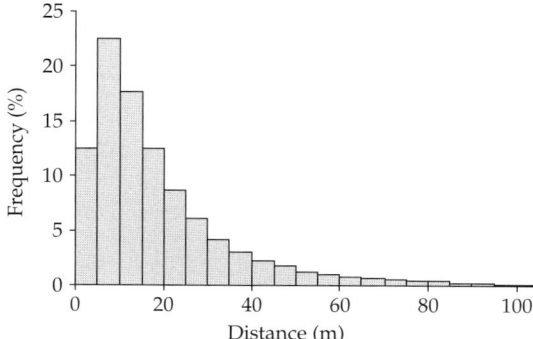

Figure 5.13 Predicted frequency distribution of dispersal distances for seeds of *Pinus halepensis* (redrawn from Nathan et al. 2001, reproduced courtesy of Ecological Society of America).

distribution of seed release heights was relatively unimportant (the model could be simplified to the release of seeds from a single mean height).

More complex models, incorporating increasingly realistic air movement (such as spatially and temporally correlated vertical wind speeds), 30-minute averaged time series of friction velocity measured throughout the dispersal season, and leaf area index canopy profiles, have been developed by Nathan and collaborators (Nathan et al. 2002; Soons et al. 2004a; Katul et al. 2005). Tackenberg (2003) has used observed high frequency wind measurements to predict seed dispersal, though Nathan and Katul (2005) argue that extrapolation to other situations may be difficult. For short grassland species, assumed to be point sources, he predicted that the great majority of propagules would land within 10 m of the parent. In weather conditions with thermal updrafts almost 20% of *Cirsium arvense* seeds were predicted to disperse more than 100 m. Most models, however, do not include a threshold wind velocity (or force) for release of seeds from the parent, so they are likely to over-estimate the frequency of very short dispersal distances in many species (Greene and Johnson 1989).

A challenge now for ecologists is to produce models that incorporate similar levels of reality for other dispersal vectors, such as animals. Although animal behaviour is highly complex and apparently erratic, many studies show that there are statistical patterns that could be quantified and built into models (e.g. Jordano and Godoy 2002; Creel et al. 2005).

Rather than assuming simple random walks or using context-specific observed displacement data, models of animal dispersal of seeds could incorporate more complex algorithms for spatially and temporally correlated movement of animals, habitat and food selection, allocation of time to different types of behaviour, and switching between behaviours. Morales and Carlo (2006) have made an excellent start in this direction.

5.4.2 Influence of multiple vectors

The models of wind dispersal assumed that there is only one dispersal vector and that further movement ceases once the propagule reaches the ground. For dispersal of wind-adapted propagules dispersing into a rough habitat, these assumptions may be considered appropriate. However, in many situations several vectors are involved in the movement of seeds from a single plant. This is particularly the case for the dispersal of fleshy fruits. Jordano and Schupp (2000), for example, recorded that 50–68% of *Prunus mahaleb* seeds were removed by seed-dispersing birds in two consecutive years. The trees were visited by 38 bird species, of which 42% were categorized as seed dispersers. In Australian tropical rainforests, 65 vertebrate species, all of them generalist feeders to a certain extent, disperse 441 plant species (Dennis and Westcott 2007). The dispersal caused by multiple dispersal vectors needs to be accounted for if we are to represent dispersal realistically from the plant's (rather than the animal's) point of view.

Ecologists often refer to the phases of movement of a propagule during its trajectory: primary dispersal moves it from the parent to the ground; secondary dispersal moves it from where it lands to another location; and so on. If we consider the population of propagules on a plant, several dispersal vectors may be involved in each phase, providing alternative pathways of movement for different proportions of the seeds (Bullock et al. 2006). For a propagule undergoing several phases of dispersal, multiple vectors (or even the same vector) act sequentially. Using an electronic circuit analogy (p. 2), vectors act either in parallel or in series and for any system we can develop a diagram showing the pathways. To determine the overall dispersal distribution of a

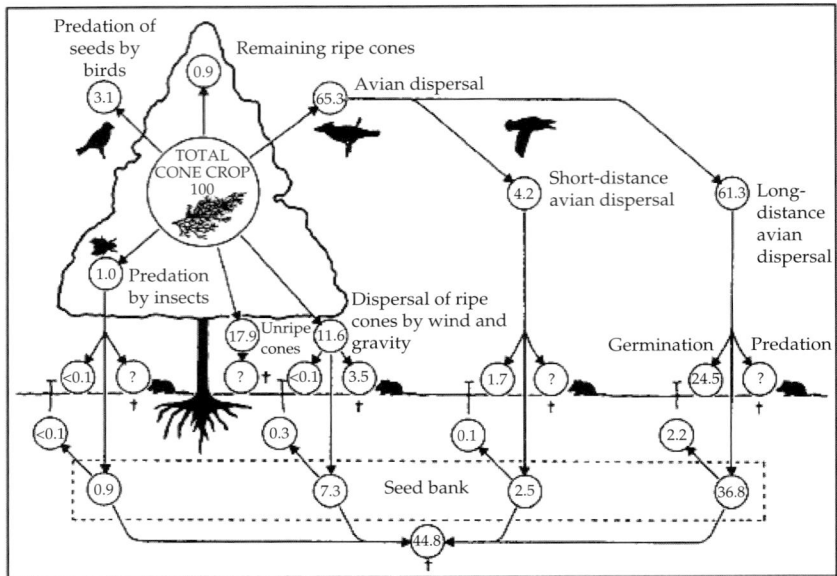

Figure 5.14 Pathways of *Juniperus virginiana* seed dispersal and predation (Holthuijzen et al. 1987, reproduced courtesy of National Research Council of Canada). Numbers show percentages.

given plant, we need to establish the probabilities of its propagules passing through each part of the diagram (e.g. Fig. 5.14; Roth and Vander Wall 2005) and the pdfs of the positions reached (relative to the position at the start of each phase). We then need to determine the dispersal distribution that results, which can be done either mathematically for simple cases or through Monte Carlo simulations of large numbers of single propagules.

If vectors act in parallel, the distances that some seeds are moved by one vector are independent of the distances that the remaining seeds are moved by the other vectors. The resulting density distribution is therefore simply the weighted sum of their density pdfs:

$$g_{comb}(r, \theta) = \sum p_i g_i(r, \theta) \qquad (5.11)$$

where p_i is the proportion dispersed by vector i (note that if there is radial symmetry $f_{comb}(r) = \sum p_i f_i(r)$). For vectors acting in series the mathematics is somewhat more difficult, since one phase of stochastic movement must finish before the next phase begins. The mathematical technique of convolution needs to be applied to combine their pdfs. In Cartesian coordinates, the convolution of two distributions

is given by

$$g_{comb}(x', y') =$$
$$\int_{-\infty}^{\infty} dx \int_{-\infty}^{\infty} dy\, g_1(x, y) g_2(x' - x, y' - y) \qquad (5.12)$$

where g_1 is the density *pdf* for the first phase of movement and g_2 is for the second phase. It is a simple matter to transform the equation to radial coordinates using the fact that $(x, y) = (r \cos\theta, r \sin\theta)$ and $\int_{-\infty}^{\infty} dx \int_{-\infty}^{\infty} dy = \int_0^{\infty} r\, dr \int_0^{2\pi} d\theta$.

Many animals can be predators, either deliberately eating seeds or inadvertently damaging them as they feed. For example, squirrels are important dispersers in temperate regions but consume far more seeds than they leave in un-recovered caches. It is thus very difficult to deal separately with dispersal by animals and mortality between the source plant and the final resting place. It is thus usual to include consumption in our circuit diagrams: the analogy might be of a resistor reducing the current passing from one step to the next. We are then dealing with realized dispersal rather than dispersal per se. It is also important to acknowledge that many propagules may not be moved at all by animals (Herrera 1984) and eventually just fall to the ground, or the flesh may be eaten and the remainder discarded close to the source (Sallabanks 1992). Clark et al. (2005), for example, found that 30–90% of the seeds of species identified as being dispersed primarily by wind, monkeys, and birds, landed directly under the source tree canopy. This short-distance un-ingested part of the population must be included as a component of the combined dispersal pdf.

In a major study of seed dispersal by frugivorous vertebrates in tropical rainforests, Dennis and Westcott (2007) reduced the complexity of their system by classifying 65 animal species into 15 functional groups based on body size, visitation behaviour, method of dispersal, predation, deposition, movement and gut throughput rates, and foraging locations. Intensive field observations were used to determine the proportions of seeds moved by each functional group. Mechanistic models for animal dispersal are only just being developed, so distance pdfs were obtained from empirical data. Gut throughput rates, taken from studies of representative frugivores in cages, and tracking data from studies of movement were used to derive a distance pdf for each functional group (see p. 60). Primary dispersal of *Elaeocarpus grandis* (Elaeocarpaceae) seeds was then estimated as in equation 5.11. The predicted frequencies of dispersal distance were greatest close to the source and declined rapidly thereafter. Seventy percent of seeds were predicted to disperse less than 100 m, while 4% dispersed more than 400 m (Fig. 5.15a). The maximum distance dispersed was 2.3 km. When converted to density vs distance, the relationship was approximately linear over the first 200 m away from the plant. It is easy to appreciate the impacts of certain species on the dispersal distribution. As Fig. 5.15b shows, most seeds travelling over 500 m were dispersed by mega-terrestrial frugivores: loss of this functional group (primarily the cassowary) from the ecosystem would reduce the maximum distance by about half. Similar results were found for dispersal of *Prunus mahaleb* in Spain (Jordano et al. 2007) using microsatellite markers of seeds to identify parental trees. Most short-distance dispersal was a result of passerine birds, while large carnivorous mammals were responsible for the longer distances.

An example of dispersal for sequential dispersal vectors comes from a study of *Bromus* spp. in agricultural fields (Howard et al. 1991). Three methods of dispersal were considered, each described by an empirical distance pdf. It was assumed that dispersal only occurs in one dimension, parallel to the direction of the machinery, and that all vehicles travel in the same direction. A proportion of seeds from each plant fall to the ground according to a normal distribution, while the rest enter the combine harvester. Some of the seeds are retained with the harvested grain (and hence are removed from the field), while the rest are distributed within the field according to an additive mixture of a normal and an exponential distance pdf (which fits observed data well: see p. 71). When all seeds have reached the ground via both pathways, they are then dispersed by a cultivator, again according to a mixture of a normal and an exponential distribution. The predicted combined outcome of all three processes, a bimodal density pdf, is shown in Fig. 5.16. One peak is close to the source plant, while the second mode is at about 2 m. The differences between the species primarily relate to the proportion of seeds entering the harvester.

5.4.3 Influence of plant architecture

So often when studying dispersal, we concentrate on those species clearly adapted to vectors that make it possible to disperse very long distances. However, there are many species that do not disperse so effectively. Dispersal is still important to these species, both in ecological and evolutionary contexts. Many temperate agricultural weed seeds, such as wild oats (*Avena* spp.; Poaceae), can travel very long distances—up to hundreds of kilometres—if they enter the harvester and the grain is sold for sowing. The great majority of seeds, however, fall directly to the ground. Trees may have very heavy propagules that, when dislodged from a tree by the wind, fall directly to the ground and perhaps roll a short distance. As we discussed earlier, even in species whose propagules are dispersed very effectively by animals, many fall or are discarded under the parental canopy. Can we predict where this component of the population falls?

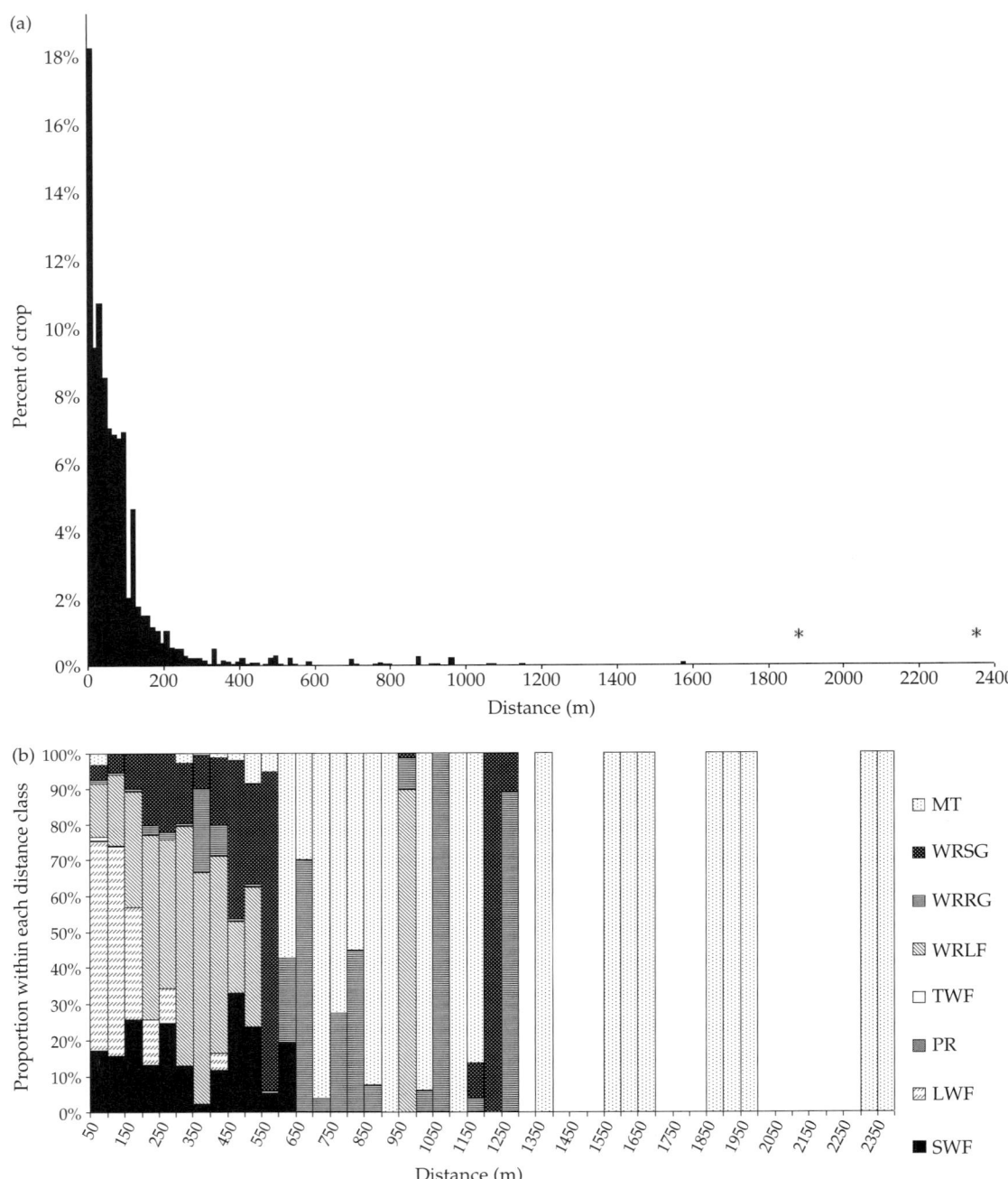

Figure 5.15 Estimated dispersal of *Elaeocarpus grandis*, based on the contributions of eight functional groups of frugivores (Dennis and Westcott 2007, reproduced courtesy of CABI). (a) Frequency distribution of dispersal distances for all dispersal vectors; (b) the contributions of each functional group. Functional groups are: MT—mega terrestrial; WRSG—wide-ranging, slow gut; WRRG—wide-ranging, rapid gut; WRLF—wide-ranging, large fruit; TWF—terrestrial within forest; LWF—large within forest; SWF—small within forest; PR—predatory rodents.

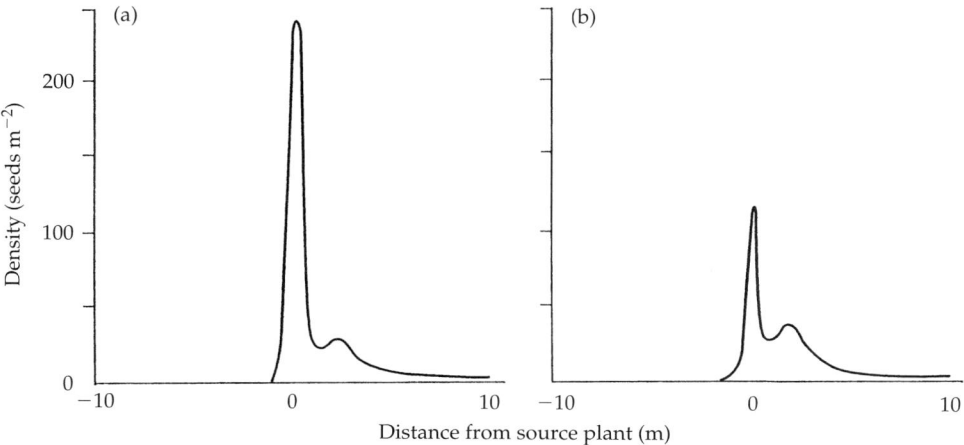

Figure 5.16 Predicted seed densities for (a) *Bromus sterilis* and (b) *B. interruptus*, resulting from the combination of passive dispersal around the parent, dispersal through a combine harvester and dispersal by a cultivator (redrawn from Howard et al. 1991, reproduced courtesy of BCPC). It has been assumed that all machinery travelled from left to right.

In the case of propagules adapted for dispersal by the wind, an assumption of the maternal parent as a point source appears quite reasonable. Even if the canopy is very broad, and clearly is a distributed source of propagules, when this is viewed against the long distances that some seeds travel, an error of a few metres in trajectory starting points will be relatively unimportant. The importance of the architecture of the parent to a dispersal distribution is thus likely to be positively correlated with terminal velocity of seeds and, less strongly, inversely correlated with plant height. Nathan et al.'s (2001) model predicted that seed distribution with height would have little effect on dispersal by *Pinus halepensis*. Trees in that study were around 9 m in height, they produced their seeds near the top of the canopy and their seeds had a mean falling velocity in still air of 0.81 ms^{-1}; Tackenberg et al.'s (2003) method would rate its Wind Dispersal Potential as moderately high (WDP 7, see Fig. 3.6). However, if the canopy dimensions of a tree are large and the falling velocity of seeds is high, the assumption of a point source may be inappropriate. Earlier in this chapter we discussed the likely shapes of dispersal distributions from broad canopies and patches of plants (that may behave like broad, single plants) and simple models that assume a population of point sources distributed in horizontal space. What effect does the three-dimensional shape of the canopy have, especially in species with restricted dispersal?

Cousens and Rawlinson (2001) modelled dispersal from artificial trees of contrasting geometric shapes: spherical, conical, and a flat disc. Propagule release points on the canopies were assumed to be spread either (1) evenly on their outer surface or (2) throughout their volume. Five hundred thousand canopy locations were selected at random. A seed was dispersed from each point in a random direction and by a random distance selected from a beta distribution. This distribution was chosen because it showed a good fit to laboratory experiments in which a range of different sizes and shapes of seeds were dropped from fixed heights and it has a finite maximum: no attempt was made to model the mechanism of movement. Distance pdfs, $f(r)$, were constructed with very narrow histogram bins; the equivalent density pdf, $g(r)$, was derived by the appropriate transformation (p. 80).

For spherical, conical, and flat canopies, whether propagules were spread throughout the canopy or just on the surface, $f(r)$ was always peaked; $g(r)$ was sigmoidal where seeds were released throughout the canopy, but peaked at the canopy edge for spherical plants where seeds were released only from the canopy surface (Fig. 5.17). The peak of $f(r)$ was closer to the stem of the plant if seed release points

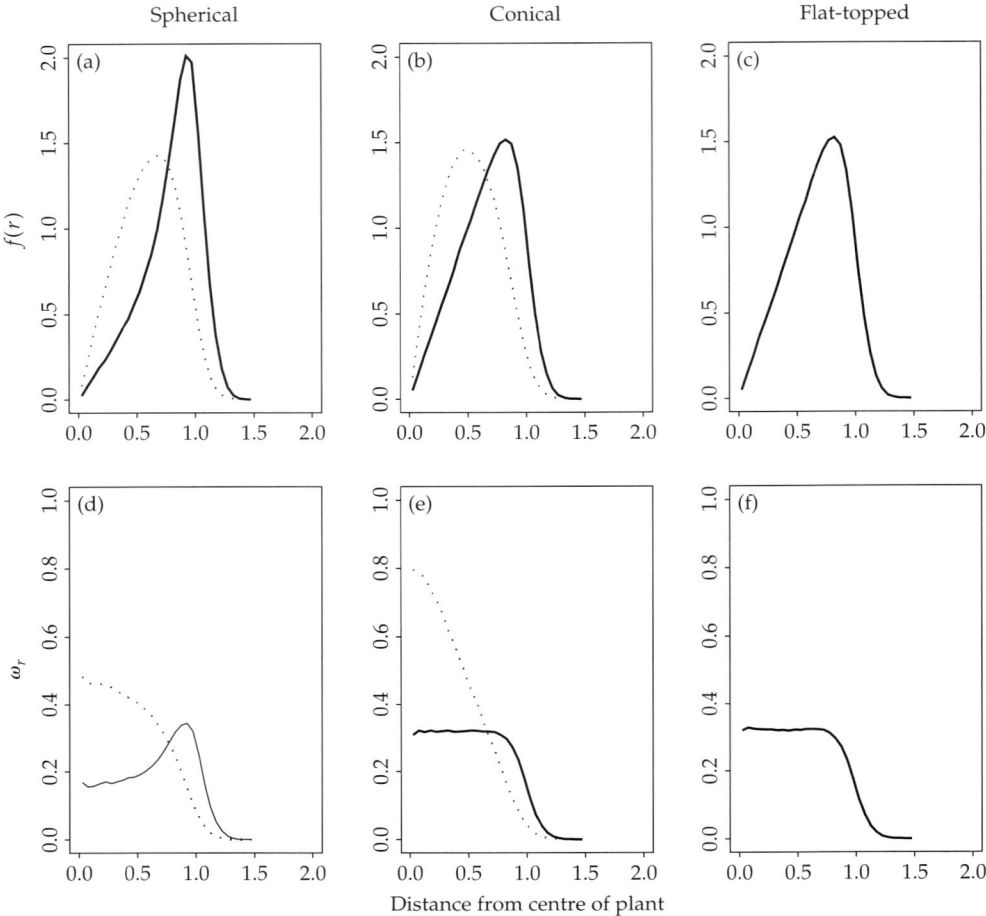

Figure 5.17 Predicted distance pdf, $f(r)$, and density vs distance (ω_r) assuming spherical, conical, and flat-topped canopies for a distributed seed source. Solid lines are where seed sources are confined to the canopy surface; dotted lines are for sources spread throughout the canopy volume (Cousens and Rawlinson 2001, reproduced courtesy of Elsevier).

were throughout the canopy. If individual propagules were dispersed by a greater mean distance, the $f(r)$ remained peaked, though less pronounced, at the canopy edge; the $g(r)$ became less peaked, and eventually sigmoidal, as mean dispersal distance increased. Hence, a peaked $g(r)$ would perhaps be more likely in species with heavy propagules than in those species adapted to wind dispersal. In no cases was $g(r)$ of an exponential or power law shape, although an exponential would have provided a reasonable fit from the canopy edge outwards.

Since density acts to modify the shape of a plant as well as its size, Cousens and Rawlinson (2001) also imitated the effects of density by changing the shape of a plant from a sphere (e.g. a tree growing without competition) to an 'umbrella' (e.g. a tree in a dense forest). Both shape and scale of dispersal distributions were predicted to be density-dependent in species with restricted dispersal potential. The distance pdf $f(r)$ was again peaked for all canopies; as the canopy came to resemble that of a plant at high density, $g(r)$ for a spherical plant with seeds on the canopy surface changed from being peaked

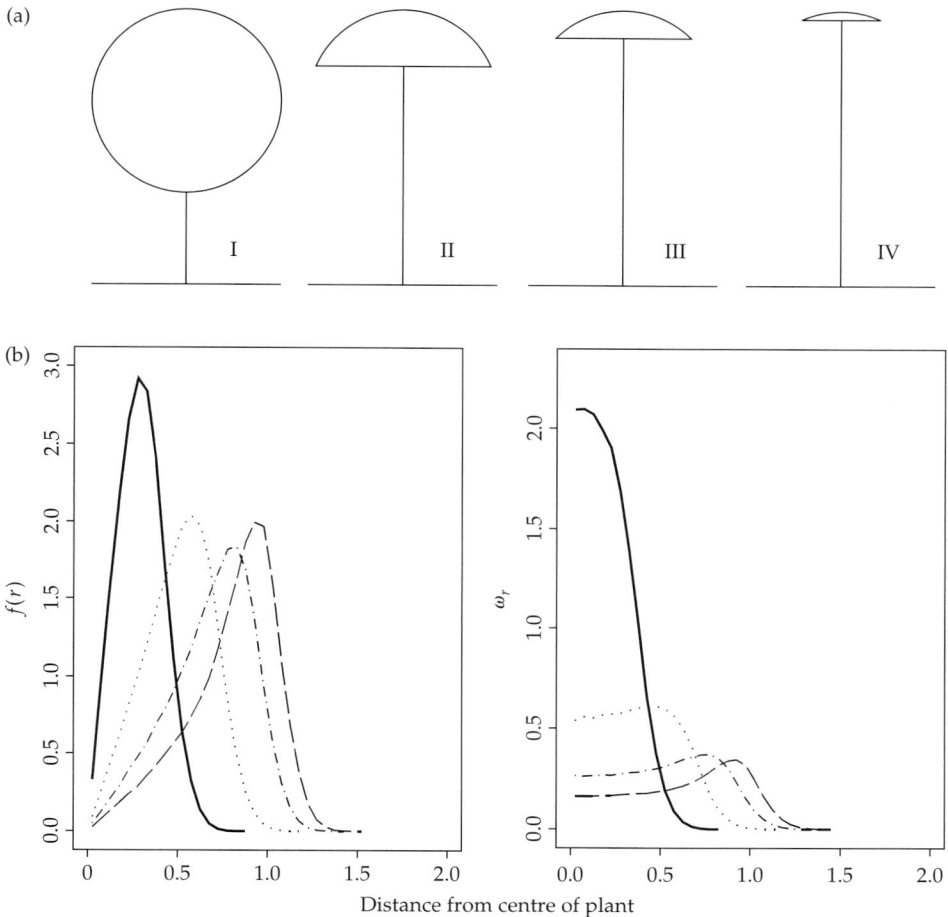

Figure 5.18 (a) Assumed changes in the shape of a spherical canopy as crowding increases; (b) effect of these changes on $f(r)$ and ω_r I (——); II (· · · · · ·); III (— · — · —·); IV (- - - - -) (Cousens and Rawlinson 2001, reproduced courtesy of Elsevier).

to being sigmoidal, with a maximum at the base of the plant (Fig. 5.18). These predictions could be tested experimentally (on small plants), for example by sowing different densities, pruning plants to different shapes, and removing fruits from some parts of a canopy. Despite many hundreds of studies that describe the effects of competition on seed production and biomass, few have examined the effects of competition on dispersal.

5.5 Sampling and experimental designs for density pdfs

We conclude our discussion of the shapes and scales of dispersal distributions by considering the design of sampling regimes, experimental designs, and methods of analysis. This critical aspect of research has been given little attention until comparatively recently (Assunção and Jacobi 1996; Skarpaas et al. 2005; Canham and Uriarte 2006). As attention has increasingly focused on the fine detail of dispersal distributions—the shapes and lengths of their tails and the equation which best fits the data—so the importance of good study design has been recognized. A poor design, no matter how much effort is expended in the field, may stop us from achieving the aims of our research. Those who have analysed published data have appreciated the inadequacies of past studies for the deductions that we now wish to make (e.g. Portnoy and Willson 1993).

There is no single best experimental design. The best design for a given circumstance depends on the aims of the study and the attributes of the system. Sampling designs need to achieve accurate estimates of the parameters of interest, with as much precision as possible for the given resources. One of the most powerful tools for exploring the effectiveness of designs is simulation. For a given 'true' density pdf and an assumed variance, the number of propagules collected in traps of different sizes and locations can be generated using a random number generator. A model is then fitted to the trap data and the performance of the design is considered against criteria such as its ability to recover the true density pdf (Skarpaas et al. 2005) and to minimize the mean squared error of the true rate of population expansion (Pielaat et al. 2006). Other aspects of design have been arrived at by researchers through common sense or experience.

The main decisions to be made in designing a sampling plan are:

- the design of traps;
- the locations to be sampled;
- the number of traps at each location;
- the number of replicate plants;
- the ways that data are to be analysed.

5.5.1 Trap design

Traps are a very convenient way to sample dispersing propagules. If there is no lateral movement after the initial contact is made with the ground, then traps will provide a good way of measuring dispersal. If there is rolling after impact, or any other secondary dispersal, any form of trap will necessarily underestimate the true distances travelled (since they will intercept them during this movement) and will artificially aggregate propagules at that point on the ground. The use of quadrats, while more realistic, will make it more difficult to recover small propagules and is inconvenient where there are large numbers of sample points and repeated sampling. The traps used most commonly are cylinders or funnels sunk into the ground, or flat surfaces covered in a sticky substance. Johnson and West (1988) suggested the use of containers of gravel, soil, or

marbles, to better imitate an intact soil surface and to allow further dispersal through the location.

A trap is assumed to estimate density at a point in space at the centre of the trap. When working along a transect, the assumption is made that the number of propagules per trap at each point represents density at a point along an infinitely thin line. This assumption is invalid at very short distances and especially with large traps, since they will sample across an arc of several degrees and over distances within which density is changing rapidly. On the other hand, small traps will almost never contain propagules at long distances, or even at short distances for plants that produce few propagules. Thus, the size of trap necessary to collect a given number of individuals will depend how far out into the density distribution tail it is and the number of propagules released from the parent: bigger is better for traps in the tail, but more systematic error will be obtained if large traps are used very close to the parent. The size of traps is usually determined by what is available, such as coffee cups, funnels, or drainage pipe, or by a subjective decision about what constitutes an appropriate scale: choice is seldom based on any scientific or statistical calculation. This size is then kept constant at all distances. However, there is no a priori reason to keep trap size constant, as unequal sizes can be allowed for at the data analysis stage. Little attention has been given to trap size in optimum design studies. However, design studies have dealt with the number of traps per location, which has the same effect on density estimates as changing trap size, and will be dealt with later.

5.5.2 Locations to be sampled

Most studies have traditionally been based on transects radiating outwards from a plant: we will consider these designs in some detail before moving on to designs for multiple plant sources.

Traps are usually spaced at either fixed distances apart, that is, in an arithmetic progression, or in a geometric progression, where the ratio of successive distances remains constant. Occasionally, however, they are arranged in a hybrid of these two designs (e.g. Bullock and Clarke 2000). The selection of distances in most studies appears to be arbitrary. Indeed, there *is* no correct number of traps and spacings, although some designs will be better than others for a particular purpose.

It is self-evident that if we wish to describe the tail of a density distribution, then we must have traps located in that region, while the same is true for the 'body' of the distribution; to describe the entire density vs distance relationship, however, we need traps spread out over a wide array of distances. Statistically, the minimum number of distances needed to fit an equation is at least one greater than there are parameters in the equation to be fitted. Theoretical studies of optimal designs (Pike and Hasted 1987) suggest that observations should be closest together where a relationship is at its steepest. For an exponential or similarly shaped density distribution, this rule-of-thumb might suggest a geometric series of distances, with spacings closest together at the shortest distances (Assunção and Jacobi 1996). The criterion used in studies of optimality of designs usually involves an assessment of how well the function fits the relationship overall (e.g. Cobby et al. 1986; Skarpas et al. 2005). An 'optimal' design may thus still give a relatively poor description of particular features, such as the position of a mode or estimates of densities in the extreme tail.

A different approach was taken by Pielaat et al. (2006). Their interest was in the use to which an equation is put, that is, spatial modelling, rather than in the fitted equation itself. Modelling (Chapter 6) has shown that the shape of the tail of the dispersal distribution is critical in determining the velocity of population expansion. Using simulations incorporating a particular spatial population model, they spaced out the initial traps at equal distances and sequentially added traps at distances which minimized the error mean square of the velocity of the population margin. They concluded that a large proportion of traps should be concentrated *towards* the tail, but not at distances so great that they do not collect any propagules (if all of them were to be empty, no information will be available to define the underlying distance distribution). For example, the area around a plant could be reconnoitred in the previous year, and the distance at which a given quadrat size would no longer regularly contain propagules could be determined from the pattern of propagules on the ground. Quadrats could then be concentrated at distances skewed towards this distance.

Optimal design calculations usually assume that the underlying relationship is described by a single function that is known. In reality, we do not know the underlying mathematical relationship before the study takes place and, indeed, many studies have the identification of an appropriate equation as an outcome. Moreover, there can be very high levels of variation in the data and it may be very difficult to determine the distances where the key features of the distribution will occur. For this reason, it may be best to 'hedge our bets', by distributing traps at more locations and to greater maximum distances than are likely to be absolutely necessary. Although some traps will contain no propagules, thus apparently wasting effort, it avoids the risk of not going to large enough distances. Similarly, there can never be too many distances included in an analysis, and separation distances can never be too small (unless they physically overlap). The logical conclusion is that you should have as many as you can possibly spare the time to operate. For example, while the number of traps and distances in Bullock and Clarke's (2000) study (Fig. 5.2) might be considered by some to be excessive, it enabled the authors to conduct the most informative study of distribution tail shape to date. One of the most common complaints of those re-analysing published data is that there were too few distances studied (e.g. Portnoy and Willson 1993), while those interested in the body of the distribution usually find insufficient traps placed at short distances to determine shape close to the parent.

Skarpaas et al. (2005) found that, in most circumstances, several transects away from a source were better than either randomly placed traps or a grid design. The main exception was when there was anisotropy in a direction that was unknown before transects were placed. However, if the direction could be anticipated (such as by knowing the prevailing wind direction), the best design involved placing transects preferentially in that direction. They did not examine the alternative of increasing the number of transects: the more directions that are sampled, the better we will be able to quantify the directionality (e.g. in a rose diagram). They also found that the ability to describe an equation relating density and distance was better if the number of propagules being dispersed was increased. With larger numbers, counts in traps will more nearly approach continuous variables, hence decreasing the error caused by the difference between discrete counts and a continuous equation. Our options for increasing source size, however, are very limited. We could select only the largest plants, but this would bias our results if larger plants have different dispersal distributions (as is the case with taller trees dispersed by wind). The authors suggest that a single plant source could be replaced by a block of plants, although they acknowledge that this could result in a different underlying shape of density distribution (see p. 94).

Principles similar to those for single sources apply to sampling designs for multiple source plants (for example to be analysed using inverse modelling). Rather than follow a rectangular grid or random location of points, it is better to consider the requirements of the model to be fitted so that good estimation of the parameters is achieved (Canham and Uriarte 2006). For each variable in the model, it is a good practice to ensure that there are observations at a wide range of values of that variable. For example, if plant size (e.g. diameter at breast height) is in the model, the sample area should include plants of a wide range of sizes, just as there should be a wide range of distances of the plants. If there are large plants outside the sampled area but close to its edge, many of their propagules will land in traps inside the area and thus their positions must be recorded and included in any analysis, otherwise this creates a strong edge effect, leading to biased estimation of model and parameters. From computer simulations, Canham and Uriarte (2006) recommend that mapping of source plants should extend outside the area in which traps are located by a distance of at least twice the mean dispersal distance.

5.5.3 Number of traps at each location

The variance of trap counts is usually greatest close to the source plant, since the mean is very much greater there than in the tail (the variance of many zeros and occasional low counts will be close to zero). If we were to aim for a constant standard deviation at all distances, as might be appropriate if we are interested only in the overall fit of an equation, we would need to locate more traps (or larger traps)

at sample distances closer to the source plant. The number of traps required at each distance can be calculated by assuming a particular density pdf, likely parameter values for the pdf, and an error distribution (we might estimate all of these from previous 'baseline' studies). However, many researchers are very interested in the tail of the distribution. In this case, perhaps it would be better to aim for a coefficient of variation (CV, standard deviation/mean) that is either constant or which even increases at greater densities. This can be achieved by placing more traps at greater distances. For example, if density vs distance (ω_r) is a negative exponential and the error distribution is binomial, the number of traps to achieve a constant CV will increase by a factor of e^k for each additional metre of distance (G Hepworth, pers. comm., where k is $1/a$ in the exponential ω_r in Table 5.2). Pielaat et al. (2006) used simulations to minimize the error mean square of the rate of population spread; they found that the optimal design for a fixed set of distances also involved larger numbers of traps towards the tail of the density distribution.

Clearly, in the presence of uncertainty about the actual density distribution, we need to use a degree of guesswork to cover various possibilities. A rule-of-thumb used by a number of researchers is to increase the number of traps proportionally to the distance from the source (e.g. Hoppes 1988). In effect, this means that for each distance the same proportion of an annulus around the parent plant will be sampled. It does not necessarily mean that CV will be constant or that the goodness of fit of an equation or estimation of its parameters will be optimal, but it does have a 'common sense' feel to it. In fact, if the proportionality is exact, the distance pdf can be described by the total number of propagules trapped at each distance, while the density pdf will be given by the mean number per trap. However, the proportionality is often not exact and the number of traps may not start increasing until some arbitrary distance from the source is reached (e.g. Clark et al. 2005; Bullock and Clarke 2000). Multiple traps at each distance will mean that sampling is spread out away from the infinitely thin transect line; however, if we ensure that the traps are evenly spread around the same angles defining an arc, we can assume that the data represent the average density for the sector.

5.5.4 Number of replicate plants

It is surely remarkable that, unlike any other branch of ecology, dispersal studies are usually unreplicated (i.e. $n = 1$). Traps are arranged along a number of transects around a single source plant (but see Thiede and Augspurger 1996, for a notable exception). No matter how many transects and how many traps at each distance, each trap constitutes a sample and not a replicate. True replication may be time consuming or require more traps, and it may not be straightforward to combine distribution pdfs from different shapes and sizes of plant, but these problems can be overcome. If we wish to have quality data and to be able to generalize from our results, the number of source plants must be replicated. Although, as Skarpaas et al. (2005) argue, time in the field and resources may be limited, much of the expense in field work often comes in getting to the site. A further day or two setting out traps or sampling will often only add marginally to the cost to the research. While recent increases in the number of sampled distances and the number of traps per distance are beneficial, an increase in replication in dispersal studies is well overdue.

5.5.5 Fitting a model

Rather than fitting equations to density, a derived variable, it is statistically more appropriate to fit models to the observed numbers of propagules in each trap, that is, as

$$n(r,\theta) = a(r,\theta)\omega_\theta(r) \qquad (5.13)$$

Trap data are integers and, especially in the tail where numbers of propagules are small, will be expected to follow non-normal error distributions. Poisson or multinomial errors are sometimes assumed in the fitting procedure. For propagules that carry multiple seeds and for those deposited in clumps by animals, however, the negative binomial may be more appropriate. Likelihoods are often higher when using the negative binomial than when using Poisson errors, but in reality the fitted curves are often almost identical (C. D. Canham, pers. comm.). Particular care should be taken when interpreting older papers written when researchers were largely confined to least squares linear regression. The assumption was often made that the variance (perhaps after log-transformation) is the same at all densities, which may not be realistic; more importantly, zero densities were frequently omitted or a constant added (since zeros cannot be transformed to logs), thus distorting the shape of the curve at longer distances.

Whatever the pdf that we decide to use, we can fit it to data as

$$n(r,\theta) = a(r,\theta)Ng(r) \qquad (5.14)$$

where N is then another parameter to be estimated. Alternatively, we might fit several equations and choose among them on the basis of their goodness of fit (e.g. using their log likelihoods if they have the same number of parameters, the Akaike Information Criterion if they differ in number of parameters, and tests for systematic lack of fit). Fitting non-linear equations can be less straightforward than linear regression and problems may be encountered more frequently with particular functions. Greene and Calogeropoulos (2002) observed that the Weibull, log-normal, and $2Dt$ can all produce ridiculous parameter estimates, depending on the attributes and quality of the data (this can also happen when assuming negative binomial rather than Poisson errors—C. D. Canham, pers. comm.). Because the parameter estimates of some models may be highly correlated (leading to large changes in parameter estimates for little change in the likelihood function), computer programs may fail to find solutions. It is thus common for those using some models, such as the $2Dt$, to have to fix some parameters or to restrict the range of parameter values allowed. However, this is inherently unsatisfying: we introduce equations with more flexibility, but then have to constrain their properties. Fixing a parameter at an incorrect value will result in incorrect values of other parameters (Snäll et al. 2007). The greater number of parameters in a mixture model, although allowing greater flexibility, also means that some parameters may be very poorly estimated if the data are limited. Indeed, Portnoy and Willson (1993) observe that many data sets are insufficient to fit even the most simple models.

5.6 Conclusions

As we have shown in this chapter, most studies of dispersal from whole plants consist of small samples from the recipient area. Although sampling protocols are improving, and methods have been developed to cope with multiple source plants, the conclusions to be reached from published studies are limited by their designs and by the species chosen for study. However, it is clear that the density of the seed rain declines rapidly with distance, being greatest close to the source plant. In contrast, the frequency distribution of dispersal distance approaches zero at the source, increases to one (or more) maximum, and declines rapidly to an (often) long tail. It is important to distinguish between these two forms of 'dispersal curve', if we are to choose the correct function to include in models of population dynamics (as in Chapters 6 and 7). The very few data sets that are sufficient to distinguish between alternative shapes of distance distribution suggest that their tails *can* be fatter than an exponential. It is not possible at this stage to determine whether the tails are more often fat than thin, whether density declines to zero close to the parent, or whether the shape of the distributions vary in some consistent way with species' dispersal syndromes.

Considering the large number of studies that have analysed dispersal from point sources, very few studies have attempted to predict, for whole plants, the distribution of dispersal locations over the entire course of the dispersal season or for all contributing dispersal vectors. We have shown examples of models that go part of the way towards this aim, predicting dispersal by wind, dispersal of a complex animal community, and dispersal of poorly adapted propagules from a distributed source. If we are to better understand the shapes of dispersal distributions, more studies are needed that incorporate the complexities of real plants, real communities, and real dispersal seasons.

Dispersal in population dynamics and evolution

The great richness of adaptations of plants and propagules to dispersal certainly has its own intrinsic fascination (Part A). This culminates in the distributions of propagules around their parents (Part B). However, to appreciate the broader consequences of dispersal, one must move on from the details of dispersal events from individual plants, to the larger-scale processes by which plant populations spread from one place to another, turn over the individuals they contain, and in so doing act selectively on the genetic variation within the species (including variation in propagule morphology), honing and changing it according to the environmental conditions which prevail. Part C of this book deals with these larger-scale processes.

The study of large-scale processes is seldom amenable to the tools of the field ecologist. The temporal scale over which we observe populations is usually considerably less than a generation in the case of long-lived trees and may only rarely exceed three generations in annuals (i.e. the length of a PhD). Moreover, as we discussed in the previous chapter, the distances over which propagules travel make it almost impossible to determine the locations of all individuals even within a single generation, except in the case of small plants with very limited dispersal. Thus, to develop an understanding of the part dispersal plays in population dynamics and for evolution, we must rely more heavily on models. Models are powerful tools for exploring the implications of biological processes. We can use them to conduct artificial (*in silico*) experiments where we ask the question: if the system works like this, what would happen?

There is a large and expanding literature exploring aspects of ecological systems that involve dispersal. Many studies do not distinguish between plants and animals and are relevant to both; a significant number are primarily relevant to animals (especially where they assume mobile parents); and others incorporate aspects of ecology that are primarily related to plants (such as a long-lived seed bank). This modelling literature is too large to review here. Instead, we will take you on a 'guided tour', picking out a selection of important topics and studies that we feel illustrate the variety of implications of dispersal for population behaviour. These methods can be applied to questions of direct relevance to particular problems. Inevitably, we have to include some mathematics, but we anticipate that the gist of the approaches, if not the detail, will be clear without delving too deeply into the maths. We have chosen to divide the text into three chapters, each examining one set of consequences of dispersal: Chapter 6 covers propagule dispersal in the context of invasions and range expansion; Chapter 7 places propagule dispersal in the context of the dynamics within populations; Chapter 8 examines the forces of natural selection which act on dispersal. It is in the light of these dynamics that our understanding of propagule dispersal can move on from the awesome diversity of individual dispersal syndromes to embrace the unifying mechanisms that determine plant dispersal.

Before embarking on our guided tour, we introduce some of the basic approaches and terminology that will be encountered in the following three chapters. Models are commonly categorized according to the ways that they handle space, the way that populations are represented, and the distribution of habitats within space.

- *Spatially-explicit* models specify the locations of the components of the population and the habitats using a coordinate system. *Spatially-implicit* models assume that exact locations are not important, although the existence of spatial separation of habitat or populations remains critical, as in metapopulation models and some models dealing with evolution.
- In Chapter 5, we contrasted the representation of dispersal data as the 'microscopic' locations of individual propagules and the 'macroscopic', spatially-distributed surface of population density. Both approaches are used in modelling: *individual-based* models (IBMs) *sensu stricto* predict the coordinates of every individual, whereas macroscopic models deal with the dynamics of *density* of individuals over space. IBMs carry more information, which means that they can give a more faithful representation of the ecological details. However, the sheer of amount of detail can make it hard to understand the results and computer-based simulations of IBMs become slow as the number of individuals becomes large. Macroscopic models of the dynamics have a simpler structure and are more amenable to formal analysis.
- Spatially-explicit models treat space as either *continuous* or discretized into a large number of cells on a lattice. *Cellular* models are sometimes preferred for

their computational simplicity and are easily linked to GIS (Geographic Information Systems). However, some care is needed when incorporating propagule dispersal which is continuous over space. Examples of cellular models are cellular automata, where each cell is potentially occupied by only a single (adult) individual (in essence individual-based models in which space is treated as discrete, and coupled map lattice models, in which the population within each cell is described by a local population size or density. Clearly, as the component cells become very small, the predictions of cellular and continuous-space models should converge.

- Simple models assume a landscape which is *homogeneous* over space and time. In reality, landscapes are typically *heterogeneous*, with continuous variation in topography, aspect, soil properties, etc. In cases such as habitat fragmentation, the inhomogeneities can be sufficient to justify a discretization of space, either into two states (suitable and unsuitable habitat) or a quantitative scale of suitability.
- Finally, we can distinguish between *analytical* models, which use formal mathematics to learn about their properties, and *simulation* models, which use computer simulations. These are extremes on a continuous spectrum. Often, to gain understanding, there is both scope for some mathematical analysis and also some need for numerical analysis. The relative emphasis on analysis and simulation depends on the complexity of the problem and the skills of the researcher.

CHAPTER 6

Invasions and range expansion

6.1 Introduction

Expansion of a population results from the combination of net reproduction (births minus deaths) and dispersal. Consider a dispersing propagule which encounters a region empty of that species. If it successfully establishes and reproduces, it will produce its own propagules, which will themselves disperse: some of these will again land in unoccupied territory, others may land next to existing plants, while others will land in gaps between plants. Thus, invasions are initiated, new patches form, expand, merge with other patches, initiate new patches further afield, and ranges are extended. The more propagules that are dispersed (the greater the level of reproduction), the more unoccupied territory will be encountered and colonized. If, instead, many of the plants are unsuccessful and fail to reproduce, or they disperse their propagules into unsuitable habitat, invasions may go into reverse: populations may contract and ultimately become extinct. A contracting population is exactly what we wish to see when managing unwanted invasions and, as we will see, an understanding of spatial pattern within populations can be important in their eradication or containment. Those people tackling active invasions will need to initiate quarantine measures to prevent further incursions, control new and existing outbreaks, and to eliminate propagule sources. However, conservationists faced with contracting species must take the opposite actions, to enhance plant survival, increase reproduction, and ensure dispersal, possibly by using re-introduction or translocation programmes.

The area occupied by a species as it invades is clearly important, in that this determines its impact on the other members of the biological community and its influence over the abiotic environment. However, the *rate* at which patches, populations, or species ranges expand can be of immense ecological and practical significance. How dynamic are species boundaries? How fast will species be able to spread in response to climate change? How much time do we have to take action to stop an invasion? What changes in land use or climate would cause a currently restricted exotic species to start expanding its range (the awakening of a 'sleeper' species), or for an endemic species to contract and become extinct? One particular question about range expansion, sometimes referred to as 'Reid's Paradox' (Clark et al. 1998a), has fuelled considerable ecological research over the past century. Clement Reid (1899) was perhaps the first to ask why species have spread so far since the last major glaciation, considering the relatively short distances over which most of their propagules are dispersed. As the ice sheets retreated, species spread hundreds of kilometres during the Holocene. Rates of spread of trees, calculated from pollen in cores taken from bogs and lakes, are as much as 150–500 m/yr (Clark et al. 1998a), despite the fact that the vast majority of their seeds are only dispersed within a few tens of metres of the parent and new trees may take over a decade before they begin to reproduce. Forest herbs have increased their ranges by similar distances, yet many of these rely on clonal growth for their spread and produce few seeds that are usually dispersed within centimetres of their parent (Cain et al. 1998).

In this chapter, we will explore the spatial dynamics of invading species, paying particular attention to the patterns formed as they spread, the rate of increase of the area invaded, and the rate at which their boundaries move.

6.2 Pattern within spreading populations

It is relatively straightforward to simulate the spread of the individuals within a population using a stochastic model and a few simple assumptions:

- Begin by placing a single plant in an otherwise unoccupied, homogeneous area and assign Cartesian coordinates to its position.
- Calculate the number of propagules that the plant produces in the season. This may be a random number selected from a discrete distribution (such as the Poisson) with a specified mean. (For perennials, the mean propagule number may be a function of plant age. No density-dependence is included in the model at this stage: plants produce the same number of propagules, no matter how many close neighbours they have.)
- Disperse each propagule by selecting a random angle and a random distance independently. For isotropic dispersal the angle is selected from a uniform distribution and the distance is selected from either an appropriate distance pdf (e.g. Table 5.1) or by converting a density vs distance regression to its distance pdf form using equation 5.7. (An alternative approach is to select random pairs of coordinates—relative to the parent—from an appropriate two-dimensional density pdf, for example Table 5.2 or, if dispersal is equal in all directions, from a distance pdf converted to it's density pdf form using equation 5.7.)
- Propagule mortality can be imitated by removing individuals at random according to a specified probability; mortality of the parent plants can be assigned in the same way (for annuals, all parents are killed at the end of the season; for perennials, we will need to keep track of their ages for calculation of their fecundity in the future).
- Whether the surviving propagules grow into plants or stay dormant at the new location for the remainder of the season can again be assigned using random numbers.
- We repeat this entire process over multiple seasons of propagule production (and hence over multiple generations) for the propagules dispersing from every plant. Each season, we can plot the positions of the plants (or the propagules) and watch as the population grows.

Fig. 6.1a shows such a simulation for an annual species. The population spreads steadily outwards over time, evenly in all directions, with the greatest density towards the centre. A series of quadrats along a transect through the centre of the population would show that the distribution of plant density in this particular case is approximately Gaussian. The rate at which the points spread outwards increases if (i) the scale parameter of the density pdf is increased or (ii) the number of propagules produced per plant is increased (if there are more individual propagules being dispersed, the rarer long-distance events will occur more often, and hence the spread will be faster).

If, instead, there are two dispersal vectors with one spreading propagules further afield than the other (often referred to as 'stratified dispersal' or 'stratified diffusion'—Hengeveld 1989), a quite different pattern of spread can emerge (Fig. 6.1b). The pattern of spread may now be very patchy, uneven in different directions, though still at greatest density towards the centre of the population. The degree of patchiness depends on the size of the difference between the scale parameters of the two component distance distributions—that is, the difference in dispersal potential of the two vectors—and the proportion of propagules being dispersed by each vector. Satellite sub-populations of plants (derived from rare, longer-distance dispersal events) are initiated and expand through more localized dispersal and eventually merge with other nearby clusters. This pattern of sporadic outbreaks and in-filling has been termed 'infiltration invasion' by Bastow Wilson and Lee 1989). The edge of the population is very difficult to define, but in any one direction range expansion occurs by a series of occasional 'jumps' followed by periods of slower spread. Similar patterns of spread occur when simulation models use a single fat-tailed density distribution (e.g. Shaw 1995), but a fat tail is not essential for this behaviour: the mixed model used in Fig. 6.1b has a tail that is (by strict mathematical definition) thin at extremely long distances.

A patchy type of invasion has considerable implications for population genetics. The satellite patches formed well away from the main invasion front are likely to be founded by just one or a small number of individuals. The majority of offspring from

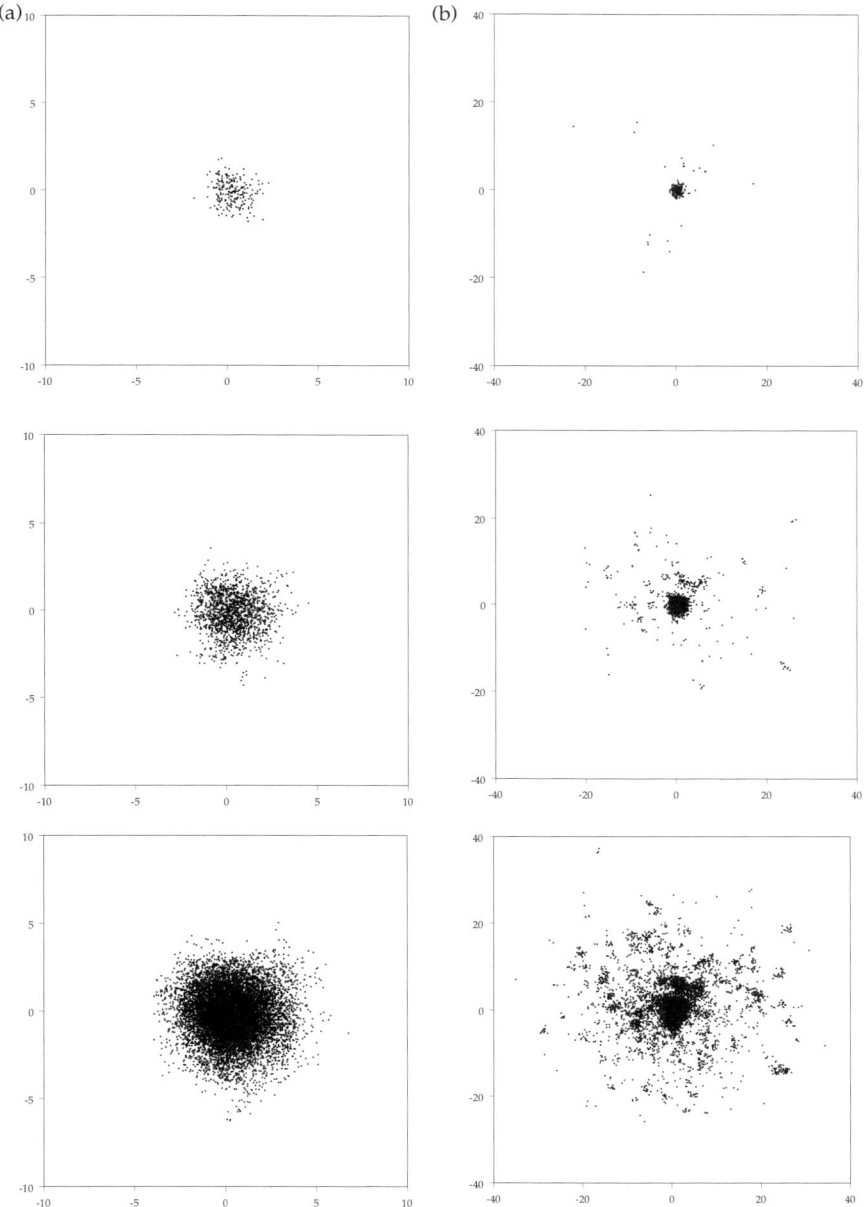

Figure 6.1 Simulation of the spread of an annual plant population over fifteen generations. Scales on maps are arbitrary units. For (a) there is a single dispersal vector, with a single distance pdf; for (b) there are two vectors, each dispersing according to a Rayleigh distance pdf (equivalent to a normal density pdf), with one having a scale parameter of 25 times the other and 1% dispersing by the long-distance kernel. Fecundity was taken at random from a Poisson distribution with a mean of 1.5 and was identical (i.e. the same random numbers) in the two simulations; all individuals die after reproduction, all seeds germinate immediately and then survive to maturity. Note that the axis for (b) is four times the length of (a).

these plants will be distributed locally, resulting in a large cluster of neighbouring plants having common maternal genes. It will be difficult for a later-arriving genotype without a selective advantage to increase in abundance within the area already occupied. Moreover, the new satellites initiated even further away are more likely to come from this new cluster than from the original source patch, since its propagules will not have so far to travel. Thus, for a population expanding in all directions, we might expect certain compass directions to become dominated by particular genotypes, purely as a result of a random process coupled with a low level of long-distance dispersal. Using a coupled map lattice model (see p. 112), Le Corre et al. (1997) showed that genotype clusters can persist for long periods of time, even after the population has spread throughout the suitable habitat (see also Ibrahim et al. 1996). When an initially diverse population spreads through a narrow corridor, a single genotype may come to dominate so effectively that the spread of other genotypes through the corridor is prevented: Bialozyt et al. (2006) refer to this as the 'embolism effect' (Fig. 6.2). This behaviour is highly sensitive to the proportion of propagules travelling long distances, with the number of 'successful' alleles and genetic diversity being greatest at an intermediate level of long distance dispersal. At very high levels of long distance dispersal, genotypes will not be able to form large patches on their own

and most will reach the end of the corridor; at very low levels of long-distance dispersal, the population spreads slowly as a wavefront, conserving more of the original genotypes.

Some researchers have commented that the patchy invasion pattern resulting from fat-tailed and stratified distance distributions is closer to reality than that predicted from a single thin-tailed distribution (see also Shaw 1995 for fungal pathogens): there are few situations, perhaps with the exception of clonal growth, where real populations expand in a continuous, steadily advancing front. Many weeds, for example, have extremely patchy distributions within arable fields (e.g. Rew et al. 1996). Although it is perhaps tempting to draw the conclusion that real distance pdfs must therefore either be fat-tailed or result from mixtures of dispersal vectors (i.e. deducing process from pattern), there are a number of reasons other than dispersal that can result in patchy populations (Cousens et al. 2006). These include spatially dependent mortality and fecundity resulting from either natural (e.g. soil, fire, grazing animals) or human (machinery, landscape modifications) causes. In addition, patches can form and persist within a homogeneous arena even when dispersal distributions have thin tails: using a cellular automaton model of an annual weed, Wallinga (1995) showed that patches may persist in the same place for many generations, under a high level of mortality and with highly restricted dispersal.

Harper (1977, based on an interpretation of a discussion by van der Plank 1963), concluded that the inverse square law is the critical density pdf at which the pattern changes from a single advancing 'wavefront' to one in which patches form and coalesce. Wallinga et al. (2002) used a cellular automaton of $0.2 \, \text{m} \times 0.2 \, \text{m}$ squares to test this. They compared the patterns produced by three distance pdfs: a negative exponential and two power curves, one declining faster than the inverse square law and one declining more slowly. Parameter values were chosen so that only 5% of seeds dispersed beyond a distance of 1 m. The exponential distance distribution produced a population with a definite edge, while both power curves produced populations that were more scattered around their periphery, with the fattest-tailed distribution producing a very scattered population.

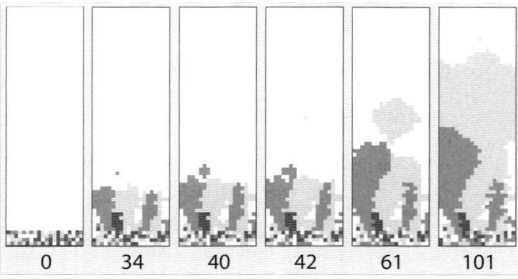

Figure 6.2 Simulation of the spread of a plant population along a corridor, where individuals differ in a maternally-inherited haploid marker and there was a low probability of a long-distance dispersal event (Bialozyt et al. 2006, reproduced courtesy of Blackwell Publishing). The model was a coupled map lattice; shading indicates the most frequent haplotype in each cell. The number of time steps is shown below each panel.

They also calculated the number of 'colonies' predicted by each pdf after 10 generations: plants were determined to be members of the same colony if they were within 5 cells of each other. The exponential had a single colony, the steep power curve 10 colonies (one primary one, surrounded by a few, small scattered ones), and the flatter power curve had 33 colonies, most of them containing a single individual. They concluded that Harper's inverse square law rule was incorrect, since both power curves resulted in patchy spread.

In our exploration of spatial models thus far, we have assumed that the environment through which the population spreads is homogeneous; however, it is relatively straightforward to incorporate habitat heterogeneity. Digital maps of the variation in key demographic parameters across the area need to be constructed: then, when we calculate plant survival or fecundity, we use spatial coordinates to extract the appropriate parameter values. An excellent illustration of this is a simple study of the clonal perennial *Trientalis europaea* (Primulaceae) in which an individual-based simulation model was used to explore the effects of habitat variation on spatial dynamics (Piqueras et al. 1999). Like many clonal plants, recruitment from seeds is relatively uncommon and local dispersal in this species is primarily through vegetative growth. Each spring, a tuber produces a single, leafy shoot; new tubers are produced at the ends of stolons towards the end of the growing

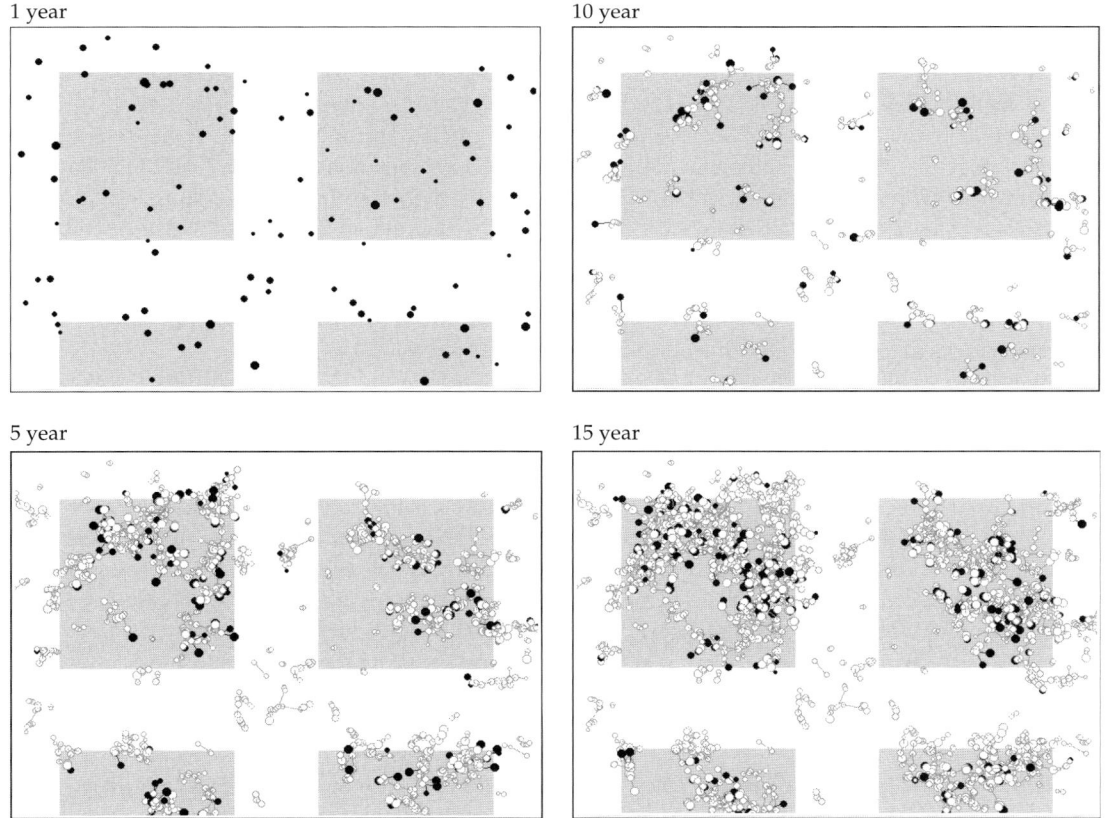

Figure 6.3 Predicted dispersal of *Trientalis europaea* through clonal growth into patches of high (grey shaded areas) or low nutrient levels (Piqueras et al. 1999, reproduced courtesy of Springer). Grey areas represent nutrient-rich patches; filled circles are living ramets, open circles are ramets produced in any of the previous years. Lines represent stolons lengths; circles indicate tuber sizes.

season, with both the original shoot and the connecting stolons decaying in the autumn. Nutrient levels in the soil affect the sizes of the tubers and the number of daughter tubers produced by a plant, but not the lengths of the stolons. In the model, values of most parameters were taken from experimental data from high or low nutrient plots; with the exception of branching angle, parameter values were size-dependent. The simulated area consisted of rectangular regions high in nutrients, surrounded by low nutrient habitat. As the shoot systems expanded, some stolons encountered high levels of nutrients (often referred to as 'foraging' in clonal species). Where the high-nutrient regions were large, ramets became locally aggregated (Fig. 6.3), since parent tubers produced more daughter tubers that were also within the region. If high nutrient regions were small, however, the daughter ramets from high-nutrient parents increasingly found themselves in low nutrient levels and shoot aggregation did not occur. Short-distance dispersal of propagules, such as tubers, facilitates proliferation of a clone in high quality habitats. Other examples of simulations of clonal shoot systems, though in homogeneous habitat, include *Alpinia speciosa* (Zingiberaceae: Bell 1979) and *Medeola virginiana* (Liliaceae: Bell 1974).

6.3 Rate of increase in area occupied

Let us return to the maps produced using our simple individual-based model (Fig. 6.1). To calculate the rate of increase in area occupied by the population, we first need an appropriate algorithm for determining the position of the population edge (which may be very irregular). We calculated the area of the population (A_t) within the convex hull drawn around the scatter of plant locations (note that we are not promoting the use of the convex hull: it is merely one of the simplest). With only short-distance dispersal, the area occupied increases steadily but slowly (Fig. 6.4a). However, even with a small proportion dispersing by long distances, the rate of occupation increases considerably: after 15 generations, the area occupied when 1% disperse long distances was 163 times the area with no long-distance dispersal. However, when the long-distance fraction is very low and population size is small, the growth in area can be very erratic, increasing by large amounts as a

result of the strong influences of single long-distance dispersal events and sometimes decreasing when these subsequently fail to reproduce. Although the model can be made biologically more realistic, its qualitative behaviour will be similar.

An alternative way of modelling the rate of change in population area is based on simple geometry. Let us assume that the net result of reproduction, mortality, and dispersal is that a patch of a species spreads outwards at a constant linear rate in all directions (we will explore the validity of this assumption using other models later) and that the density within the occupied area is homogeneous. Since the area of a circle is πr^2, the square root of the area occupied by the patch increases linearly over time. If, instead of having a single expanding patch, the same initial area is divided into many smaller patches that do not overlap as they expand, the area occupied at any given time will increase with the number of initial patches (Mack 1985). Whereas a single circular patch with initial radius r_0 and rate of radial expansion Δr will have area $\pi(r_0 + t\Delta r)^2$ after time t, n patches starting with the same total area will have a combined area $\pi(r_0 + \sqrt{n}t\Delta r)^2$. It has therefore been argued that a population expanding in a patchy manner, initiating many new small patches (these are referred to by Moody and Mack (1988) as 'nascent foci' or 'satellite foci'), will increase in total area faster than a population expanding as a single patch.

Shigesada et al. (1995) further explored the dynamics of a system in which an initial patch spawns new satellite patches, each growing at a constant radial expansion rate. The additional assumption was made that none of these patches would overlap (as in the early phases of an invasion and in which satellites form a long way from their parent patches). Initially, the square root of the area of the single patch increased linearly (i.e. area vs time was quadratic); however, as new satellites were initiated, expanded, and then initiated satellites of their own, the rate of increase of area became exponential. This was the case regardless of whether the rate of initiation of new patches by a parent patch was constant, proportional to patch circumference or proportional to patch area. (It should be noted that many economic models of the management of invasive species assume a constant proportional rate

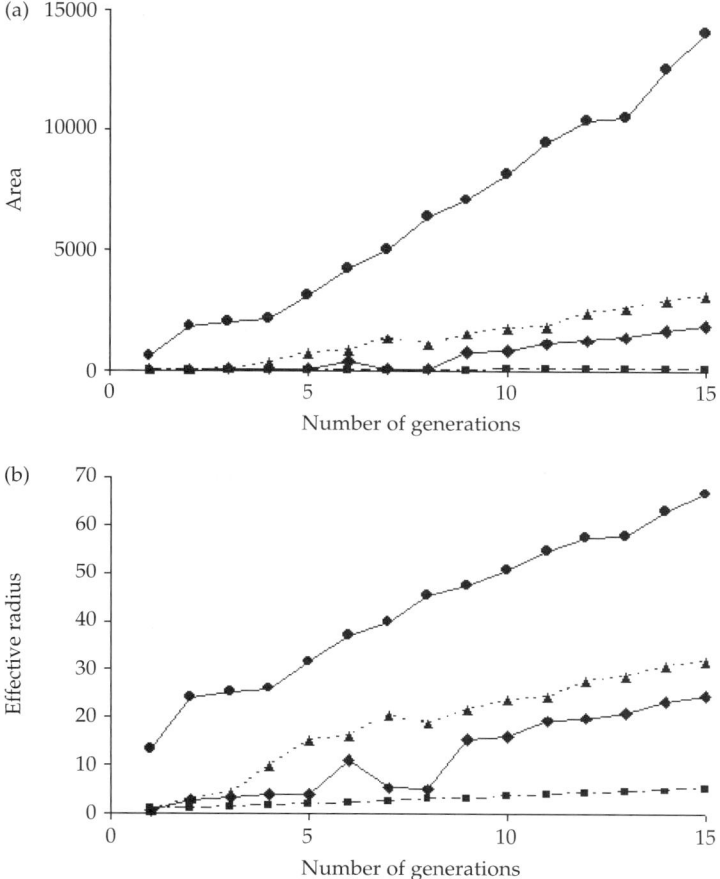

Figure 6.4 Increase in (a) area occupied by the expanding populations shown in Figure 6.1 and (b) effective population radius (see below). The four lines are: ■ all seeds dispersed by a single short-distance Rayleigh distribution; ♦ 0.01%; ▲ 0.1%; and ● 1% dispersed by a long-distance Rayleigh distribution.

of increase— i.e. x% per year—in the area infested, equivalent to an exponential rate of increase: see Auld et al. 1987.) When they allowed patches to overlap, in a 'coalescing colony' model (Fig. 6.5), with all patches being initiated a given distance ahead of the advancing parent patch, the rate of increase of area was reduced. The effective range radius increased linearly when either the satellite initiation rate was constant or when the initiation rate was proportional to the circumference; for an initiation rate proportional to patch area, the radius accelerated steadily.

It has been shown for field data of a number of organisms, even where there are multiple and irregular patches, that graphs of the square root of area against time are often well described by a straight line (Hengeveld 1989). These relationships can be used to estimate the 'effective range radius' (Shigesada et al. 1995): the radius of a circle whose area is equal to the combined area of all patches, i.e. $\sqrt{\Sigma(\text{area})/\pi}$ and its rate of increase. As Lonsdale (1993) showed for *Mimosa pigra* (Mimosaceae), although the fit of data to a quadratic may be very good, the fit of the same data to an exponential may be better. For an exponential relationship, the percentage increase in area is constant (Cousens and Mortimer 1995), whereas for a quadratic the percentage increase in area will

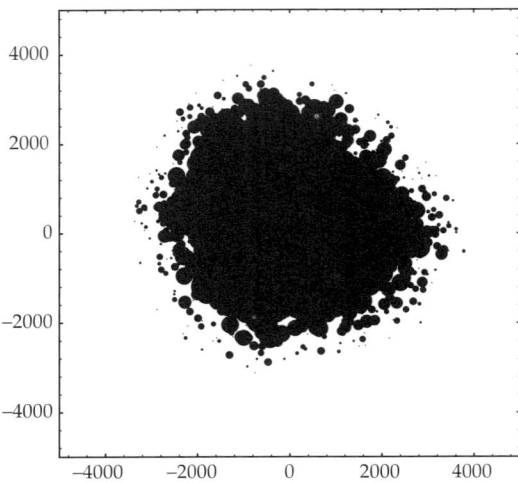

Figure 6.5 Simulation from the Shigesada et al. (1995) coalescing colony model (reproduced courtesy of University of Chicago Press). New satellite patches are initiated ahead of the advancing main colony and then merge with it.

decline over time even in an unlimited space. Eventually, of course, as the suitable habitat becomes fully occupied the population's area will approach its upper limit and the rate of increase will level off (see Shigesada et al. 1995).

The initiation of new patches has important consequences for the control of invasive species. Moody and Mack (1988) examined the control of a population consisting of one very large and many small non-overlapping patches, each with the same radial rate of expansion. It was shown that a single control event which eliminates all satellites will have a much greater effect on the future infested area than controlling an annulus around the edge of the main patch. They elaborated their model by allowing new satellites to be initiated in each time step, with the rate of initiation increasing as a logistic function of the total area. Patches had threshold ages below which they could not be detected (and hence could not be targeted for control); radial expansion rates could differ between the main patch and the satellites and according to whether satellites exceeded their threshold age. Simulations showed that it is important to both detect and destroy a large proportion of the satellite patches early in their development. Reduction in the initiation of new satellites, by reducing the area of the main patch,

was generally less effective as a strategy for curbing the rate of population growth than eliminating the satellite patches as they were detected. This result has had considerable impact on the development of management strategies for invading species; moreover, it agrees with rules of thumb developed by field practitioners, such as the 'Bradley method' of ecological restoration, which target satellite populations as well as working inwards from the margins of large patches rather than putting all resources into reducing the densest parts of the largest patches (Bradley 1971). However, two other modelling studies reached the opposite conclusion: an unpublished study using control theory, where large numbers of fruits cause birds to aggregate in the core population (Shea et al. 2002); and a cellular model of two species where satellites form as a result of very long-distance dispersal events, predominantly from the core population (Wadsworth et al. 2000). Perhaps there is no single optimal management strategy: the appropriate action may depend on the population biology of the weed, the behaviour of the vector and, as found by Taylor and Hastings (2004), the size of the budget available for weed control. Further research is clearly needed.

Simple geometric models are useful in a conceptual exercise to consider general principles, but they are simplistic in the way that they incorporate both population growth and dispersal. Cellular models allow simulations to be based on more complex, realistic biological information. Two examples of simulations from cellular (coupled map lattice) models are shown in Figs. 6.6 and 6.7. In the first, seeds of the annual weed *Datura ferox* are dispersed by a combine harvester within an homogeneous field sub-divided into 0.7 m × 0.7 m squares. Seed production by each cell was a non-linear function of density within the cell, while dispersal was determined by a two-dimensional table of probabilities allocated to cells within a certain distance of each source cell. The population increased rapidly in the direction of the harvester. The relationship between the square root of the area occupied by the population and time was almost linear (although the limited number of generations means that initial model instabilities may partly obscure the true long-term trend). The second example shows the predicted migration of a long-lived deciduous tree (based on *Quercus*) at

Box 6.1 Cellular population models

Cellular models have been extremely popular amongst ecologists for simulating the area occupied by an invading species: it is easy to separate out the effects of dispersal, reproduction, and mortality and to allow density to vary spatially; they are based on the types of demographic data collected routinely by most field ecologists; they are straightforward to program, without any advanced mathematical knowledge; and their cellular structure means that they are easy to link to Geographic Information Systems (GIS).

To produce a coupled map lattice model, begin by considering an 'arena' divided into discrete cells of equal size. Cells are usually square, as this is easy to handle in programming languages. However, the disadvantage of squares is that the centres of neighbouring cells are not all the same distance away from the cell that they enclose: the centres of diagonal cells are further away than cells with adjacent sides. Hexagonal cells (González-Andujar and Perry 1995; Tilman et al. 1997) have the advantage that every neighbouring cell is the same distance away from the cell that they surround. The population must be initiated by assigning a number of individuals to particular cells. Within each generation, we apply the equation

$$N_{t+1} = N_t + B - D + I - E \qquad \text{(B6.1.1)}$$

to every cell in turn (using a 'do loop' in a computer program), calculating the number of plants (or propagules) in the cell in the next generation (N_{t+1}), that is, the local population density, from the number present in the previous generation (N_t), the number of births (B) and deaths (D), which can be calculated from a simple non-spatial population model, immigration (I, dispersal in) and emigration (E, dispersal out). Any of these parameters can be replaced by a density-dependent function. Spatial variation in habitat can be incorporated by designating particular demographic parameter values to the cells within particular regions of the arena. The progress of population expansion can be followed by mapping the population density; the area occupied by the population at a given time is obtained simply by counting how many cells are occupied (or perhaps those with a density above some minimum level); and the total number of individuals can be found by summing the contents of the cells. Cellular automaton models differ from coupled map lattice models in that density-dependence is incorporated by specifying that each cell can contain only a single adult plant.

Many cellular models treat their cells in the same way as the patches in metapopulation models (see p. 151): dispersal into a cell is considered to be constant within a given neighbourhood distance (e.g. McInerny et al. 2007. Thus the dispersal pdf away from a source plant is a step function, with a high proportion remaining in the source cell, a constant lower probability in receptor cells, then zero at greater distances. Clearly, in comparison with the distributions presented in Chapter 5, this is a crude representation of dispersal. However, for illustrative purposes it can be sufficient and its simplicity ensures that complex landscape scenarios can be modelled with a minimum of computing power.

Emigration into one cell from another can be calculated from a more realistic density pdf (Table 5.2), using the number of propagules produced by the plants and the coordinates of the cell margins. However, the discrete nature of cells can still lead to significant errors if cells are large relative to the steepness of the pdf. If all individuals are assumed to be located in the centre of a cell then short-range dispersal on a coarse grid will result in no dispersal at all. However, if the incoming propagules are assumed to be evenly re-distributed within the receptor cell, for calculating events in the next generation, then some dispersal distances will be inflated. Propagules only just managing to reach the cell may be moved further within the cell as a result of this assumption, and this will be more important when modelling invasions and range expansions. One of the key decisions to be made is therefore the size of the cells: the smaller the cells, the more realistically the discrete spatial structure can be described by continuous mathematical functions and the smaller the errors. How should we choose the size of a cell?

- In cellular automata the cells can be defined as the size that can be occupied by one adult individual. Given the great plasticity of plant growth in relation to density, this is clearly a difficult concept to apply: plants at high density will occupy less space per plant than they will at low density. The law of constant final yield (Kira et al. 1953), however, dictates that the fecundity of a cell will be the same regardless of whether one adult occupies it or several, and thus an error in the choice of cell size may have little consequence to the predicted spatial dynamics, at least with respect to fecundity.

- In many coupled map lattice models, where each cell can contain any number of plants (the state variable for the cell is the local population density, which is not necessarily a round number), the cell size is often chosen to be similar to the sizes of quadrats used to collect the field data from

continues

Box 6.1 continued

which the demographic parameters of the model were derived (which themselves are usually arbitrary). However, if the dispersal of the species is very limited in extent, this may mean that the entire seed shadow will fall within just a few neighbouring cells and thus dispersal will be effectively a crude step function poorly reflecting our understanding of the underlying dispersal process.

- Ideally, we should base our (compromise) choice of cell size on the results of model simulations run at a range of cell sizes. For example, Collingham et al. (1996) found that their coupled map lattice model with a Gaussian density distribution, lost accuracy if the cell widths were larger than half the root mean square displacement of propagules.

---

a geographic scale through an heterogeneous environment. Cells were 1 km square and varied in their carrying capacity according to the proportion of the cell occupied by deciduous woodland. Seeds were dispersed according to a compound dispersal function, with 95% landing in the parent cell and 5% dispersing according to a bivariate normal pdf with root mean square displacement of 6 km (this value was necessary to produce rates of spread comparable to those estimated from palaeoecological data) truncated at a maximum distance of 20 km.

A more complex cellular model was used by Schippers et al. (1993) to simulate the spread of the clonal weed *Cyperus esculentus* (Cyperaceae) in a cultivated field. Dispersal was stratified, with some tubers being spread locally by soil tillage and others being moved over longer distances as a result of adhering to farm machinery. There was also vertical soil movement, with position in the soil profile determining how many tubers would produce shoots. The rate of increase in area of the population depended on the use of herbicides and their effectiveness in killing tubers. One important prediction was that there could be a period of time when patch size was still increasing even though the overall plant number was declining.

6.4 Rate at which population boundaries move

Once again, simulations using individual-based models *can* be used to predict the rate of spread of the edge of a population. Returning to the model used in Fig. 6.1a, we have calculated the edge of the population by taking the area (A_t) of the minimum

convex polygon (the convex hull: see p. 78) around the scatter of plants and calculating the effective range radius as $\sqrt{A_t/\pi}$. Fig. 6.4b shows that, after some initial instabilities, the increase in radius was approximately linear from generation nine onwards. The rate of increase in the radius was 0.34 units per generation when all dispersal was by short distances, rising to 3.1 units per generation when 1% disperse by long distances. We could also use the output from simulations of cellular models in the same way, using the number of occupied cells to estimate A_t.

However, much more powerful *analytical* methods can be used to explore the rate of population range expansion. These can divided into two groups: those that consider both space and time as continuous, referred to as *reaction-diffusion* models; and those that consider space as continuous but time as discrete, referred to as *integro-difference* models (p. 126). Both consist of two mathematical components, one describing movement of individuals (often referred to as the dispersal *kernel*, or redistribution kernel), the other describing population growth rate. We will discuss these two types of model in turn and the results that have emerged from their application. However, as we begin to consider dynamics in heterogeneous or fragmented landscapes, the use of continuous-space models becomes limited and we will return to cellular simulation models.

6.4.1 Reaction-diffusion models

Reaction-diffusion models relate the change in state (in this case population size) at a point in space to diffusion (of members of the population) and a

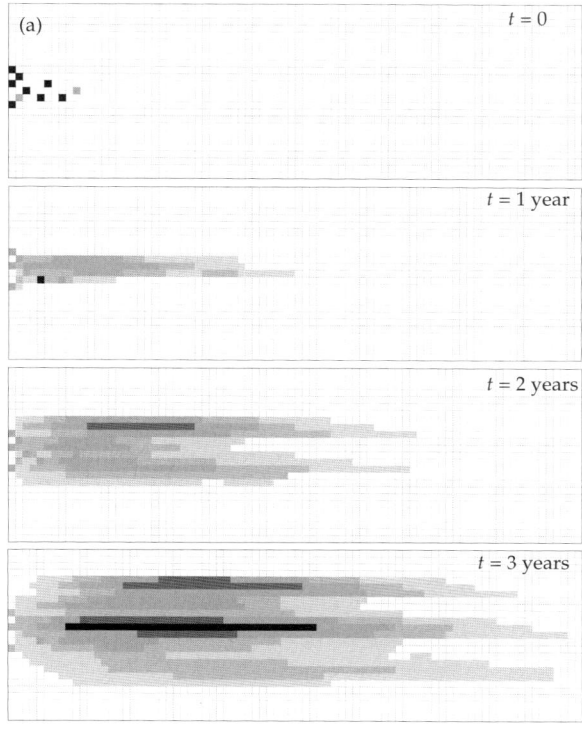

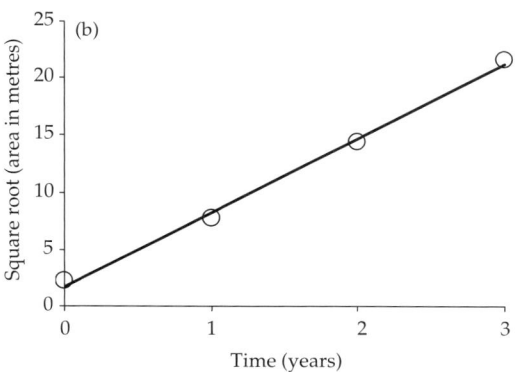

Figure 6.6 Simulation of population spread of the annual weed *Datura ferox* by a combine harvester in an homogeneous field, using a coupled map lattice model. (redrawn from Ballaré et al. 1987). In (a), cells are 0.7 m × 0.7 m; black indicates the highest density within a cell. (b) shows the change in the square root of the total area occupied.

reaction (population growth rate):

$$\text{Rate of change of density} = \text{dispersal rate} + \text{growth rate}$$

(Lewis 1997). The reaction-diffusion model used most commonly for single species populations is

$$\frac{\partial u}{\partial t} = D\left(\frac{\partial^2 u}{\partial x^2} + \frac{\partial^2 u}{\partial y^2}\right) + F(u) \qquad (6.1)$$

where D is the diffusion coefficient (with units distance^2/time) and u is the local population density at spatial coordinates x, y and time t (Holmes et al. 1994). $F(u)$ is the population growth function: for exponential growth, $F(u) = \alpha u$, where α is the intrinsic growth rate, while for logistic population growth $F(u) = \alpha u(1 - u/K)$, where K is the population carrying capacity (maximum local population density). The diffusion term $D\left(\frac{\partial^2 u}{\partial x^2} + \frac{\partial^2 u}{\partial y^2}\right)$

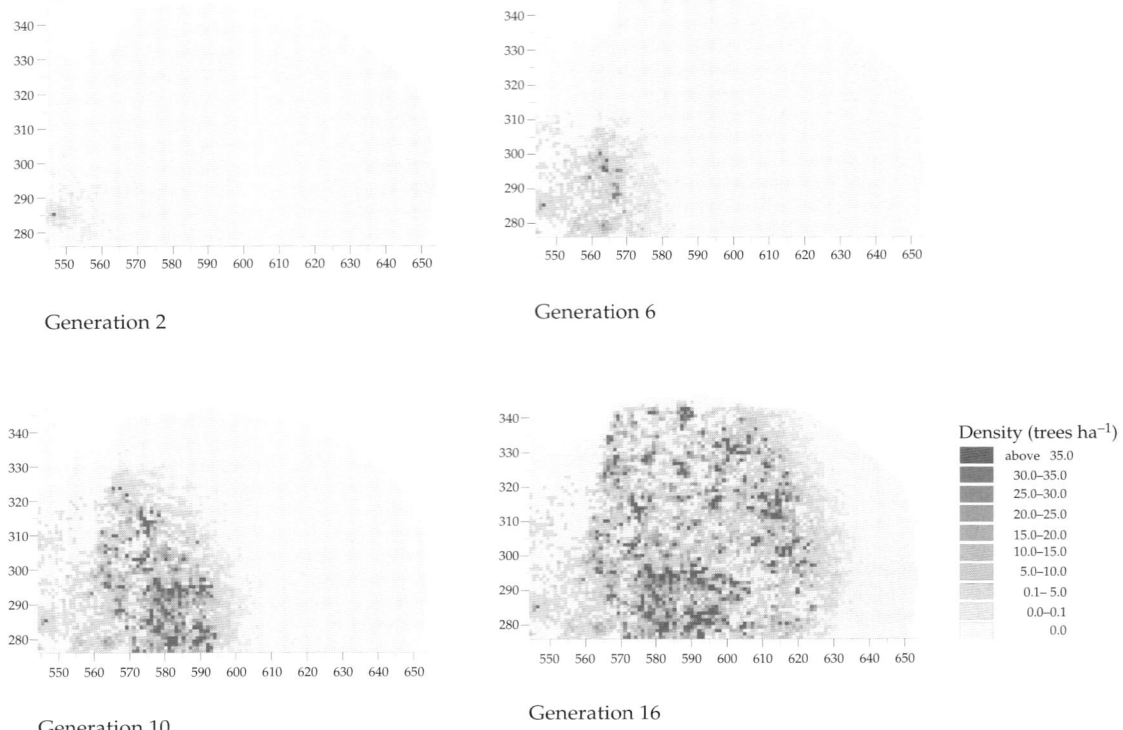

Figure 6.7 Simulation of the spread of a long-lived deciduous tree (based on *Quercus*) through an heterogeneous environment at a geographic scale, using a coupled map lattice model (Collingham et al. 1996, reproduced courtesy of Opulus Press). The grey area indicates the county of Norfolk, UK; cells are 1 km square.

is derived from the assumption of a random walk and in the absence of population growth results in a two-dimensional Gaussian (normal) distribution of population density which becomes wider and flatter over time (the mean squared displacement equals $4Dt$). Equation 6.1 has a long history and has been central to the development of ecological invasion theory. Key papers were by Fisher (1937), who developed the model for the movement of a novel gene in a habitat occupying a single spatial dimension and with logistic population growth, and Skellam (1951), who considered two dimensions with exponential population growth. Excellent discussions of reaction-diffusion population models and their assumptions are given by Holmes et al. (1994) and Shigesada and Kawasaki (1997).

The most important prediction that emerges from a mathematical analysis of equation 6.2 is that populations (those for which the model assumptions are appropriate) will spread equally in all directions, advancing as a wave of population density (Fig. 6.8). Contours of population density spread steadily outwards. The velocity of the leading edge of the population wavefront approaches and then maintains a constant value of $\sqrt{4\alpha D}$ (e.g. Fisher 1937; Kolmogoroff et al. 1937; Skellam 1951). This is the case for any growth function ($F(u)$) in which (a) the maximum population growth rate is greatest when the population density is small (i.e. there is no Allee effect) and (b) the growth rate is always positive when the population density is below the carrying capacity of the environment (Hadeler and Rothe 1975). Since the rate of increase in the radius of the population is constant, then the square root of the area within a given density contour will increase linearly with time (see p. 119), as found in many

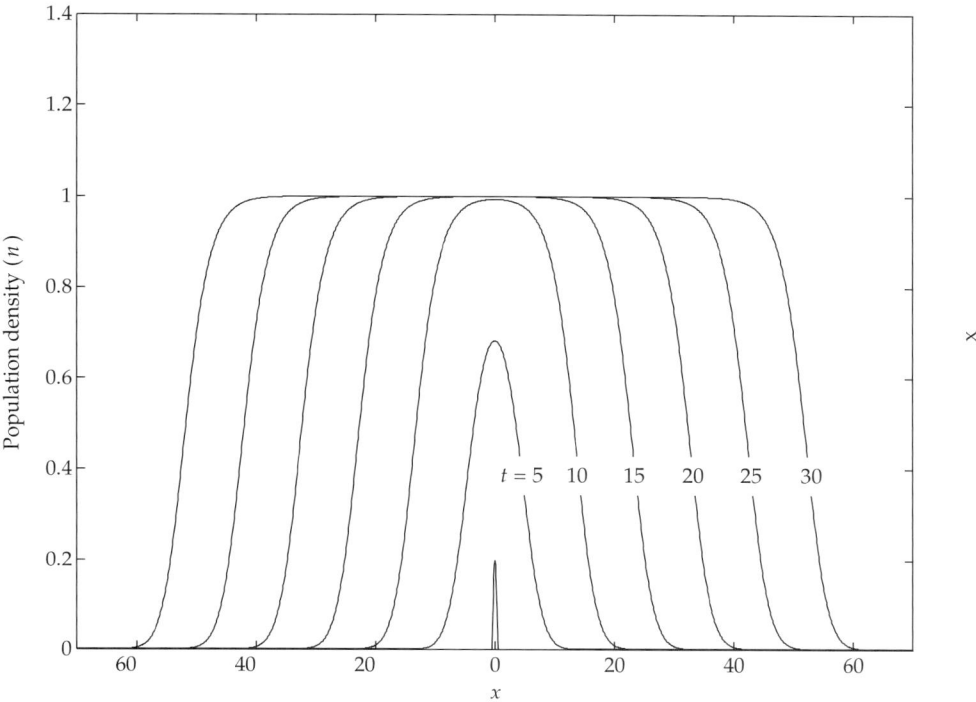

Figure 6.8 Waves of population expansion, predicted by a reaction-diffusion model incorporating logistic population growth. The population begins at the centre and steadily moves outwards, building in density as it spreads. Each line represents a cross-section through the population density profile at successive, equal time intervals. (Lewis et al. 2006, reproduced courtesy of Springer).

real populations (Hengeveld 1989). Clearly, anything that increases or decreases the coefficient of diffusion (rate of emigration) or the rate of population growth (i.e. reproduction and survival) will respectively increase or decrease the rate of spread.

What happens if we vary some of the assumptions of the model? If we replace simple diffusion, with its infinite-tailed dispersal function, by a function having a flatter peak and truncated tail (based on a correlated random walk), we obtain the reaction-telegraph model (Holmes 1993). For this model, the population still spreads out in a wave, the velocity of the wave still reaches a constant value, but the rate of spread is slower and the leading edge of the wave is steeper (the size of the differences between the two models depends on the parameter values used). If we allow the rate of population growth to become negative at low population density, for example due to reduced pollination success and inbreeding depression (the 'Allee' effect), the rate of spread can

be considerably reduced (see also Kot et al. 1996). Using a cubic population growth equation for $F(u)$, Lewis and Kareiva (1993) found that populations will spread (the wavefront velocity can be positive), provided that certain thresholds for population density and area are exceeded. This is because although the population density at the leading edge would not be sufficient to overcome the Allee effect on its own, dispersal from the heart of the population ensures that density at the wavefront increases. The larger the diffusion constant or the density at which growth rate becomes positive, the larger the initial population size required for the population to spread. Travis and Dytham (2002), using a cellular IBM, also found that a mild Allee effect had a huge effect on range expansion but, because they allowed evolution of the dispersal strategy, the evolved strategy in the core population was lower and this reduced the push from the centre and further slowed the rate of range expansion.

It is implicit in equation 6.2 that the arena within which the population is spreading is homogeneous (since its parameters are not a function of x or y). What happens in a heterogeneous arena, where there may be regions that are good or bad for reproduction, mortality, or dispersal? Clearly, an increase in either D or α will result in an increase in rate of spread. Thus, as the population moves through a landscape, its rate of spread will increase and decrease according to the conditions that it encounters. In a one-dimensional habitat consisting of repeated bands of high and low α, Shigesada et al. (1986) found that the population developed into a travelling periodic wave. The average rate of spread was found to be approximately $\sqrt{4\alpha_a D_h}$ for narrow bands of habitat, where α_a is the arithmetic mean intrinsic rate of growth and D_h is the harmonic mean diffusion coefficient (the harmonic mean is calculated from the reciprocals of the values). Since the harmonic mean can be much lower than the arithmetic mean if variation is considerable, this means that the average velocity in a heterogeneous area may be much less than the average of the velocities that would be achieved in large areas of homogeneous habitat (Holmes et al. 1994). Thus the grain size of the variation in habitat may be a factor in determining the rate of population spread. (Kinezaki et al. [2003] have extended this model into a two-dimensional striped habitat, while Weinberger [2002] gives a detailed mathematical treatment.)

This exploration of reaction-diffusion models may be interesting from a mathematical perspective, but is it relevant to *plant* populations? Plants do not undergo continuous random walks: not even their seeds do this while they are in motion. They do not reproduce continuously through time (although they might reproduce over many years in the case of perennials, seed production is highly seasonal). However, as we will see when discussing integro-difference models in the following section, many of the qualitative predictions of reaction-diffusion models are unchanged when we allow time to be discrete and when we allow more realistic dispersal frequency distributions. Importantly, from an historical perspective, a reaction-diffusion model was the first model to be used to examine the implications of a species' population biology to the rate

of invasion (specifically Reid's paradox concerning plant migration after the last ice age—see p. 113). Skellam (1951) made the assumptions that: the post-glaciation expansion took place over 18,000 years; the distance travelled by the species' geographic limit in that time was 600 miles; the generation time of oaks is around 60 years; over its lifetime an oak at the edge of the population produces 9 million mature offspring (net reproductive rate = R_0). This gives the duration of the range expansion as 300 generations. The intrinsic rate of increase $(\alpha) = \log_e(R_0) = 16.01$; the Mean Square Displacement over one time unit for a random walk in two dimensions is $4D$ (Holmes et al. 1994). Since distance = velocity × time = $t\sqrt{4\alpha D}$, we obtain the root mean square distance as $600/(300 \times \sqrt{16.01}) = 0.5$ miles (800 m). Even allowing for considerable error in the estimates of the parameters, this calculation indicates that most acorns resulting in new plants must travel further than they would achieve through passive dispersal. This led Skellam to agree with Reid, that animals capable of moving seeds over long distances, such as rooks, must have played a major role in oak dispersal. This conclusion has had a lasting impact on ecologists' understanding of the processes involved in dispersal on a landscape and geographic scale. Clark et al. (1998a) also argue that the conclusion rules out diffusion as an appropriate process for dispersal of such species, requiring that we consider qualitatively different processes—and hence different types of model.

6.4.2 Integro-difference models

It is clear that several of the key assumptions of reaction-diffusion models are inappropriate for plant populations. Recruitment from the seed bank exhibits marked periodicity in predictable (e.g. temperate) environments or may be episodic in unpredictable (e.g. desert) environments. For most of the year, plants grow vegetatively, with only short periods when reproduction occurs. Although the propagules of some species remain on the plant for some weeks, months, or even years, the primary dispersal event is usually far from being a continuous process. Thus, for both population growth and dispersal, it is more appropriate to consider time as

Box 6.2 Modelling in one dimension and the selection of dispersal kernels

Almost all modelling of the rate of population spread using reaction-diffusion and integro-difference models is done using one-dimensional models. This is because a single spatial dimension makes the mathematics much simpler, in many cases analytical solutions can be found, and simulations are easier to program. It is assumed that all population growth and dispersal occur along an infinitely thin line. But is this assumption reasonable from a biological point of view? What are the implications for the choice of an appropriate equation for the dispersal kernel? What does this mean for ecologists wishing to see whether dispersal data from a 2-D field study conform to the conditions under which a 1-D model would predict a particular type of dynamics?

In a number of circumstances adult populations are confined to narrow habitat strips. For example, Fisher (1937) begins his classic paper with the statement 'Consider a population distributed in a linear habitat, such as a shore line'. Higher plants on sandy beaches, as well as algae on rocky shores, are limited to narrow bands in the transition zone between salt and fresh water, aquatic and aerial habitats, through a combination of their physiological tolerance, physical disturbance, and interactions with other species (Lewis 1964). Other obvious examples are lake margins, rivers, river-banks, waddies, road verges, and railway lines. Dispersal, however, may not be so confined and a large proportion of propagules may be lost laterally. Keddy (1980), for example, considered dispersal perpendicular to the shore to be sufficient to determine the population dynamics of *Cakile edentula* within sand dunes (i.e. away from its preferred strandline habitat). If, in modelling in one dimension, we intend to imitate real population dynamics in one-dimensional habitats, we need to ensure that the way we distribute propagules is appropriate. The most direct way that we can do this is by ensuring that our choice of kernel is based on dispersal observations collected from that same habitat (e.g. Donohue 1998 for *C. edentula* along a beach). Rather than estimate the number of seeds dispersing along the habitat strip by counting the number of seeds on plants (i.e. plant fecundity), we would need to base our estimates on the numbers of seeds collected in traps (or seedlings counted in quadrats), to allow for the proportion of seeds lost from the strip (i.e. realized dispersal—see p. 92). In the case of rivers or coastlines with strong currents, the kernel would be markedly asymmetric. Neubert and Parker (2004) show how plausible kernels for a one-dimensional habitat can be derived from simple assumptions about propagule trajectories. If, however, we base our choice of kernel in the one-dimensional habitat on field data collected in a two-dimensional habitat, we must make a number of assumptions to re-create what we would expect such a species would do in a one-dimensional habitat. For example, we might funnel all propagules along the one-dimensional habitat, obtaining the kernel by integrating a density vs distance (ω_r) transect study through 360° (i.e. using the distance pdf as the kernel in the model).

Rather than viewing the habitat as one-dimensional, an alternative is to think of modelling in one-dimension as taking a slice through a two-dimensional patch that is large enough for the edges to be straight. Thus, Holmes (1993) states: 'When the invasion advances across a two-dimensional landscape in a straight wave front, the invasion velocity is the one-dimensional invasion velocity. When the front is curved outward, the invasion velocity is slower at the leading edge, so that the front gradually becomes straight until it again has the one-dimensional invasion velocity.' Effectively, the linear edge means that sideways movement cancels out and we only need to consider dynamics perpendicular to the wavefront. The appropriate kernel to use in this case is the *marginal* pdf of the two-dimensional density pdf. If we express the density pdf in Cartesian coordinates (rather than the radial coordinates used throughout Chapter 5), then the kernel in our one-dimensional model should be the marginal probability

$$k(x) = \int_{-\infty}^{\infty} g(x, y)\, dy \tag{B.6.2.1}$$

(Caswell et al. 2003; Neubert and Parker 2004; Lewis et al. 2006; Pielaat et al. 2006). If $g(x, y)$ is a Gaussian, then the marginal probability is also Gaussian (though with different parameter values). If $g(x,y)$ is a Bessel function (as predicted by equation 4.6), the marginal distribution is an exponential in both positive and negative directions (Van den Bosch et al. 1990); Clark et al. (2001) give an equation for the marginal pdf of the $2Dt$ distribution. Lewis et al. (2006) found that the use of a regression curve for density vs distance along a transect (ω_r) as the kernel in a 1-D model, rather than deriving the marginal *pdf* from the 2-D density pdf, can result in the under-estimation of wave speed.

In many instances, theoretical ecologists choose a convenient equation with broadly the right shape as their 1-D kernel, usually one in which the mathematics is straightforward. They are concerned with the general behaviour of the system, rather than with particular species.

continues

Box 6.2 continued

After exploring the predictions of the model, it is usual to consider the impact of relaxing its core assumptions. If it is found that two classes of kernel lead to different types of spatial dynamics (e.g. constant vs accelerating velocities from thin and fat-tailed kernels respectively), it is natural for a field ecologist to ask which of the kernels is supported by data from the species that they study. Unfortunately, the appropriate comparison between the 1-D model kernel and dispersal data collected from a 2-D arena is often not straightforward and there may be problems when

extrapolating from a 1-D model to a 2-D model (Fort 2007). Where the 1-D model is considered to be modelling a slice through a large, straight-edged patch, then the model kernels must be converted to two dimensions before they are compared with field transect density vs distance data. If the limiting case for a constant velocity is an exponential, for example, then assuming isotropic dispersal and working in the opposite direction to the previous paragraph, field data need to be compared with a Bessel ω_r and not an exponential.

---

discrete. Further, density *pdfs* of dispersed propagules on the ground are often more leptokurtic than the normal distribution that arises from diffusion (Chapter 5). Thus, we need a way of modelling populations when those two assumptions do not apply.

Integro-difference models allow us to predict the spatial population dynamics of a species, by combining a discrete model for population growth with an appropriate function that describes dispersal around a parent plant. Let us begin by considering population growth in the absence of spatial movement. We can relate the population density at time $t+1$ to the density one time unit earlier, using a difference equation:

$$u_{t+1} = F(u_t) \qquad (6.2)$$

where $F(u_t)$ is some function of population density (u) at time t. $F(u_t)$ may be a simple mathematical function, such as the discrete version of the exponential, $F(u_t) = \alpha u_t$, or logistic, $F(u_t) = \alpha u_t/(1 + au_t)$, where α is the finite rate of increase and $a = (\alpha-1)/K$ where K is the carrying capacity. The dynamics of population density can be predicted from applying equation 6.2 iteratively for successive time intervals. For a good introduction to such models, see Begon et al. (1996b, Chapter 3). Other possible functions for $F(u_t)$ include the Beverton-Holt (1957) stock-recruitment curve, the Ricker (1954) curve, and the Hassell (1975) curve. Alternatively, $F(u_t)$ may be a more complex expression, based on a flow chart of the plant's life history (e.g. Sagar and Mortimer 1976; Cousens et al. 1986; Allen et al. 1996) and including

a seed bank and density-dependent survival and fecundity.

In each time step, however, propagules could potentially move from their parent to every other point in space. To define the resulting population density in space, we must integrate the movements from all possible parents to all possible finishing positions. Thus, we obtain the *integro-difference* equation

$$u_{t+1}(x,y) = \int \int k(x,y\,|x',y')\, F(u_t(x',y'))dx'dy'$$

$$(6.3)$$

where the dispersal kernel $k(A|B)$ is the pdf of a propagule dispersing from a parent at point B to point A. If dispersal is isotropic and the kernel is invariant among plants, the pdf depends only on the distance from the parent, that is,

$$k(x,y\,|x',y') = k(x - x', y - y') \qquad (6.4)$$

and thus the kernel is equivalent to the radial coordinate 'density pdf' $g(r) = g\left(\sqrt{(x - x')^2 + (y - y')^2}\right)$ in Chapter 5. Although we have expressed equation 6.3 in two dimensions, theoretical ecologists almost invariably assume that space is a single dimension (see Box 6.2). In which case, the integro-difference model becomes

$$u_{t+1}(x) = \int k(x - x')F(u_t(x'))dx' \qquad (6.5)$$

The choice of a kernel function to use in the model is also dealt with in Box 6.2.

Simulations using the one-dimensional integro-difference model with a normal dispersal kernel and logistic population growth, show qualitatively similar dynamics to the reaction-diffusion model (Fig. 6.8). Populations again expand as a wavefront at a constant velocity $\sqrt{4\alpha D}$ (Weinberger 1982; Kot 1992; Kot et al. 1996). Other shapes of thin-tailed kernels (with a steeper or as steep decline as an exponential: see p. 81) result in different, constant, velocities (e.g. Kot et al. 1996). In general, the speed (c) of the wavefront is given by

$$c = \min_{s>0} \frac{1}{s} \log_e [\alpha M(s)] \qquad (6.6)$$

(Weinberger 1982) where M() is the moment generating function of the dispersal kernel (see p. 84); this assumes no Allee effects and no overcompensation in the population dynamics (M. R. Lewis pers. comm.). Lewis et al. (2006) show how a non-parametric kernel, based on a histogram or a non-parametric kernel density estimate (see p. 78) rather than a fitted equation, can also be used to estimate the speed of a population wavefront. Non-parametric kernels have a finite tail, since they are based on a finite number of individuals, and predict a finite wavefront speed.

Wavefronts resulting from dispersal kernel equations with tails 'fatter' than an exponential are predicted to accelerate: their velocity will increase steadily over time (Kot et al. 1996; Fig. 6.9). There is no finite moment generating function, so the velocity must be estimated by simulation using equation 6.5. While populations clearly cannot continue to accelerate indefinitely (Kot et al. 1996), species having such fat-tailed dispersal distributions will increase their ranges more rapidly than those with thin tails. Constant rates of spread that are much faster than predicted from a single, thin-tailed distribution can also be achieved by a compound kernel incorporating a small proportion of long-distance propagules (Clark et al. 1998a). As was the case for reaction-diffusion models, an Allee effect in the population growth component of the model will again slow down the rate of spread: if the tail of the kernel is fat, an accelerating wavefront can be converted to a constant, but still very high, velocity by an Allee effect (Kot et al. 1996).

There is now a large, and ever increasing, body of literature on integro-difference models, far too big to attempt to summarize here. Most does not differentiate between plants and animals, exploring the dynamics of populations in general. We will give just four examples that discuss models specifically with respect to population dynamics of plants.

6.4.2.1 Sensitivity of the rate of population expansion to dispersal and reproduction

Clark (1998) explored the dynamics of a population of trees. His one-dimensional model incorporated

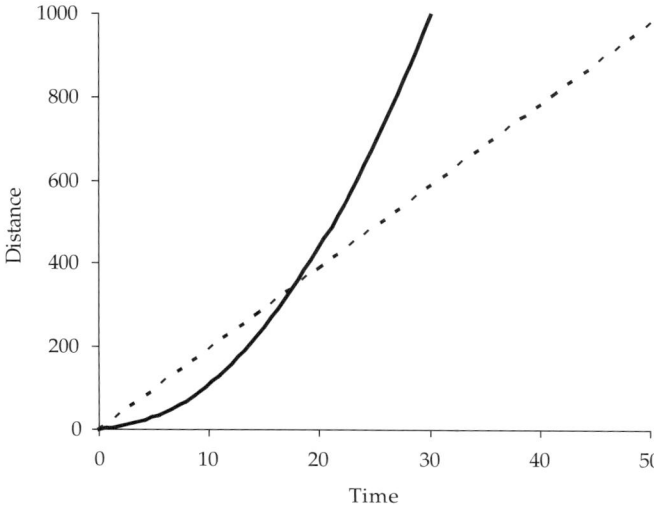

Figure 6.9 Rates of spread of populations predicted by integro-difference models in one dimension, using either a 'thin-tailed' kernel (the normal distribution, $k(x) = \frac{1}{\sqrt{4\pi D}} \exp\left(-\frac{x^2}{4D}\right)$) or a 'fat-tailed' kernel $\left(k(x) = 0.25\alpha^2 \exp\left(-\alpha\sqrt{|x|}\right)\right)$ (modified from Lewis 1997).

discrete time steps of one generation (T, calculated using life tables) and a net reproductive success R_0 over the tree's entire life. Dispersal was given by the density distribution

$$k(x) = \frac{c}{2\alpha\Gamma(1/c)} \exp\left(-\left|\frac{x}{\alpha}\right|^c\right) \qquad (6.7)$$

where α and c are distance and shape parameters respectively; $c = 2$ for the normal distribution, $c = 1$ for the exponential distribution (and thus for $c < 1$ the tail is fat). For a normal density distribution, the velocity of spread was a constant $\frac{\alpha}{T}\sqrt{\log_e R_0}$; sensitivity analysis showed that dispersal and generation time have proportionally more influence on rate of spread than fecundity. The influence of fecundity increased if R_0 was very low. Generation time and R_0 became increasingly important as the tail of the density distribution became fatter.

6.4.2.2 Effect of life history in structured populations on rate of expansion

Perennials characteristically grow in unevenly aged stands, with their reproduction and survival being a function of their size. Neubert and Caswell (2000) extended the integro-difference modelling approach to allow the inclusion of a stage-specific matrix of life history parameters; the model is in one dimension and assumes a Laplacian dispersal kernel (an exponential in both directions). The matrix approach also allowed the calculation of sensitivities and elasticities of rate of population spread to the fecundity and survival of the different stages. The model was parameterized for *Dipsacus sylvestris* (Dipsacaceae) and *Calathea ovandensis* (Marantaceae). For both species, the velocity of the population wavefront was approximately linearly related to the finite population growth rate, compared with a curvilinear relationship predicted for an unstructured population. For *C. ovandensis*, dispersal was assumed to be by four ant species, each moving a different proportion of the seeds. The rate of population expansion was primarily determined by the longest-dispersing ant species, even when that species moves only a small proportion of seeds. A similar conclusion has been reached by Woolcock and Cousens (2000) modelling dispersal of weed seeds and by Clark (1998) for stratified dispersal by unspecified vectors. Neubert and Caswell (2000) concluded that estimating

the distance that propagules are moved by the long-distance vector is more important than knowing what proportion of seeds dispersed in that way, provided that the proportion is small. Buckley et al. (2005) adapted the Neubert and Caswell model to analyse *Pinus nigra* invasions, including a Gaussian kernel in two dimensions for short-distance dispersal and a one-directional exponential for long-distance dispersal. They found that in an ungrazed shrubland and a pasture, elasticity of habitat invasion speed was most sensitive to the probability of establishment and the long distance dispersal parameter. In grazed habitats, both elasticity and sensitivity of invasion speed were greatest for the severity of grazing.

6.4.2.3 Mortality required to stop population expansion

To control weeds in cropping systems, we have the option of using herbicides and/or other methods of increasing mortality. Using models, we can determine the threshold level of mortality required to stop a population from spreading and thus to control an invasion. Using a model in which dispersal was by a combination of three mechanisms: a combine harvester, tillage implements, and passive dispersal around the parent, Woolcock and Cousens (2000) found that the level of mortality from a herbicide required to stop a weed patch from spreading was the same as that required to cause population decline in a non-spatial model. This contrasts with Schippers et al.'s (1993) prediction from a cellular model that *Cyperus esculentus* patches can still be spreading while overall abundance decreases. Allen et al. (1996) also used an integro-difference model to derive an equation for the threshold mortality level for preventing spread. The threshold depended on life history parameters but not on parameters related to dispersal.

6.4.2.4 Influence of model assumptions on rate of invasion

Conventional reaction-diffusion and integro-difference models have infinite tails to their dispersal distributions and, though the probability is very low, predict that some seeds will disperse to very great distances. However, the number of seeds produced by a plant is finite and the edge of a real population is the result of discrete events. Thus, a smooth line fitted to frequency data may provide a good interpolation, but it does not truly express the qualities of real populations. Clark et al. (2001) therefore calculated the probability of the furthest-dispersing individual landing within a given distance interval, using a stochastic one-dimensional model (see also Kinlan et al. 2005). They assumed that the rate of population spread is driven by seed production by the individuals dispersing the most extreme distances, that is, as a series of 'leaps' forward by discrete individuals, with subsequent infilling of the gaps left behind. A constant net reproductive rate R_0 had a strong effect on the distance of the most extreme disperser. The rate of spread, approximately $\frac{1}{T}\sqrt{(\pi b R_0/2)}$ if R_0 is large, where b is the distance parameter from Clark's 2Dt pdf (the shape parameter was 1) and T is the generation time, was greater than predicted by conventional integro-difference models using normal or exponential kernels. However, it was less than that obtained using a fat-tailed kernel with conventional integro-difference models. Significantly, the model did not predict accelerating wavefronts.

6.4.3 Cellular models

While the continuous-space models can be elegant in their ability to generate analytical solutions, their use is limited. As we extend them to include more aspects of the species' biology, competing species, herbivores and diseases, temporal variation in weather, habitat conditions and land management, and so on, mathematical solutions become beyond our reach. Simulations using cellular models are often more amenable to the investigation of these more complex systems.

As an example, Higgins et al. (1996) used a cellular automaton model to investigate the sensitivity of rate of population spread in *Pinus* spp. to the return interval of fires, life history parameters of the pines, and their mean dispersal distance. The starting population was a single line of trees, one cell wide, with cells being 10 m square. Recruitment only occurred following fires, which were assumed to cover the entire landscape; adults were less likely to be killed by fire. Dispersal was determined by an exponential distance pdf with a mean distance of either 20 m or 70 m and truncated at 1 km; dispersal directions were selected at random. Simulations were conducted as a factorial experiment, with all possible combinations of two levels of each parameter being varied and replicated ten times. The rate of spread was calculated from the position of the furthest advanced individual at each time step. They found that the rate of spread was greater when early age of reproductive maturity and high fecundity were combined with longer mean dispersal distance, thus supporting a similar conclusion reached from an analysis of species invasiveness within the genus (Rejmánek and Richardson 1996). Not surprisingly, the most important factor regulating the rate of spread was found to be dispersal ability, while adult mortality had the least effect. There was also a large interaction between the minimum age of reproduction and mean dispersal distance. In a modification of the pine model, Higgins and Richardson (1999) compared the rate of spread using a single Weibull distribution for dispersal distance with the spread resulting from a combination of two Weibull distributions that had been fitted to the same data. They found that the rate of spread under the stratified dispersal distribution increased the rate of spread by a factor of 4.5. Allowing 0.1% of seeds to move 1–10 km resulted in an order of magnitude increase in the rate of spread. Qualitatively similar results have been found by Clark (1998) in a one-dimensional cellular model of tree populations parameterized for several tree species.

Cellular models really come into their own when we want to investigate the implications of spatial heterogeneity in habitat conditions for range expansion (such as habitat fragmentation from land clearing). We simply allocate habitat qualities to particular cells and then simulate the spread of a population through the arena. Habitat types may be simply suitable vs unsuitable habitat, but it is straightforward to allow habitat quality to vary on a quantitative scale. We can compare the effects of patches arranged at random, according to particular statistical qualities (such as fractals, with different degrees of contagion), on the basis of field surveys or from satellite images; we can also examine the impacts of allocating this heterogeneity at different scales ('grain' size) (Gardner et al. 1987; Hiebeler 2000; Collingham and Huntley 2000).

Perhaps not surprisingly, one of the main conclusions to be reached from such models is that the effect of landscape fragmentation on the rate of colonization depends on the grain of the landscape (the sizes of landscape units and the distances between them) relative to the scale of the dispersal pdf. With (2004) has proposed that lacunarity, a measure of the variability in inter-patch distances, is an important measure of landscape pattern. At low levels of suitable habitat in the landscape, there is a lacunarity threshold that coincides with the threshold in colonization success (With and King 1999). Thus, they hypothesized that perhaps the size and distribution of gaps between suitable habitat patches is more important than the connectivity between suitable habitat patches. There is some empirical support for this idea from work on insects (With et al. 2002) but so far not for plants.

Adjoining patches of suitable habitat act as pathways through which the species can spread to other patches. Patches do not have to be physically connected, since dispersal can take the species across gaps, depending on how large these are. As the patches become further apart (i.e. the proportion of suitable habitat decreases), it becomes more difficult to find a pathway through the landscape and the velocity decreases rapidly after a critical point of habitat availability is reached. Eventually, spread becomes impossible. This type of process is referred to as *percolation* (see With 2002 for a general discussion of percolation theory and its implications for landscape ecology). The most basic percolation models have patches assigned at random and dispersal is assumed to be only between neighbouring non-diagonal cells (i.e. four nearest neighbours: a Neumann neighbourhood). The *percolation threshold* (the critical proportion of suitable habitat for population spread) in this simple model is 59%. Above this level of suitable habitat, there is a high probability that there will be at least one continuous pathway (the percolation cluster) that spans the entire landscape. As the level of suitable habitat falls below 59%, for example as a result of forest clearance, then the probability that the percolation cluster becomes broken increases rapidly. However, if spread also occurs into adjacent diagonal cells (i.e. eight nearest neighbours: a Moore neighbourhood), if dispersal can occur across unsuitable cells, or if the landscape

is fractal rather than random, the critical proportion may be very much lower (Wiens 1997). However, the amount of available habitat required to allow range shifting to occur for a short-range disperser in a fragmented habitat may be even higher than 59% (McInerny et al. 2007).

Where a species spreads through a narrow corridor surrounded on both sides by totally unsuitable habitat, there will be fewer pathways for the species to pass through. Tilman et al. (1997) showed that reducing corridor width and, to a much lesser extent length, reduced the critical proportion of remaining good habitat at which spread ceases. Clearly, the critical level will depend on the same species attributes as in a homogeneous area of suitable habitat: life history attributes, contributing to rate of population increase, and dispersal frequency distribution. Because fragmentation makes it harder to spread through a narrow corridor, Tilman et al. (1997) drew the intriguing conclusion that a habitat must be more pristine (fewer patches of unsuitable habitat) to serve as an effective corridor than to serve as a viable long term refuge. Unfortunately, because of edge effects corridors are usually *more* degraded than large areas containing suitable habitat.

There have been a number of studies of the spread of species through cellular landscapes consisting of habitat mosaics. In one example using parameter values based on *Tilia cordata* (Tiliaceae), Collingham and Huntley (2000) dispersed seeds by a combination of two bivariate normal distributions, one for short distances and one for longer dispersal distances. The landscapes comprised cells of suitable and unsuitable habitat, generated from an algorithm in which cell types were randomized at three hierarchical grain sizes. As the proportion of suitable habitat decreased, velocity through the landscape declined. Velocity changed only slowly when suitable habitat was relatively abundant, but then beyond a critical proportion of good habitat the velocity declined rapidly to zero (Fig. 6.10). If patches were large, this decline occurred at a higher proportion of good habitat than if the patches were small.

Higgins et al. (2003c) used a different algorithm to create a fractal landscape of suitable and unsuitable habitat cells, as well as a random distribution of cells.

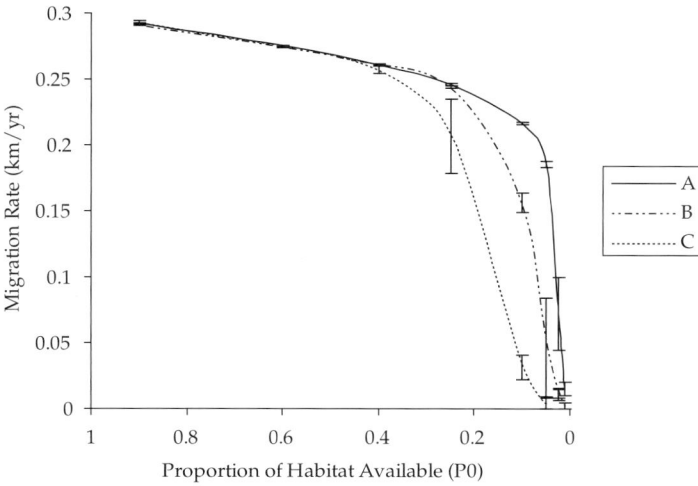

Figure 6.10 Predicted rates of spread of *Tilia cordata* in fractal landscapes of suitable and unsuitable habitat patches: A is the finest grain of subdivision, C is the coarsest (Collingham and Huntley 2000, reproduced courtesy of Ecological Society of America).

Their density distribution for dispersal was a composite of a stratified dispersal function (comprising exponentials for short and long-distance dispersal) and a stopping distance function, mimicking the effect of barriers to secondary dispersal. As expected, rates of spread were again decreased by a reduction in the proportion of suitable habitat. Depending on the type of habitat pattern (fractal or random) and parameter values for dispersal and stopping distance, migration rates were either linearly related to the proportion of good habitat or they followed a sigmoidal relationship. In the latter case, although there was a critical level of good habitat below which velocity decreased rapidly, a slow rate of spread was still maintained when little good habitat remained. Species were predicted to be insensitive to habitat loss only if they had very high fecundity and a high capacity for long-distance dispersal. They found that their results fitted the empirical model

$$v = \frac{\alpha}{1 + \exp\left(\beta - \frac{P}{\varepsilon}\right)} \tag{6.8}$$

where v is migration rate, β is the critical threshold proportion of suitable cells, P is the actual proportion of suitable cells, and ε is a shape parameter describing the rate of decline in velocity after the critical point.

6.4.4 From a paradox to a predicament

Many ecologists would now consider that we have solved Reid's paradox: rare, long distance dispersal events do indeed occur (Chapter 5), we know some of the mechanisms that cause them, and they are sufficient to predict the rates of spread after the last ice age estimated from pollen cores. As a result, there is now a considerable interest in the shapes of the tails of real density distributions and the shapes of tails predicted by mechanistic models of dispersal (Chapter 5). To be able to estimate rates of spread in a range of ecological contexts, and to compare the likely rates of spread of different species, we need to be able to measure the tails of dispersal distributions. Unfortunately, fat tails are almost indistinguishable from thin tails in most empirical studies (e.g. Wallinga et al. 2002), though more intensive sampling regimes may help: see Bullock and Clarke, 2000). Clark et al. (1998a) showed that the statistical fit of a thin-tailed pdf, predicting a rate of spread of just a few metres per year, to tree dispersal data was almost the same as for a compound pdf in which 5% of propagules were dispersed according to a fat-tailed pdf, and which predicted rates of spread compatible with estimates of around 200 m per year for trees in the early Holocene. An incorrect assumption of a thin-tailed dispersal kernel in a model, where in reality the dispersal distribution is (partly) fat-tailed, will therefore considerably underestimate the rate of invasion.

Thus, we have swapped a paradox for a predicament: although we know that the tails of dispersal distributions are of major significance to population dynamics, it is almost impossible to measure them. However, Clark et al. (2003) argue that we are also very uncertain about most of the other measurements that go into calculating rate of spread, such as net reproductive rate and generation time. We usually make complete guesses at the lifetime production of seeds, the number of these producing reproductive adults and the generation time. How accurate is Skellam's estimate of 9 million reproductive offspring per oak tree, for example, or his generation time of 60 years? The crucial data concern life history parameters at the edge of an actively spreading population and yet most of our data come from within long-established communities. Thus, even if we spend considerable time and effort improving our estimates of long distance dispersal, in particular the shape of the tail, our ability to predict rates of invasion will still be highly uncertain. Our methodological predicament therefore extends to the difficulty of estimating any demographic or dispersal parameters with sufficient accuracy.

6.5 Conclusions

As we have shown in this chapter, since the 1990s there has been a major ecological focus on the expansion of populations, fuelled by an interest in invasions and range expansions past, present, and future. Models, comprising a range of structures and assumptions both simple and complex, have been employed to great effect. We have firmly established the importance of (often very rare) long distance dispersal events in dictating the rate of spread the patchiness of population margins. However, the difficult task of estimating the extreme tails of distance distributions, and indeed of the various other parameters in the models, will mean that our ability to predict expected rates of spread will remain, frustratingly, elusive.

Although attention has been focused firmly on the part played by dispersal, we must not forget that the rate of spread is dependent on *both* the dispersal distribution and the fecundity (or more precisely the number of propagules successfully reaching adulthood) of the species. Greater numbers of propagules produced by a plant will, on average, produce more long-distance dispersers with a greater likelihood that at least one of them will establish.

The fragmentation of native habitat remnants in many parts of the globe, largely due to agriculture, will certainly make it difficult for species to spread in response to future climates and in the face of other global and local changes. Modelling is hardly needed to determine that without positive human intervention many species will be unable to spread and will thus face extinction. However, models are still needed to better evaluate the protocols for the management of these species, as well as for the invasive species that threaten both endemic communities and agricultural production. In Chapter 7, we turn our attention to the influence of dispersal on the dynamics within populations.

CHAPTER 7

Propagule dispersal and the spatial dynamics of populations and communities

7.1 Introduction

Plant ecology is inherently spatial. More so than for most animals, the location of a plant is a matter of life and death. Propagule dispersal is clearly the core determinant of where plants come to be located, and is therefore central to population dynamics. This simple observation motivates this chapter on dispersal and dynamics of spatial patterns (Levine and Murrell 2003 give a detailed review of the literature).

First we consider the signal (or lack of signal) of local dispersal which sits inside maps of the locations of individuals (spatial patterns), and introduce some spatial statistics as measures of spatial structure. Second, we show how fundamental the part is that dispersal plays in local population dynamics by examining spatial patterns within populations as they unfold over time, given assumptions about dispersal. Third, we illustrate the key role that dispersal plays in the outcome of local spatial competition between two species. Fourth, because dispersal also couples local populations at larger spatial scales, we consider the part it plays in the dynamics of metapopulations. Fifth, and more speculatively, dispersal is likely to affect the properties of species living together in multispecies communities, so we examine some possible effects it has on the structure of plant communities.

7.2 Dispersal and spatial patterns in nature

Spatial patterns of individuals are a good place to start the study of spatial aspects of population dynamics. The patterns carry information about dispersal, because the pattern is an outcome of processes in which dispersal is likely to have played a prominent part. The patterns are also templates upon which processes including dispersal will be built in the future. It has to be understood from the start that patterns and processes are intricately interwoven, and that isolating the precise role of dispersal in the emerging patterns of populations and communities is not a trivial task. Nonetheless, the spatial pattern does give a foundation on which to build population and community dynamics, and has been the motivation for many ecological studies. Sometimes this work is at a seemingly superhuman scale. For instance, over the last 25 years, tropical forest ecologists have been mapping the position of all trees in large natural forest plots. At the time of writing, this amounts to approximately 3 million trees of about 6000 species, approximately 10 % of the known tropical tree flora of the world.

Here we use two contrasting spatial patterns of tree species to see what inferences about dispersal might be made from them. One is from a diverse tropical forest, and the other is from a species-poor temperate forest. To make the measure of relative spatial locations precise, we introduce some second-order spatial statistics along the way.

7.2.1 Spatial patterns

7.2.1.1 A tropical example
Figure 7.1a shows the locations of individuals of a species, *Shorea leprosula* (Dipterocarpaceae), in a 50 ha plot in the Pasoh Forest Reserve in Malaysia

censused in 1987. In this plot, all trees 1 cm or more in diameter at breast height (dbh) have been mapped to gain understanding of the biodiversity and dynamics of dipterocarp rainforests. *S. leprosula* is itself one of the large dipterocarp species, and grows up to 60 m tall.

Scattered across the plot in Fig. 7.1a, there are about 3000 stems of *S. leprosula*. A visual inspection of the map suggests that the trees are not distributed independently at random across the plot: there are signs of some clustering of the trees, a feature quite widely observed in spatial patterns of tree species in tropical forests (Hubbell 1979; Condit et al. 2000). We could speculate on the reasons for clustering. First, limited dispersal of fruits is obviously a prime candidate. Although the fruits have wings which aid wind dispersal, the nut is quite large. Moreover, dipterocarps tend to have quite short dispersal distances, most fruits falling within 40 m of the parent tree (Appanah and Rasol 1995; Blundell and Peart 2004). Second, variation in the physical environment is a possibility, as there could be some areas more suitable for *S. leprosula* than others. However, the Pasoh plot has relatively little variation in elevation and, over much of the plot, the soils are alluvial. There remains sufficient variation in soil texture to generate different distributions of certain species (Debski et al. 2002; Plotkin et al. 2002), but the spatial pattern of *S. leprosula* is not obviously correlated with features of the environment. Third, interactions among neighbours could contribute to non-randomness, athough in plant communities dominated by competitive interactions, the outcome is more likely to be fewer individuals close together (local inhibition) rather than more (local aggregation). Actually, the impression created by Fig. 7.1a is a bit misleading, because it suggests there are empty regions of the plot. In reality the rainforest is continuous. *S. leprosula* is just one of about 800 tree species, contributing less than 1% of the total number of stems. A better impression is obtained from Fig. 7.1b, which gives a map of all the stems in the small boxed region, 1 ha in size, near the origin of Fig. 7.1a, with the small number of *S. leprosula* stems highlighted. Seen in this light, *S. leprosula* is locally rare; the cluster of points near the bottom left are small individuals within about 20 m of a large one (Fig. 7.1c).

7.2.1.2 A temperate example

The rarity of *S. leprosula* in the plot above, and the relatively short distances over which its fruits disperse, means there is some chance of picking up a signal of its local dispersal in the spatial pattern of small trees relative to large ones. In other circumstances we should not hope to learn too much about propagule dispersal from locations of established plants. The seed shadows of individual plants may overlap a lot, and a great deal can happen between the time of dispersal and the time of becoming established. Schupp and Fuentes (1995), for instance, were quite pessimistic about the information which could be garnered from locations of adults.

To illustrate these difficulties, the location of beech trees (*Fagus sylvatica*) in a 1 ha plot of mixed beech-spruce (*Picea abies*) forest in Rothwald, Austria is shown in Fig. 7.2a. This plot was censused in 2001. Rothwald is remarkable as one of the few fragments of old-growth forest remaining in Europe, having escaped exploitation in part because of the difficulties in extracting timber from it and more recently through being maintained for hunting by the Rothschild family. As in the case of *S. leprosula*, it appears as though there is clustering in the map of beech stems, particularly among small individuals. However, the average density of beech individuals is about 20 times greater, large individuals being common through the plot, often with a few metres of each other. Here the seed shadows from different parent trees overlap greatly (Chapter 5.3.2 p. 80), and it would be surprising if a strong signal of propagule dispersal could be picked up in the spatial pattern of small and large individuals.

7.2.2 Statistics of spatial patterns

Visual inspection of spatial patterns is a weak basis on which to make inferences, and it is useful to put in place some precise second-order spatial statistics. This will also help later, when a spatial extension to population dynamics is developed, and a statistic of spatial structure is needed as a state variable. A lot of thought has gone into statistical measures of spatial point patterns by mathematicians and ecologists; some recent reviews of the subject area are given by Stoyan and Penttinen (2000), Diggle (2003), and Wiegand and Moloney (2004). Below are some

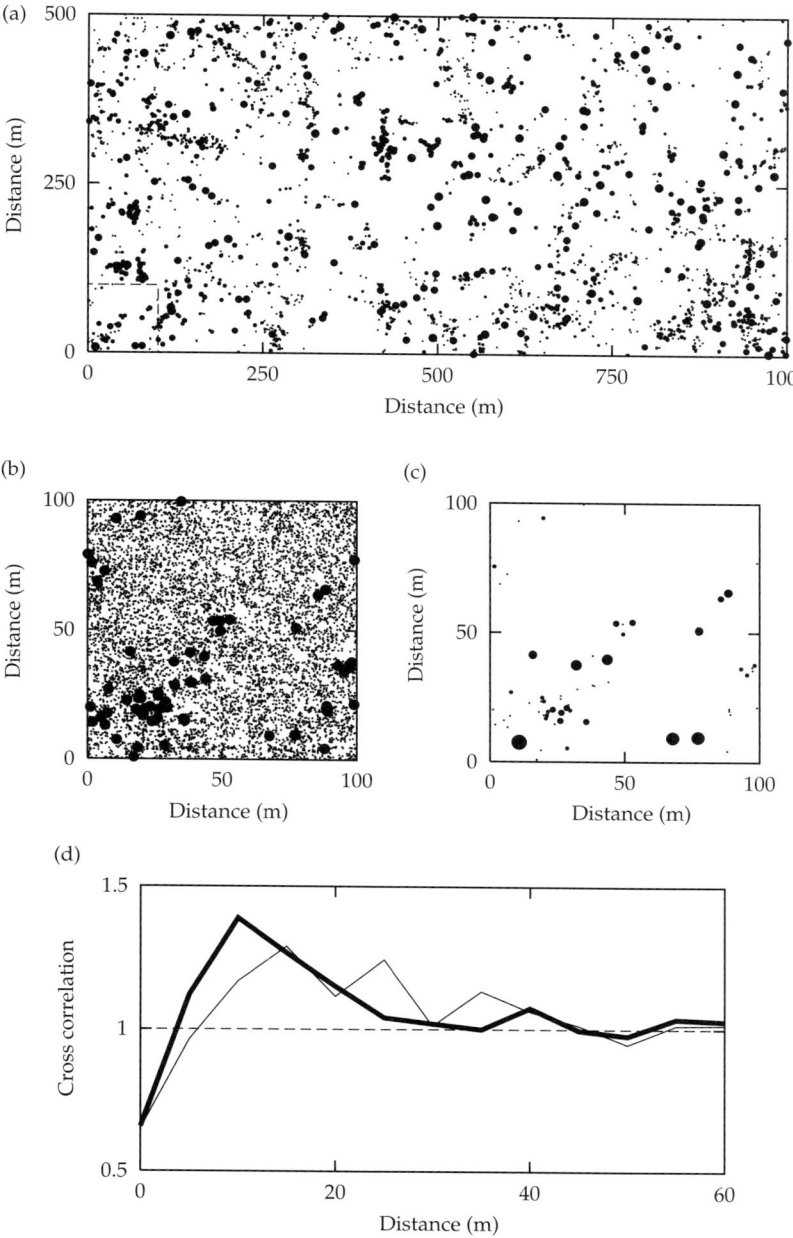

Figure 7.1 Spatial information on *Shorea leprosula* in a 50 ha plot at Pasoh, Malaysia (first census). (a) Locations of each stem; diameter of points is proportional to $\log_2$ diameter at breast height (dbh). (b) Locations of all stems in the small boxed region of map a; the locations of *S. leprosula* are are shown as the larger circles. (c) Locations of *S. leprosula* in the small boxed region of map a, the diameter of points is proportional to $\log_2$ dbh. (d) Cross correlation functions of large stems (dbh > 32 cm) with small stems: dbh 1 to 2 cm (heavy line) dbh 2 to 4 cm (thin line). The large-scale forest plot at Pasoh Forest Reserve (Chapter 7) is an ongoing project of the Malaysian Government, initiated by the Forest Research Institute Malaysia through its Director General, and under the leadership of N. Manokaran, Peter S. Ashton, and Stephen P. Hubbell. Supplemental funds come from: the National Science Foundation (USA) BSR Grant No. INT-84-12201 to Harvard University through P. S. Ashton and S. P. Hubbell; Conservation, Food and Health Foundation, Inc. (USA); United Nations, through the Man and the Biosphere programme, UNESCO-MAB grants 217.651.5, 217.652.5, 243.027.6, 213.164.4, and also UNESCO-ROSTSEA grant No. 243.170.6; and the continuing support of the Smithsonian Tropical Research Institute (USA), Barro Colorado Island, Panama. P. F. Chong assisted in structuring the Pasoh demography data, enabling us to analyse the data in the present way.

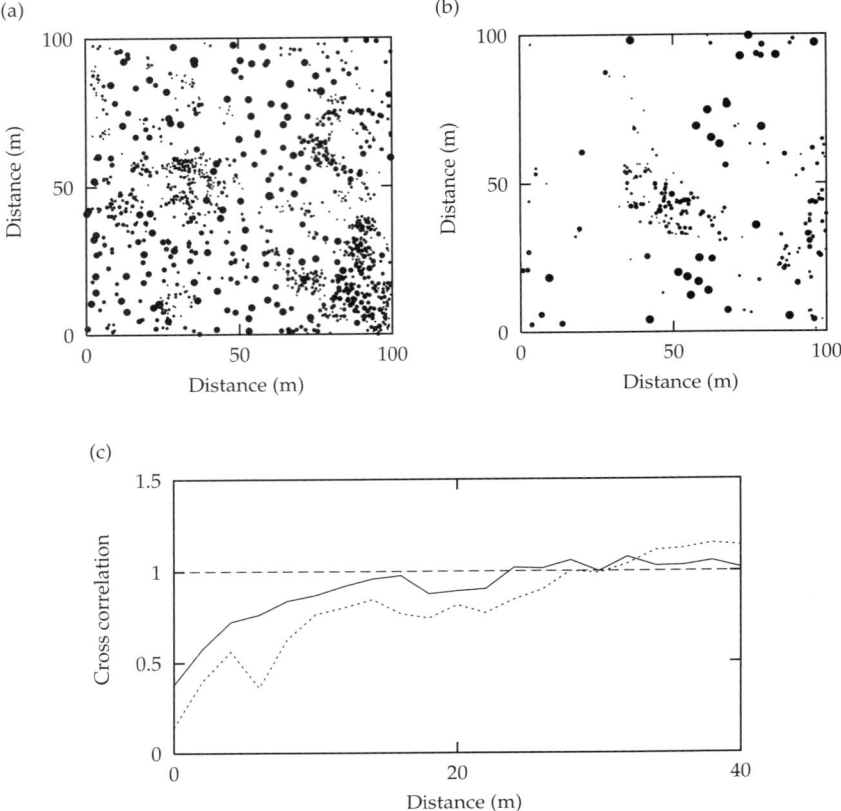

Figure 7.2 Spatial information on beech and spruce trees in a 1 ha plot at Rothwald, Austria. Locations of beech stems (a) and spruce stems (b); diameter of points is proportional to $\log_2$ dbh. (c) Cross correlation functions of beech stems (dbh 1 to 4 cm) and large stems (dbh >32 cm) of beech (continuous line) and spruce (dotted line). The data for this illustration were kindly made available by Dr Georg Gratzer, Department für Wald- und Bodenwissenschaften, Universität für Bodenkultur, Wien, funded by Project P14583 of the Austrian Science Fund.

statistics of a spatial pattern of individuals in a finite two-dimensional space, using the notation of Diggle (2003), illustrated in the context of the patterns in Figs. 7.1 and 7.2.

7.2.2.1 First spatial moment

An easy way to get started is with the familiar notion of the density of individuals, or in statistical terms, the intensity of points, at position $x = (x_1, x_2)$:

$$N(x) = \lim_{|dx| \to 0} \left\{ \frac{E[n(dx)]}{|dx|} \right\} \quad (7.1)$$

Although this measure may seem a bit obscure, it is essentially the average density at location x. It takes a small piece $|dx|$ of the whole area (which contains

the point x), and counts the number $n(dx)$ of individuals in it. The area is small enough so that it contains at most one individual; the expected value E is used because the spatial pattern is generated by a random process and would not put individuals at exactly the same locations if the dice were thrown again. Often it is assumed that the density does not vary over the region (i.e. the process is stationary) in which case we replace $N(x)$ with N, getting the spatial average density typically used in ecology.

7.2.2.2 Second spatial moment

It is not a big step to go from the density of single individuals, to the density of pairs, one individual

---

Box 7.1 Some technical details about second-order spatial statistics

We assume here that the spatial pattern on which the statistics are measured is stationary and isotropic, so the expressions defined in equations 7.1 and 7.2 are written as:

N density of individuals
$\tilde{C}(r)$ density of pairs of individuals displaced by distance r

$\tilde{C}(r)$ has dimensions $(\text{Length})^{-4}$, and it can help to have a normalized, dimensionless measure

$$C(r) = \tilde{C}(r)/N^2 \qquad (B7.1.1)$$

sometimes called the pair correlation function (Stoyan and Penttinen, 2000). This function is readily interpreted: if the plants are independently and randomly distributed in the region, $C(r) = 1$ for all r; if $C(r) > 1$ at a distance r, then the plants are aggregated above the mean value N at this distance; if $C(r) < 1$ at a distance r, then there is inhibition of plants at this distance. Another normalization is

$$O(r) = \tilde{C}(r)/N \qquad (B7.1.2)$$

referred to as the O-ring statistic (Wiegand and Moloney 2004).

An alternative, cumulative, second-order statistic, more widely used, especially by statisticians, is Ripley's (1976) K function:

$$K(r) = 2\pi \int_0^r \rho \, C(\rho) \, d\rho \qquad (B7.1.3)$$

with dimensions $(\text{length})^2$. It differs from the pair correlation function in measuring of the pair density from a distance 0 up to a distance r, as opposed to the pair density at a distance r. On one hand this cumulative measure is preferred over the pair correlation function for statistical inference because it can be computed without any assumption about how pairs of individuals are binned according to distance r. On the other hand, from an

ecological perspective, the pair correlation function has the merit of being a rather direct measure of how a plant 'sees' the community in which it is embedded (the 'plant's-eye' view of the community: Turkington and Harper 1979; Mahdi and Law 1987). In any event, it is easy enough to transform between the pair density and cumulative pair density as necessary.

It is straightforward to extend these measures to spatial patterns of two (or more) classes of points, such as individuals of different sizes or of different species. The two types, say 1 and 2, are distinguished in the definition of the pair density (equation 7.2) by the indexing $n_1(dx) \, n_2(dy)$, with the pair density written as $\tilde{C}_{12}(r)$, and similarly for the other second-order statistics. The pair cross-correlation function defined in this way is used in this chapter to investigate the spatial structure of small saplings relative to large trees responsible for propagule production. In this way, we can look for for evidence of local propagule dispersal from large trees and other neighbourhood effects.

Sometimes it is thought that second-order statistics like Ripley's K function are only applicable to spatial patterns which are homogeneous in space. This is a misapprehension. Although we do not use it here, there is an inhomogeneous version of the K function which can be used if there are grounds to suspect inhomogeneities, such as might be caused by variations in the physical environment in which the plants live (Baddeley et al. 2000; Waagepetersen 2007).

In carrying out computations of second-order spatial statistics on a spatial pattern, some assumption has to be made about pairs near the boundary of the region. In this chapter, the simplest assumption of toroidal boundary conditions is made. Often one would want a more realistic edge correction. For this, and for most other calculations on spatial point patterns, functions are available in the packages *Splancs* and *Spatstat*, in the statistical programming language *R*.

---

being at location x and the other at x':

$$\tilde{C}(x, x') = \lim_{|dx| \to 0, |dx'| \to 0} \left\{ \frac{E[n(dx) \, n(dx')]}{|dx| \, |dx'|} \right\} \qquad (7.2)$$

If the assumption of stationarity is made, the pair density is determined simply by the spatial displacement between the individuals $\xi = x - x'$; if in addition there is no directionality (i.e. the process is isotropic), all that matters is the Euclidean distance $r = \|x - x'\|$

between them. It is important to understand that second spatial moments are functions of a spatial displacement or distance, rather than scalar quantities. More details of the functions used are given in Box 7.1. What this measure of second-order spatial structure means probably seems a bit obscure right now. In fact, it provides a measure of the 'plant's-eye' view of the community, best introduced in the context of the explicit examples below.

7.2.2.3 Higher-order spatial moments

We could in principle continue to the density of triplets of three individuals, to get a third spatial moment, and so on. However, almost all research at present concentrates on the spatial structure as detected by the second spatial moment.

With the background about spatial statistics in place (Box 7.1), we can revisit the spatial maps in Figs. 7.1 and 7.2 to look for a signal of propagule dispersal, using the relative locations of the smallest stems and the large trees responsible for most fruit production, as measured by cross-correlation functions. The functions are normalized so that a value of unity corresponds to the average pair density, as described in Box 7.1.

7.2.2.4 Cross-correlation functions of S. leprosula

The way to interpret the cross-correlation functions of *S. leprosula* (Fig. 7.1d) is to think of a large tree at the origin 'looking' around it into the forest at small saplings of the same species. At the shortest distances, these conspecific saplings are relatively scarce on the average. At intermediate distances there is an excess of conspecifics on the average. As the distance gets still greater, the density of conspecific saplings relaxes, eventually becoming close to the spatial average. The simplest explanation for the excess of conspecific saplings is the dispersal of propagules, fruits tending to come to rest quite close to the adult tree from which they originate. How typical such cross-correlation functions are in general is an open issue at present. Results obtained by Plotkin et al. (2002), using a quite different statistical technique to detect clusters of conspecific trees at Pasoh, suggest we should be cautious in generalizing. In their analysis, it was relatively unusual to find species with large trees near the centre of clusters.

That there is a relative scarcity of young trees of *S. leprosula* at the shortest distances in Fig. 7.1 is also of interest. One interpretation is that survival from germination to reach a dbh of 1 cm is especially low near adults, an important theme in tropical forest ecology (Janzen 1970; Connell 1971; Nathan and Casagrandi 2004). This will be considered further in Chapter 7.6.4 (p. 154).

7.2.2.5 Cross-correlation functions of F. sylvatica

The message which emerges from the cross-correlation functions of beech (Fig. 7.2c) is quite different. There is no inflation of the density of small saplings at intermediate distances from the large trees; all that can be seen is a relative scarcity of small saplings close to the large trees. Arguably this is because the distances between trees are too short to enable the seed shadow from individual trees to be clearly distinguished. The main signal which remains is of some local inhibition of small saplings in the vicinity of large trees. This inhibition is not unique to large beech trees, because it occurs in much the same way in the vicinity of large spruce trees as well. It would seem that small beech saplings simply do not 'like' to be close to large trees.

7.2.3 Limits to inference from spatial patterns

You could object to these attempts to draw inferences about dispersal on the grounds that the dispersal characteristics of each parent are different, depending on the size of the tree, the shape of its canopy, properties of its neighbourhood, and so on (Cousens and Rawlinson 2001; Chapter 5.4.3 p. 101). At best, all we can hope for is an average over the effects of a great many different parents. Also, the locations of germinating propagules, rather than small saplings, would give more direct information on dispersal; much important biology may happen between the time when seeds germinate and the time when they are first recorded at a size of 1 cm dbh (Wills and Condit 1999; Harms et al. 2000; Blundell and Peart 2004). The longer the delay from the time of seed fall to measurement of positions of plants, the more time there is for intervening processes involving mortality to leave their mark on the pattern. It may be no more than a 'ghost' of the seed shadow which remains.

More generally, you might wonder to what extent spatial patterns such as those in Fig. 7.1a reflect dispersal as opposed to other processes such as local competition, predation, and environmental heterogeneity. In this context, a study by Seidler and Plotkin (2006) on the spatial patterns of 561 tree species at Pasoh is of particular interest. They asked whether any signal of different modes of dispersal could be detected in the patterns. Despite all the caveats above, they found a strong signal, with spatial clustering ranked by dispersal syndrome: ballistic > gravity > gyration > wind > animal (<2 cm) > animal (2 to 5 cm) > animal (>5 cm). The implication from this is that dispersal is an important determinant of spatial pattern in the field, which in turn implies that dispersal is an important process in plant population and community dynamics.

Spatial patterns like the ones shown here tend to generate as many questions as answers about ecological processes. Different dynamical processes may give rise to indistinguishable spatial patterns, so it would be misleading to read too much about dispersal from them. Limited dispersal of fruits is an obvious possibility for the spatial structure of *S. leprosula* and one which explains some features of the relative location of large and small trees. However, we cannot exclude other effects of the environment, such as the existence of gaps suitable for growth having been present at times in the past and the existence of special kinds of interactions, such as shared mutualists, which might lead to clusters of individuals. Ultimately, there is no substitute for direct knowledge of the spatial dynamics.

7.3 Dispersal and local spatial dynamics of single species

Data from natural communities give snapshots of spatial patterns at fixed points in time. In reality, spatial patterns continually unfold through time as propagules develop, disperse, grow, and die. To understand how propagule dispersal affects population dynamics requires knowledge of these spatial aspects of population dynamics. We will look at the effects here in the form of a simple, individual-based model (IBM), essentially a stochastic version of logistic population growth, continuous in time and in a two-dimensional space (Bolker and Pacala 1997; Law et al. 2003). This is a 'toy' model intended to get some understanding of the effects of certain spatial processes, rather than to grapple directly with the dynamics of systems as complex as those in Figs. 7.1 and 7.2.

7.3.1 A stochastic model

First, we give a little background on the modelling framework. The IBM deals with birth, death, and movement events of individuals; these are in general contingent upon the current spatial pattern of the population, given as a function $p(x, t)$, the density of individuals at position x at time t. (Technically, $p(x, t)$ can be thought of as a sum of a set of Dirac delta functions, there being one such function for each individual; see Dieckmann and Law 2000: Box 21.2). Time and space are continuous. The region is finite and homogeneous (variation in environmental conditions is not incorporated) and, for simplicity, thought of as being on a torus, so that there are no boundaries.

Second, the birth process is defined in terms of the probability per unit time $B(x, x')$ at which an individual positioned at x produces an offspring at x'

$$B(x, x') = b \, m(x' - x) \tag{7.3}$$

Here b is an intrinsic per capita birth rate, and movement is coupled to birth by means of a function $m(x' - x)$, given here in a two-dimensional space $x = (x_1, x_2)$, describing the probability density with which the offspring finally comes to rest at a location $x' - x$ relative to the parent. It is the effects of this function which is of interest here. In the spirit of Box 5.1 (p. 85), we refer to this as the dispersal kernel.

Third, the death process is defined in terms of the probability per unit time $D(x, p)$ at which an individual positioned at x in the spatial pattern p dies

$$D(x, p) = d + d' \int w(x' - x) \left[p(x', t) - \delta_x(x') \right] \, dx' \tag{7.4}$$

The term d in equation 7.4 is an intrinsic death rate. The rest of the right-hand side is a neighbourhood-dependent death rate which makes it more likely for an individual with many neighbours to die; it is this which gives the dynamics logistic-like properties. The function $w(x' - x)$ is an interaction kernel which weights neighbours according to how close they are to x, normalized so that the volume under the function is 1. This is multiplied by the density $p(x', t)$ of neighbours at x' at the current time t. The integral then sums over all the individuals, and the

Dirac delta function $\delta_x(x')$ removes the individual at x, which cannot be part of its own neighbourhood. Finally the parameter d' scales the overall effect of the neighbours.

Notice that there are two components to these processes. First there are non-spatial parts which would have to be present in any logistic birth-death process, namely b, d, and d'. Second there are parts which deal specifically with development of spatial structure: the dispersal and interaction kernels. It helps to understand that these two kernels typically have an opposing effect on spatial pattern. Dispersal over short distances leads to local aggregation of individuals whereas competitive interactions concentrated over short distances lead to local inhibition (matters would be different if interactions were mutualistic). The dynamical outcome depends on how these two processes play out against one another, through their effect on the spatial pattern.

The framework used here is one of a number of ways in which a stochastic, spatial, birth-death process could be written down. An alternative is a lattice in which space is discrete rather than continuous (e.g. Ellner 2001; Chapter 6 Box 6.1 p. 121). On the one hand lattices have the advantage of being easier to deal with computationally. On the other hand the discretization of space usually means less detailed information on dispersal kernels. Time could be discrete rather than continuous, and there are good ecological reasons for adopting this approach in seasonal environments. Modelling the process in continuous time requires care to ensure the time scale is correct. The Gillespie algorithm is a good way of achieving this (Gillespie 1976), and proceeds by summing the total probability per unit time of an event, the sum being over all types of event and all individuals. The algorithm then takes the time to the next event from an exponential distribution of waiting times, the sum being the parameter of this distribution.

7.3.2 Dynamics in a homogeneous environment

To show how important an effect dispersal can have, Fig. 7.3 gives some examples involving different assumptions about the kernel, while holding everything else constant. The dispersal kernel is a Gaussian function (in two dimensions), with a single parameter $s^{(m)}$ which defines how far offspring are dispersed (column 1 of Fig. 7.3)

$$m(x' - x) = \frac{1}{M} \cdot \exp\left(-\frac{1}{2} \frac{\|x' - x\|^2}{(s^{(m)})^2} \right) \qquad (7.5)$$

M being a normalization constant. The use of a Gaussian function is just for the sake of simplicity, and could be changed to match specific shapes thought to apply in reality (Chapter 5). The interaction kernel is also assumed to be Gaussian, but is held constant and is assumed to have a parameter $s^{(w)} = 0.02$, which causes strong local inhibition. The spatial patterns in the second column of Fig. 7.3 were recorded after 30 time units had elapsed, starting with 20 individuals distributed independently at random across the region. Unsurprisingly, the space is less fully filled after a fixed period of time has elapsed as a result of shorter dispersal distances—an important effect of local dispersal is simply to make changes in populations take place more slowly. Also, there is much more evidence of clumping when the dispersal parameter becomes smaller, as the pair correlation functions computed for the patterns in column 3 of Fig. 7.3 show (see also Chave et al. 2002; Levine and Murell 2003).

The increasing clumping with decreasing dispersal in Fig. 7.3 appears to be more a consequence of empty areas than a consequence of nearby individuals being closer together. Local competition in the examples shown is strong enough to counter, at least in part, the tendency for tighter clusters to develop. It is important to understand that individuals may appear to be more clustered in the pair correlation function simply as a consequence of greater rarity over the region as a whole, rather than through some special mechanism which brings them physically closer together.

It is not just that it takes longer to fill the space as dispersal distances are made shorter: the average density over space also remains lower in the long term. This becomes clearer from looking at these densities (an average over 20 realizations) over time (Fig. 7.4). For comparison with the spatial dynamics, the population density from the familiar non-spatial

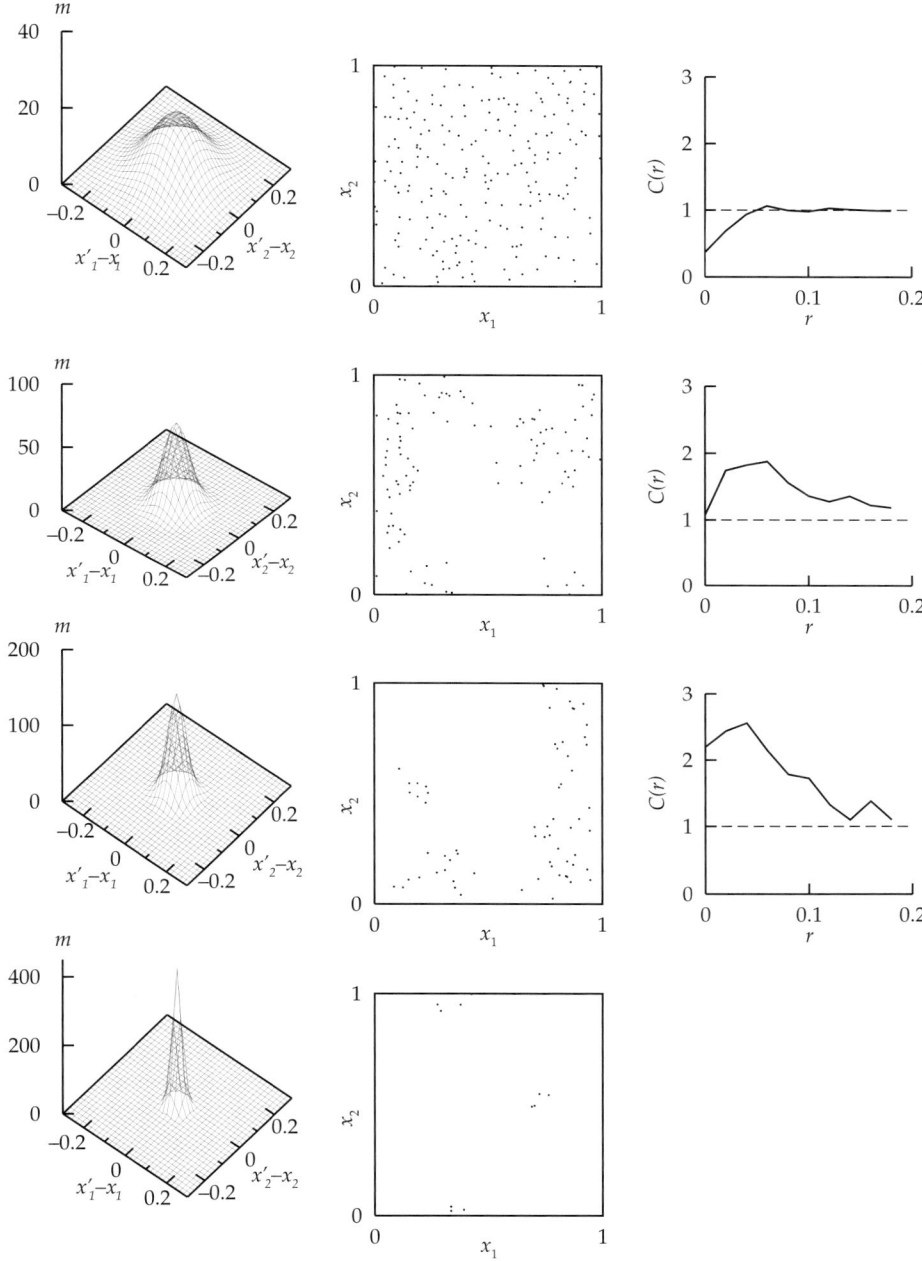

Figure 7.3 Effects of different dispersal kernels on the spatial structure of simulated populations. All dispersal kernels are Gaussian functions, with a parameter s_m describing the width of the kernel: row 1, $s^{(m)} = 0.1$; row 2, $s^{(m)} = 0.05$; row 3, $s^{(m)} = 0.035$, row 4, $s^{(m)} = 0.02$. First column shows the dispersal kernels. Second column shows a spatial pattern of a stochastic realization of a spatial logistic IBM after 30 time units have elapsed. Third column shows the pair correlation function $C(r)$ for the spatial pattern illustrated; there are too few points to compute this in the fourth row. All parameter values apart from $s^{(m)}$ are kept constant: $b = 0.4$, $d = 0.2$, $d' = 0.001$, $s^{(w)} = 0.02$.

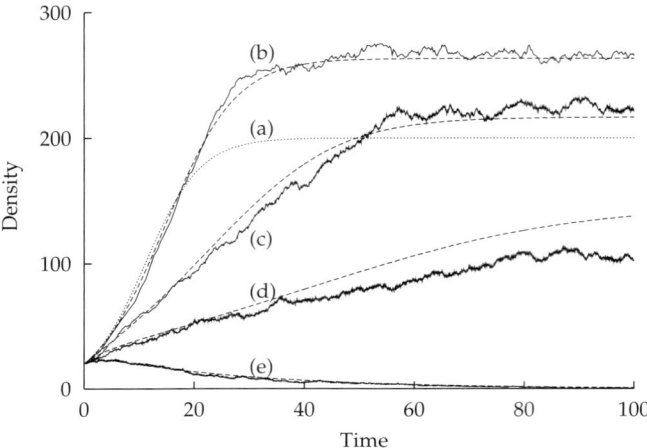

Figure 7.4 Average densities over time of populations with spatial logistic dynamics. The lines show the effect of different Gaussian dispersal kernels, all other parameters being kept constant as in Figure 7.3. (a) Density of a non-spatial logistic equation; (b) $s^{(m)} = 0.1$; (c) $s^{(m)} = 0.05$; (d) $s^{(m)} = 0.035$, (e) $s^{(m)} = 0.02$. Jagged lines are averages over 20 realizations of the stochastic IBM. Smooth lines are from a moment approximation to the stochastic process.

logistic equation is shown. With long dispersal distances, the population density actually reaches values substantially greater than would be expected from the non-spatial model. As dispersal distances are made shorter, the densities reached become smaller. It does not matter that the average density is low over the whole region; the crucial point is that the density is high in the small neighbourhood in which individuals are competing. Eventually dispersal distances become so short that the population simply implodes on itself and vanishes altogether (see also Etheridge 2004).

These results demonstrate in a graphic way how important propagule dispersal is in spatial population dynamics. In particular, the results suggest that dispersal over short distances is likely to be relatively disadvantageous, because shorter dispersal distances lead to slower population growth, and ultimately to lower population densities.

Clear though the results are from this, there is field evidence both for and against a positive relationship between dispersal and local abundance. In a broad survey of British plants, Hodgson and Grime (1990) found that species associated with arable land, spoil, and wetlands had features facilitating dispersal through space. Mabry (2004) compared seven pairs of herbaceous woodland species, each pair containing one more common and one more restricted species from a single genus or family. The more common species were characterized by smaller and more numerous seeds. An obvious

interpretation of this is that abundance is greater in species which produce more seeds, but an additional effect could be that smaller seeds are dispersed over longer distances contributing to the success of these species. One effect of poor dispersal is a complete absence of individuals at small spatial scales: you can see these empty areas in Fig. 7.3. Local absence of species has been documented by Zobel et al. (2000). They showed that some calcareous grassland species would establish themselves at small spatial scales (10×10 cm) if introduced as seeds, implying that their absence was due to the limited dispersal rather than an inability to grow at the site. In contrast to results of this kind, Rabinowitz and Rapp (1981) found a negative relationship between dispersal ability and abundance in seven prairie grass species spanning a range of different abundances. A possible explanation is that, because most space would typically already be filled by other species, plants dispersing in the field would not be filling up empty space; in practice the capacity to disperse may be less important than the capacity to get established in competition with other plants (Tilman 1994). Dispersal is only part of the story, albeit an important part.

7.3.3 Dynamics in a spatially heterogeneous environment

You should bear in mind that the results above apply to an environment which is homogeneous in space. In environments which vary over space, parents are likely to become concentrated in locations which are favourable for survival. In this case, the disadvantage of local dispersal seen above has to be weighed against the disadvantage of wider dispersal into less suitable habitats. On one hand, dispersal over too short a distance would leave offspring in strong competition with sibs and parents. On the other hand, dispersal over too long a distance would often mean that offspring end up in locations which are intrinsically poor for growth and survival. Dispersal over distances both too short and too long is disadvantageous. Just what the best compromise is depends on the spatial structure of the environment in which the population lives.

Snyder and Chesson (2003) argued that the key point is the combined effect that the dispersal kernel and spatial structure of the environment have on the covariance over space of the per capita rate of population growth and relative population density (the 'growth-density' covariance). If you find the meaning and implications of this not immediately evident, the first point to note is that the density of a population at the next census depends not only on the average over space of the per capita growth ratio in the intervening period of time, but also on the growth-density covariance. This covariance is a key quantity for measuring spatial effects. For instance, if individuals are concentrated in patches with a large growth ratio, the overall growth of the population is greater than it would have been for other arrangements of individuals. Second, Snyder and Chesson defined a dispersal function and a function describing the population growth ratio at each location. Third, they took the Fourier transforms of these two functions which decompose them into signals at frequencies corresponding to spatial lags. They then showed that the covariance is the integral of a function of these Fourier transforms, in which large contributions to the integral correspond to frequencies at which the dispersal kernel and growth function are simultaneously large. The crucial point is that the covariance (and hence the population growth) is greater the more synchronized the spatial scale of dispersal and environmental structure are.

As a simple illustration of the benefit to a population of having a dispersal kernel with a spatial scale similar to that of its environment we show results of simulating some IBMs in Fig. 7.5. The first column has three contrasting dispersal kernels, and it is the intermediate one in the second row which best matches spatial scale of the environment. The simulations all start with 100 individuals at random locations. As time progresses, the first thing that happens is that the numbers fall, because most individuals are not in favourable habitats. But this decline is reversed as new individuals come to be located in high-quality patches (see the second column for the association between individuals and environment at time 15). At this stage, the dispersal kernels have major effects on the populations. Intermediate dispersal distances allow fast population growth, because offspring often end up in the same favourable habitat as the parent, but not too close to the parent. A small proportion move further away, mostly into poor habitat and have relatively short lives. In contrast, large dispersal distances result in slower population growth, because more individuals are dispersed beyond the favourable patches, with the result that a larger proportion of individuals are in low-quality locations, short though their lives tend to be. In further contrast, low dispersal distances result in still slower population growth, because offspring suffer the same disadvantage as in a homogeneous environment of not being able to escape from the neighbourhood of parents and siblings.

Although there are large differences in the transient dynamics of populations with high and intermediate dispersal, differences between them in the long term are less easy to see. This is no doubt in part because of stochastic variation. In addition it seems that most of the favourable space is eventually occupied, even when dispersal is over long distances. Two further studies, which go much deeper than this, showed that the benefits of a dispersal scale similar to that of the environment can extend to the equilibrium population density in the long term. First, Bolker (2003 Fig. 1) obtained the greatest equilibrium density when the two scales (and the scale of competition) were the same. He observed this in a stochastic model, but not in a moment approximation (see Box 7.2). Second, North and Ovaskainen (2007 Fig. 3) found a clear maximum

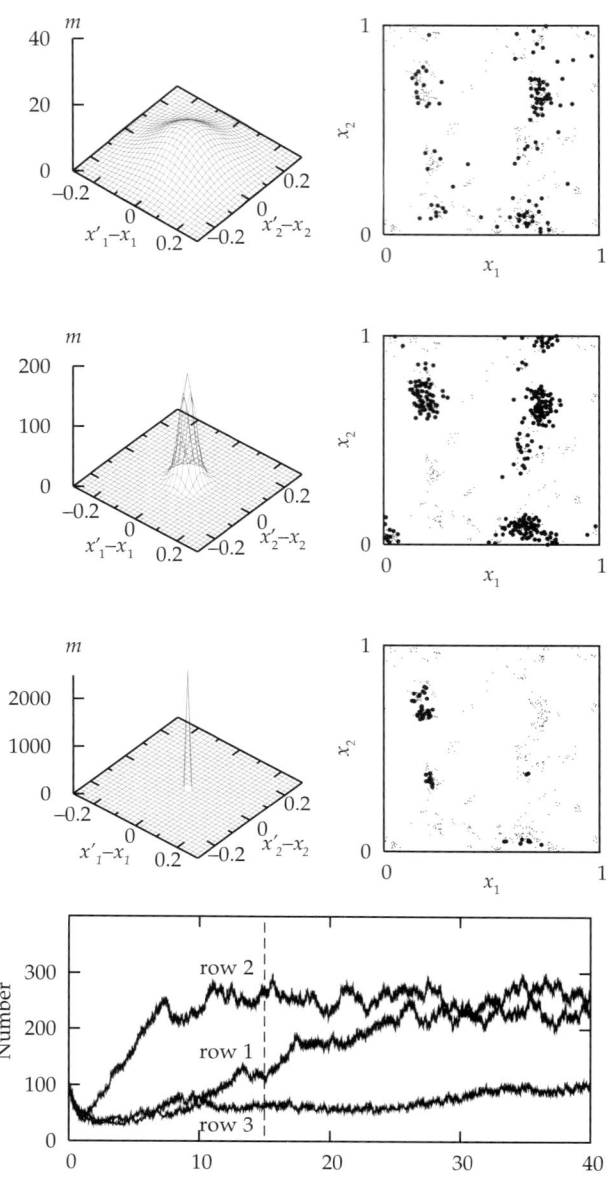

Figure 7.5 Effect of dispersal kernels on growth of a population in a fixed heterogeneous landscape. The first column shows the dispersal kernels, and the second column shows a corresponding pattern of individuals (circles) at time 15. The best habitats are those in which the quality of the habitat (density of points in the second column) is greatest. The graph at the bottom shows the population size of realizations of the IBM using the kernels in previous rows. The spatial IBM is the same as that used in Figs. 7.3 and 7.4, with the addition of habitat-dependent mortality. Dispersal kernel widths: row 1, $s^{(m)} = 0.12$; row 2, $s^{(m)} = 0.03$; row 3, $s^{(m)} = 0.005$. In addition, there is a low probability (0.01) of a dispersal event to any location in the area. All parameter values apart from $s^{(m)}$ are kept constant: $b = 1.0$, $d = 0.4$, $d' = 0.001$, $s^{(w)} = 0.01$. The contribution of habitat to the death rate of individual i is given by the number of habitat points h_i in a region of radius 0.05 around i, and takes the form $-A(h_i - B)$, where $A = 0.2$ and $B = 12$. A minimum mortality rate is set as 0.1, to ensure turnover of individuals; values smaller than this are replaced by 0.1.

in the equilibrium density when the habitat and dispersal scales were similar and the scale of competition was different. They used a perturbation expansion to obtain this (see Chapter 7.4.1 p. 148).

7.4 Dispersal and local spatial dynamics of two competing species

The effect of dispersal on population growth in the presence of another competing species, could be

---

Box 7.2 Some background about dynamics of spatial moments

The moment approximation takes both the density of individuals and the density of pairs (equations 7.1 and 7.2) as state variables in a system of differential equations. By using the second moment, the pair density, as a state variable, we hold in place some information on how the spatial pattern changes over time. Moreover, the first and second spatial moments are coupled so that the spatial structure affects the density of individuals, and vice versa. The dynamical system for a single species has the general form:

$$\frac{d}{dt} N = F_N \left(N, \tilde{C}(\xi) \right) \tag{B7.2.1}$$

$$\frac{d}{dt} \tilde{C}(\xi) = F_C \left(N, \tilde{C}(\xi) \right) \tag{B7.2.2}$$

assuming homogeneous two-dimensional space, with $\xi = x' - x$ (Law et al. 2003). This is a slight simplification because, when you derive the moment dynamics from the stochastic process, the equation describing the dynamics of the second moment contains an expression with the third moment. This means that the third moment must be

replaced by some function of the first and second moments to achieve a closed dynamical system of the kind above; see Murrell et al. (2004) for a discussion about moment closures.

As an illustration that the moment approximation works quite well (but is still no more than an approximation), numerical solutions of the moment dynamics of the spatial logistic equation are given in Fig. 7.4, for parameter values corresponding to each set of stochastic realizations. The main properties of the time series, such as the asymptotic value they approach, and the rates at which they approach these values, match the average of the stochastic realizations quite well, although there is some discrepancy in simulation (d).

The general form of a two-species moment approximation is a bit more complicated than that of the single-species approximation. Clearly there are two first moments which are free to change: N_1, N_2. In addition there are three second moments to deal with: $\tilde{C}_{11}(\xi)$, $\tilde{C}_{12}(\xi) = \tilde{C}_{21}(\xi)$, $\tilde{C}_{22}(\xi)$.

---

investigated in a two-species version of the the IBM above. However it is easier to see what happens from a deterministic approximation derived from the stochastic process. We take this approach here, using moment approximations in continuous space which can be derived from the IBM above (Bolker and Pacala 1999; Law and Dieckmann 2000; Murrell and Law 2003). Some background about the moment approximation is given in Box 7.2.

7.4.1 A model for dynamics of spatial moments

The notion of modelling the dynamics of densities N_1, N_2 of individuals in two competing species, indexed 1 and 2, is familiar in theoretical ecology in, for instance, the famous (or some would say infamous) Lotka-Volterra competition equations:

$$\frac{d}{dt} N_1 = (b_1 - d_1) N_1 - d'_{11} N_1^2 - d'_{12} N_1 N_2 \tag{7.6}$$

$$\frac{d}{dt} N_2 = (b_2 - d_2) N_2 - d'_{21} N_1 N_2 - d'_{22} N_2^2 \tag{7.7}$$

The dynamical system is shown here in a slightly unusual form to keep in place the separation

between births and deaths. The parameters b_i, d_i, d'_{ij} are as defined in equations 7.3 and 7.4, except that the dynamics of species indexed i are now affected by the other species j.

Equations 7.6 and 7.7 assume a well-mixed community in which individuals interact in accordance with their average density over space. This assumption has to be abandoned if there is spatial structure and interactions are local: here dynamics are likely to be strongly affected by the spatial structure. It is intuitive, for instance, that, if both species are independently clustered, individuals of the different species, are less likely to encounter one another. It can be shown that the Lotka-Volterra competition equations, modified to deal with the effects of spatial structure, have first moment dynamics:

$$\frac{d}{dt} N_1 = (b_1 - d_1) N_1 - d'_{11} I_{11} - d'_{12} I_{12} \tag{7.8}$$

$$\frac{d}{dt} N_2 = (b_2 - d_2) N_2 - d'_{21} I_{21} - d'_{22} I_{22} \tag{7.9}$$

where $I_{ij} = \int w_{ij}(\xi) \, \tilde{C}_{ij}(\xi) \, d\xi$ and takes care of the effect of type j on type i in the neighbourhood

(Murrell and Law 2003). Notice that, in a well-mixed community, you could replace the pair density $\tilde{C}(\xi)$ with $N_i N_j$, thereby recovering equations 7.6 and 7.7. You may wonder where the dispersal kernel for each species has got to in equations 7.8 and 7.9. In fact the kernel does not directly affect the density of individuals, because it simply moves them from one location to another. Therefore it does not appear in these equations. But the dispersal kernel clearly does affect the pair densities: for instance, small dispersal distances would tend to inflate pair densities at short distances. So the dispersal kernel does appear in the dynamics of the second moments, and the effects of this feed through to the dynamics of the first moments because of the coupling of the differential equations. Flux terms in the dynamics of second moments are quite intricate, because they have to deal with processes operating on both individuals in the pair. For this reason we do not write them out here; they can be found in the appendix of Murrell and Law (2003).

You should be aware that there are other ways of constructing deterministic models for the dynamics. For instance, if we were to begin with a stochastic model in a discrete lattice space, a system of ordinary differential equations for the first and second moments could be derived. Such models are referred to as pair approximations; examples, though not for the specific ecological system above, are Hiebeler (2000) and Ellner (2001). Such models are likely to be more tractable mathematically, but they carry over the same drawbacks as stochastic lattice models in the discretization of space. Another recent development is a systematic expansion around the mean-field dynamics, a perturbation expansion, which avoids the use of a moment closure (Ovaskainen and Cornell 2006; North and Ovaskainen 2007).

7.4.2 Dynamics in a homogeneous environment

What effect does dispersal have on coexistence of the species? The answer is that high dispersal helps a species in competition, all other things being equal (Fig. 7.6). In Fig. 7.6 the dispersal kernel of species 1 is held constant throughout, and the distance over which species 2 disperses is varied. As shown in the first column, dispersal distances of species 2 are

raised above those of species 1 (first row), kept the same as species 1 (second row), and dropped below species 1 (third row). Other parameters are set to allow coexistence when the dispersal distances of the two species are similar (second row). The second and third columns show the paths of trajectories in phase planes of population density. The paths are a good deal more complicated than those of the standard model (equations 7.6 and 7.7), because they are projections from a higher-dimensional phase space involving also the second moments. But the picture is still clear. When species 2 has dispersal sufficiently greater than species 1, it is able to eliminate species 1, and, when it has dispersal sufficiently smaller, species 1 is able to eliminate it. Evidently the benefits of dispersal apply just as much in the case of competing species as they did in the case of a single species living in isolation (Figs. 7.3 and 7.4).

Although columns 2 and 3 of Fig. 7.6 look superficially similar, they achieve coexistence in quite different ways. In the second column, deleterious effects of each species on the other are weaker than the deleterious effects on themselves, a scenario leading to coexistence in non-spatial models of competition. This operates through the density-dependent death rates d'_{ij} in equations 7.8 and 7.9, and is often interpreted as an outcome of niche differentiation. In the third column coexistence comes about from neighbourhood interactions acting over shorter distances between species than within species. This operates through the interaction kernels $w_{ij}(\xi)$ in the integral expressions in equations 7.8 and 7.9, and is termed 'heteromyopia' (Murrell and Law 2003). Heteromyopia is a good deal more intricate than 'niche' coexistence, and comes about because the interaction kernels cause a particular kind of spatial structure to develop which permits coexistence.

Overall, so far as we can tell, a species with greater dispersal can destroy coexistence, but cannot generate coexistence in the long term, all other things being equal. If the only difference between the two species, living together in a homogeneous environment, is that one has greater dispersal than the other, then this species ultimately wins. Nonetheless, local dispersal can potentially make the rate of replacement slow (Levine and Murrell 2003), because it reduces the rate at which a species fills the region

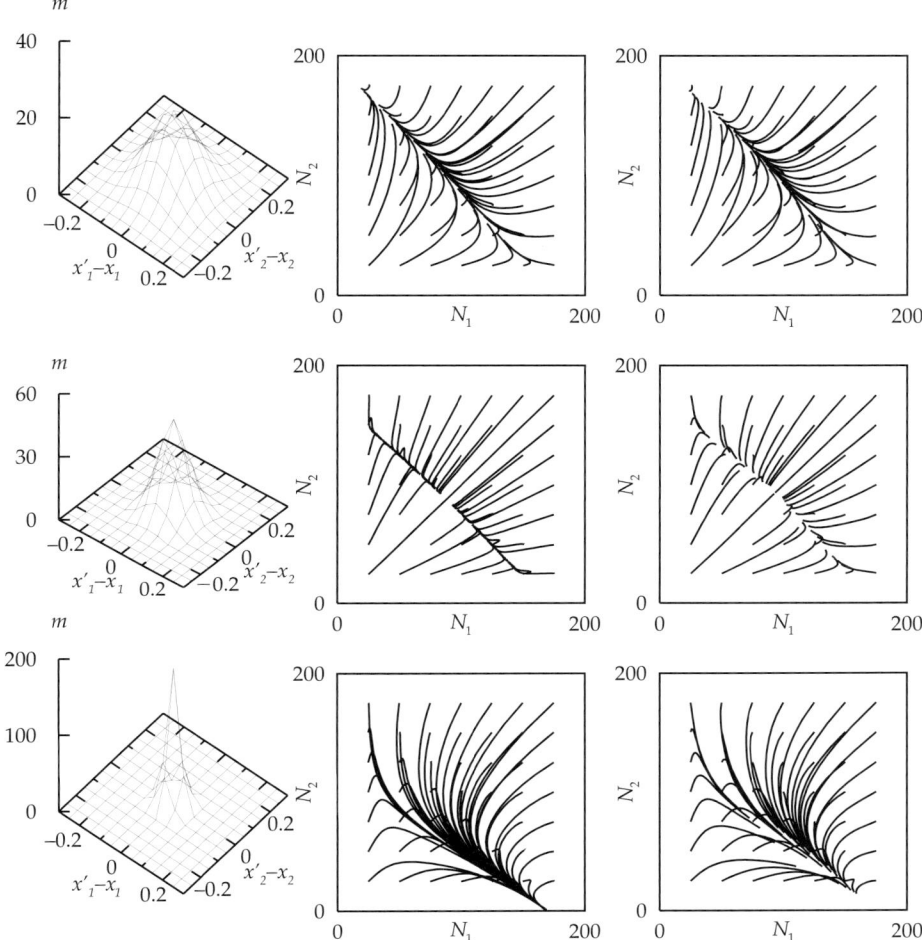

Figure 7.6 Differences in dispersal kernels of two competing species destroy coexistence. Each row refers to a different dispersal kernel for species 2, as shown in the first column: row 1, $s_2^{(m)} = 0.09$; row 2, $s_2^{(m)} = 0.06$; row 3: $s_2^{(m)} = 0.03$; the dispersal kernel of species 1 is held at $s_1^{(m)} = 0.06$ throughout. Columns 2 and 3 show projections of the phase portrait of moment dynamics on to the plane of population densities N_1, N_2. In column 2, parameters of the dynamics are chosen to achieve coexistence in row 2 through weaker competition between than within species: $d'_{11} = d'_{22} = 0.001$; $d'_{12} = d'_{21} = 0.0009$; $s_{11}^{(w)} = s_{12}^{(w)} = s_{21}^{(w)} = s_{22}^{(w)} = 0.06$. In column 3, parameters are chosen to achieve coexistence through heteromyopia: $d'_{11} = d'_{22} = d'_{12} = d'_{21} = 0.001$; $s_{11}^{(w)} = s_{22}^{(w)} = 0.06$; $s_{12}^{(w)} = s_{21}^{(w)} = 0.03$. Other parameters are held constant at: $b_1 = b_2 = 0.4$; $d_1 = d_2 = 0.2$.

(see Figs. 7.3 and 7.4). A consequence of slower replacement would be to allow more species to coexist at a given time in a community in which resident species die out and new species arrive from outside. In keeping with this, Adler and Muller-Landau (2005) found species richness increased with shorter dispersal distances in a non-equilibrium simulation model. In this instance, mortality was generated by host-specific enemies to mimic the Janzen–Connell hypothesis (Chapter 7.6.4 p. 154), and it could be said that the species therefore occupy different niches. However, we would expect to see a similar non-equilibrium result for coexistence of species without niche differentiation.

Results on the effects of dispersal on competition available from experiments and the field are as yet equivocal, although there are now a small number of competition experiments in which the effect of different dispersal patterns has been mimicked by different spatial arrangements of planting. For instance, Stoll and Prati (2001) found that weaker competitors produced more above-ground biomass in competition with other species when arranged in spatial clumps, than when distributed at random. The opposite was true for the most competitive species.

7.4.3 Dynamics in a spatially heterogeneous environment

Although a species with greater dispersal wins in competition in a homogeneous environment, it does not necessarily help a species living in a heterogeneous environment, as we noted in the single-species case.

Suppose we have two species living in a heterogeneous environment, each with a preferred environmental state in which it can out-compete the other, that is, some niche differentiation. Suppose also that we hold one species (A) with an intermediate dispersal kernel which concentrates it in the environmental states it prefers, and that we vary the dispersal kernel of the other species (B). If propagules of B disperse over long distances, they gain the advantage of less competition with other offspring and parents. At the same time they suffer two disadvantages: they are more likely to find themselves at locations they do not favour, and these locations are already likely to be occupied by species A which out-competes them. This suggests that long-distance dispersal will prevent coexistence of B with A. As the dispersal distance is made shorter, we should reach a point at which species B can coexist with A. If the dispersal distance of B is made still shorter, there comes a point at which again it cannot persist, but the reason in this case is that competition with conspecifics becomes too great. The message is that, in contrast to the homogeneous environment, a species with intermediate dispersal is at an advantage over one with longer dispersal, and coexistence is promoted by shortening the dispersal distance.

Snyder and Chesson (2003, 2004) analysed the outcome of competition between species in a spatially heterogeneous environment, continuing with the decomposition of spatial inhomogeneities into Fourier transforms mentioned in Chapter 7.3.3 (p. 145). Remember that they introduced the notion of the growth-density covariance (called $\Delta \kappa$ below), and showed a simple algebraic dependence of this on the dispersal kernel and the environmental heterogeneity. We now add two more concepts, a spatial storage effect ΔI and a nonlinear competitive variance ΔN which also contribute to the initial growth ratio $\tilde{\lambda}_i$ of a competing species i:

$$\tilde{\lambda}_i = \tilde{\lambda}_i' + \Delta I - \Delta N + \Delta \kappa \qquad (7.10)$$

where $\tilde{\lambda}_i'$ is the contribution to the growth ratio of non-spatial processes. The outcome of competition can be understood in terms of $\tilde{\lambda}_i$ when a competing species first enters an inhomogeneous arena with a resident species at a stationary state. Coexistence of the species would be implied by mutual invasibility, that is, a positive growth ratio for each species when the other is at a stationary state. (It is as well to bear in mind that the kernel describing within-species processes now contains effects of both dispersal and intraspecific competition, so direct effects of dispersal are less easy to isolate.)

Snyder and Chesson's storage effect describes the manner in which the quality of the environment and strength of competition covary over space. The storage effect of the introduced species is measured as the difference between the resident's covariance and its own covariance. You might imagine, for instance, that the resident and introduced species have different environmental preferences, and the resident is crowded into its preferred locations (so the resident's environment-competition covariance is large). If interspecific competition is over small distances so that the resident has little effect on the introduced species, the benefit to the growth of the introduced species through the storage effect is large. (A possible problem is that local dispersal in the introduced species could quickly generate clusters, so that its own environment-competition covariance becomes large.) The nonlinear competitive variance comes about from the fact that, if a function $f(x)$ is nonlinear for some argument x, the average value of the function $\overline{f(x)}$ is not equal to the value of the function computed for the average value of the argument

$f(\bar{x})$. Here the functions in question are the variances of the resident and introduced species and the argument is competition, which are combined into a single measure as the difference between variance in the resident and that in the introduced species. The nonlinear competitive variance contributes to the growth ratio of the introduced species in a positive way if the variance is greater for the introduced species than for the resident.

If there seems to be an embarrassment of spatial variances and covariances here, you might bear in mind that the formal moment-dynamic model would have the dynamics of five kinds of second moments: $\tilde{C}_{11}(\xi)$, $\tilde{C}_{22}(\xi)$, $\tilde{C}_{12}(\xi) = \tilde{C}_{21}(\xi)$, $\tilde{C}_{1e}(\xi)$, $\tilde{C}_{2e}(\xi)$, where e denotes the environment, as well as the dynamics of two first moments N_1, N_2. We should not be too optimistic about finding a simple overarching effect of dispersal in these circumstances. Its effects are likely to be contingent on the nature of the heterogeneous environment, the competitive interactions within species, and the competitive interactions between species.

7.5 Dispersal and dynamics of metapopulations

The population dynamics considered above are best thought of as operating within a locality small enough for dispersed propagules to reach all locations on a timescale comparable to that of births and deaths. In addition, ecologists have a strong interest in dynamics in which timescales are separated into a slow process of dispersal from one favourable habitat (patch) to another, and a fast process of local births and deaths within patches. This separation comes about when patches are spread out into larger total areas such that the chance of dispersing from one patch to another becomes small but not negligible. Fig. 7.5 illustrates the transition. (It is achieved in this instance by reducing the distance over which propagules disperse, rather than by increasing the total area, but the overall effect is much the same.) As dispersal distance is reduced, newborn individuals become less and less likely to move from one patch to another, making such dispersal events rare relative to births and deaths within patches. This separates the time-scale of dispersal events between

patches from the shorter timescale of birth and death events within patches.

The timescale separation leads to the notion of a metapopulation as a set of local populations linked by dispersal, and a metacommunity as the equivalent of this with more than one species (Hanski and Gilpin 1997; Holyoak et al. 2005). The environment is usually thought of as binary, containing patches which can support a local population, embedded in a matrix of unsuitable habitat. Metapopulations are of interest because their dynamics might be given in terms of rates of colonization of patches and extinction of local populations within patches, and thereby avoid having to deal with the complicated details of births and deaths. In addition, they are a natural framework in which to investigate population dynamics in landscapes increasingly fragmented by human activity. Dispersal plays a key part in the dynamics (Fig. 7.5). It has to be low enough to prevent the whole system from being homogenized and, at the same time, not so low that the local populations are effectively isolated. Somewhere between these extremes the metapopulation takes on dynamics of its own.

The original model of metapopulation dynamics introduced by Levins (1969) took the proportion p of suitable patches containing a given species as the state variable. The flux of this quantity in continuous time through the rate of colonization and extinction of local populations was defined by the differential equation:

$$\frac{dp}{dt} = cp(1-p) - ep \tag{7.11}$$

where t is time, c is the per patch rate at which propagules disperse and colonize other patches, and e is the per patch extinction rate. The equilibrium proportion of occupied patches in the model is $\hat{p} = (c - e)/c$. Dispersal plays a part in these dynamics through its effect on the colonization parameter, and it is clear from the equilibrium value that the greater the rate of dispersal beyond the boundaries of individual patches, the greater the proportion of patches containing the species will eventually be. You can see the effect of the dispersal kernel on patch occupancy in Fig. 7.5. The example with the shortest dispersal distances (row 3) leaves many more patches unoccupied than the example with intermediate dispersal (row 2), notwithstanding the fact that both have a low background probability of dispersal over the whole area.

Although Levins' model has the virtue of simplicity, it does miss some important features of the examples in Fig. 7.5. For instance, it lacks any notion of physical location of patches, so dispersal events are distributed over the whole metapopulation. In practice, the proximity of a patch relative to other occupied patches affects the rate of colonization. In practice, environmental quality is often continuously varying rather than binary (good, bad). In practice, patch size affects local population size and hence the rate of extinction. In recognition of such limitations, more recent studies on metapopulation dynamics respect patch locations and size, and metapopulations are investigated as spatially-explicit stochastic processes on mosaic landscapes.

We might debate whether groups of local populations in the field really meet the conditions for being metapopulations (Ehrlén and Eriksson 2003; Freckleton and Watkinson 2003). However, irrespective of models, it is reasonable to expect that the lower the dispersal rate, the greater should be the proportion of suitable patches unoccupied at any time. Evidence from the field supports this conjecture. For instance, dispersal limitation has been observed in several field experiments involving a pulse input of seeds into natural plant communities. This includes a multipecies pulse into an oak savanna grassland which resulted in increased diversity and altered the abundance of species in ways still detectable eight years later (Tilman 1997; Foster and Tilman 2003). In a similar experiment on a grassland undergoing succession, Foster (2001) also observed that addition of seeds increased the species richness; interestingly this effect was at its strongest in sites of low productivity, suggesting that in more productive environments resident species are better able to exclude newcomers. Evidence of dispersal limitation can also be detected during the natural dispersal of propagules. McEuen and Curran (2004), in a study of seed rain in four northern hardwood forests, noticed that only 4 out of 17 species showed evidence of exchange across forest fragments at the landscape level, these being the species associated with disturbance. The message from studies of the kind above is that local communities are often not saturated with species, and a corollary is that limited dispersal potentially plays an important part in community structure.

7.6 Dispersal and community structure

The arguments above suggest that, all other things being equal, greater dispersal is beneficial both for a single species, and also when a species is in competition with another species. Of course, it is often the case that all other things are not equal. We have already seen that, in the presence of spatial heterogeneity in the environment, intermediate dispersal distances can increase the growth rate of single-species populations and enable coexistence with a competitor. There are many other ways in which other things are not equal, some of which we describe here in the context of dispersal in multispecies communities.

7.6.1 Dispersal and the colonization-competition trade-off

An obvious way in which dispersal can be coupled to other ecological processes is through propagule size. Seeds containing more resources at the time of germination are more likely to survive in competition, but parents can produce fewer of them for a given allocation to reproduction. If dispersal is passive, their greater mass may result in them being dispersed over shorter distances. From this we get the general notion of a trade-off between colonization and competition. In the present context, a distinction should be made between number and dispersal of seeds which both separately contribute to colonization of new locations (Holmes and Wilson 1998; Bolker et al. 2003).

The colonization-competition trade-off has the interesting feature of enabling species to coexist (Hastings 1980; Tilman 1994). This is because a species with smaller propagules is more likely to end up in gaps, but less likely to survive when in places already occupied by a more competitive species. It has been suggested that an unlimited number of species could be packed into a community along this trade-off (Tilman 1994), but this seems to be a consequence of an assumption made about asymmetric competition (Adler and Mosquera 2000), and more moderate assumptions lead to a relatively small number of coexisting species.

The picture suggested by theoretical studies is therefore of a community of coexisting species arranged along a continuum from low to high colonizing ability, and at the same time from high to low competitive ability. Although there is empirical evidence for colonization-competition tradeoffs, for instance during succession in old fields (Gleeson and Tilman 1990), the evidence for a specific role of dispersal is less clear (Holmes and Wilson 1998; Bolker et al. 2003). Among the small number of studies which deal specifically with dispersal, Yeaton and Bond (1991) found that a dominant shrub in the South African fynbos was maintained in a community by virtue of specialized ant dispersal, despite having a low competitive ability against the other main dominant shrub species. But Brewer et al. (1998) found only rather weak signs of such ranking among clonal species in a salt marsh. In practice, it is not easy to separate the effects of dispersal and reproduction on colonization. This is because prolific reproducers tend to have small seeds, and small seeds in turn promote passive dispersal.

7.6.2 Dispersal, competition, and biodiversity

How does dispersal affect local (α) biodiversity, regional (γ) biodiversity, and the turnover of species from one area to another (β-diversity)? These are questions which can be considered in the context of metacommunities, allowing for the likely possibility that some species are able to out-compete others (Mouquet and Loreau 2002, 2003; Chase et al. 2005). When dispersal between local communities is low, it is argued that α-diversity should be be low and β-diversity high, in other words that most of the γ-diversity stems from different sets of species in different local communities. As rates of dispersal between local communities are made greater, there is more mixing across the metacommunity as a whole, increasing α-diversity and decreasing β-diversity, so that

more of the γ-diversity is attributable to diversity within local communities. Still higher rates of dispersal allow the most competitive species to dominate the metcommunity, so that α-diversity falls again, with an accompanying decline in γ-diversity.

It might seem difficult to test these arguments in the field, because species differ in their capacity for dispersal. However, you could look for metacommunities in which patches of suitable habitat are more isolated (having lower influx of species), or less isolated (having greater influx of species). Harrison (1997) did just this in a study on patches of serpentine soils which have a rather distinct flora associated with them. The patches studied were either from within large continuous areas of serpentine, or from small, isolated fragments. Consistent with the argument above, the small isolated patches had lower α-diversity, greater β-diversity, and greater γ-diversity than the patches in large continuous areas.

7.6.3 Dispersal, productivity, and biodiversity

An interesting related phenomenon is the humped-back relationship in which biodiversity is greatest in environments with intermediate productivity. Although not ubiquitous, this phenomenon is well-documented in plant and animal communities (Waide et al. 1999). Pärtel and Zobel (2007) suggested that, if dispersal was also greatest at intermediate levels of productivity, it could generate this pattern. (By dispersal they meant the combined effects of numbers of propagules and the distance they disperse, which they referred to as 'relative dispersal probability'). Their argument was based on the premise that the species pools and degree of competitive exclusion would be similar at different productivities.

With this in mind, Pärtel and Zobel (2007) took a large number of European plant species, and assigned to each a dispersal syndrome, an expected seed output, and a measure of the productivity of their typical environments. The relative dispersal probability was greatest for species associated with intermediate productivities, in keeping with their hypothesis. As they pointed out, this is not to suggest that other explanations for the productivity-biodiversity relationship are incorrect. The finding of Foster (2001) that dispersal limitation goes down as productivity increases is interesting in this context. It suggests that species in productive environments are better able to exclude newcomers than species in unproductive environments, and this would also cause a reduction in biodiversity as productivity increases at high enough productivities.

7.6.4 Dispersal and the Janzen–Connell hypothesis

The famous Janzen–Connell (J–C) hypothesis (Janzen 1970; Connell 1971) can be thought of as a theory about the relationship between at least three kinds of kernel. These are the dispersal kernel, the kernel describing effects of conspecific neighbours on survival (mediated both through direct interactions and also through host-specific enemies), and the corresponding kernel(s) describing the effect of heterospecific neighbours. Other things are not equal in the J–C hypothesis because the concentration of mortality of propagules around parents is host specific: different species have different enemies.

The J–C hypothesis argues that, because dispersal is local, most propagules come to rest close to parent plants. At the same time, host-specific enemies (pathogens, seed predators, and herbivores) are also likely to be concentrated close to the parents, with the result that propagules should be more likely to survive at greater distances from the parents. Experimental results supporting the J–C hypothesis have been observed in a number of studies. For instance, Hood et al. (2004) found seedlings of a West African timber tree had a greater survival rate when grown in soil taken from locations remote from the species than when grown in soil from under its canopy. The J–C hypothesis is of great interest to plant ecologists: it suggests species-rich communities could be maintained because rare species would have an advantage over common species.

A priori, one might expect a J–C process to leave its mark on a spatial pattern, with highest densities of surviving offspring at intermediate distances from the parents. The pair correlation functions shown for *S. leprosula* in Fig. 7.1d illustrate this property.

However, such a straightforward expectation can be upset by the quantitative details of the kernels involved. Nathan and Casagrandi (2004) investigated models in which both the dispersal and the density of seed predators decayed exponentially with distance from the source (with parameters D and q respectively). They showed that a necessary condition for the survivors to have a maximum density at an intermediate distance is that $q < D$, that is, that dispersal should extend over longer distances than the deleterious effects of predators. This is illustrated in Fig. 7.7a, which shows the spatial pattern as expected from the J–C hypothesis with $q < D$ (J–C pattern). Fig. 7.7b shows the spatial pattern with $q = D$, referred to as the Hubbell pattern because the increase in survival is not able to compensate sufficiently for the decline in dispersal with distance (Hubbell 1980). Fig. 7.7c shows the spatial pattern with $q > D$, called the McCanny pattern (McCanny 1985). Clearly the J–C pattern is just one of a number of possible distributions of survivors.

It is also important to realize that processes other than J–C could cause survivors at intermediate distances to be at the highest densities. Consider, for instance, the demonstration in Fig. 7.1d that small individuals of *S. leprosula* are underrepresented in the neighbourhood of large individuals. This might be expected just on the grounds that young individuals are less likely to survive where large trees have created a canopy through which little light can travel, irrespective of the identity of the large trees. An inference about a J–C effect would be stronger if we knew local inhibition to be a special feature of conspecific interactions. In fact, it can be shown that small *S. leprosula* is underrepresented in neighbourhoods of large individuals of other species as well as in those of conspecifics, which leaves a J–C interpretation in doubt. Even if we were confident that the lack of small individuals was unique to the neigbourhood of large conspecifics, we ought to bear in mind that the shape of the canopies of these large individuals could be responsible (Chapter 5.4.3 p. 101).

Another point to keep in mind is that the existence of J–C patterns in the short term is not sufficient to demonstrate coexistence of species in the long term. The latter requires knowledge of dynamics over time. Chapter 7.4.2 (p. 148) suggests that the

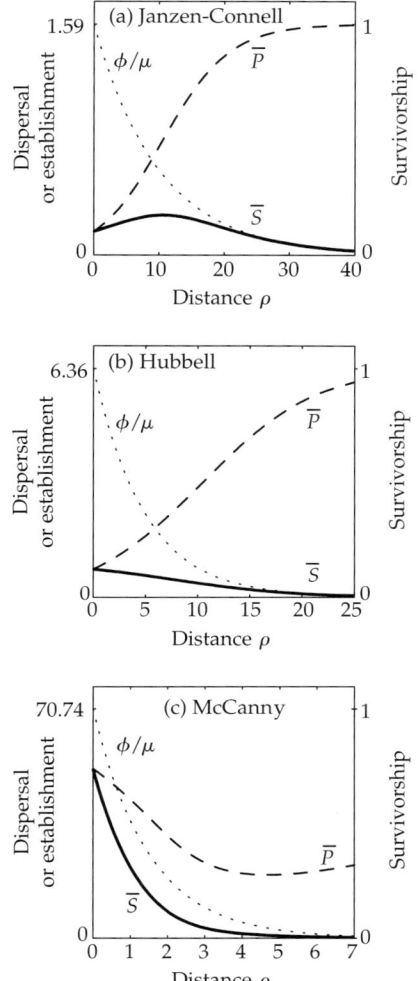

Figure 7.7 Examples of distributions of surviving seeds as a function of distance from source and local seed predation (Nathan and Casagrande 2004, reproduced courtesy of Blackwell Publishing). The term ϕ/μ is the dispersal kernel $\phi(\rho)$, scaled by the distance-independent sources of mortality μ; $\bar{P}$ is the proportion of seeds surviving at equilibrium; $\bar{S}$ is the density of seeds surviving at equilibrium; the quantities are related by $\bar{S} = \bar{P}\phi/\mu$.

combined effects of the conspecific and heterospecific interaction kernels, given as $d'_{11}I_{11}$, $d'_{22}I_{22}$, $d'_{12}I_{12}$, and $d'_{21}I_{21}$ in equations 7.8 and 7.9, are critical for coexistence. In this formalism what matters is that interspecific terms should be smaller than intraspecific ones, which can be achieved by specialist enemies acting on the mortality rates and/or

on the shape of the interaction kernels. The dispersal kernel plays a part in these expressions through its effect on the pair density inside the integrals, although the kernel is not itself a component of the first-order dynamics.

One way to examine effects of natural enemies on coexistence of species in the long term is through a non-equilibrium stochastic birth-death process which allows immigration and extinctions of species in some locality. Adler and Muller-Landau (2005) did simulations of such a process to investigate the effect on species richness of the distance over which propagules and enemies disperse. Surprisingly, they found that, if dispersal distances were long, reducing the distances over which enemies dispersed had the effect of reducing species richness on the average. It was by making the dispersal distances short that the species richness was increased. An interpretation of this would be that short-distance dispersal leaves more gaps in the spatial pattern of the community open for colonization by species from outside (see for example Fig. 7.3).

7.6.5 Dispersal and the neutral model

The neutral theory is a null model of plant community dynamics which describes the number of species which would be present in some region under the assumptions that all species are the same, that all space is occupied, that individuals die at random, and that they are replaced immediately by new individuals (Hubbell 2001; Alonso et al. 2006). A new individual can be of the same species, or of another species in the local community, or of another species from elsewhere in the metacommunity (how likely this is depends on the immigration rate set by a parameter m), or even of an altogether new species.

An interesting development of this theory has been to equate the immigration parameter m with the mean dispersal distance of all species (Condit et al. 2002). In this way it is possible to predict the degree to which communities should differ in species composition (β-diversity) according to the distance they are apart. The finding from tropical rainforest communities is that, with a plausible value for mean dispersal distance, the β-diversity is in the correct region over intermediate distances, but not over short or long distances (Condit et al.

2002). An extension to this is to use the patterns of abundance of species to estimate the immigration parameter m. Latimer et al. (2005) compared South African fynbos communities with tropical rainforest communities. Both types of community are very diverse at a regional scale, but in the fynbos the β-diversity is especially high, and in the rainforest communities the α-diversity is especially high. The values of m estimated from the abundance data are over two orders of magnitude lower in the fynbos, which they argued is consistent with their specialized modes of dispersal.

Interesting though such analyses are, the evidence from section 7.4 is that neutrality of competing species would require them to have the same dispersal kernels. It seems insufficient therefore to envisage a neutral model based on an *average* dispersal distance of the species because, unless they all had the same dispersal kernels, the species would not be neutral. As noted by Fuentes (2004), there is concern about the robustness of the neutral model to small departures from neutrality.

7.7 Conclusions

This chapter has brought propagule dispersal into the context of population dynamics. Because dispersal is a spatial process, this inevitably takes us into spatial aspects of these dynamics. The starting point is a map giving locations of individuals. Maps of this kind are of great interest in their own right, and no doubt carry a lot of ecological information in them. Some of this information can be distilled into the form of second-order statistics, from which inferences about dispersal kernels may sometimes be drawn.

The consequences of local propagule dispersal for population dynamics are profound. Local dispersal causes clumping, slows down changes in density, and affects the asymptotic state ultimately achieved by a population. When different species compete, dispersal over greater distances gives a species an advantage which can eliminate competitive coexistence which might otherwise occur. These effects of propagule dispersal all point to an advantage associated with greater dispersal in a homogeneous environment. However, matters are different in the presence of spatial environmental heterogeneity, where too much dispersal can move offspring away from favourable parts of the environment. In this case, the greatest success may be achieved with an intermediate distance of dispersal.

The arguments in this chapter have been framed in the context of population dynamics of interacting species. However, it is a small step from the competition described here to a simple process of mutation and selection operating on dispersal of propagules. In this case one type becomes the mutant, differing in some feature of its dispersal, and the other type becomes the resident. The indications from this chapter are that a mutant with increased dispersal distances would spread to fixation in a homogeneous environment: other things being equal, greater dispersal appears always to be advantageous. However, the outcome would be less clear cut in a heterogeneous environment, because of the potential disadvantage of dispersal over long distances. Seen in the light of natural selection operating on genetic variation in dispersal, the myriad of morphological variants of propagules becomes more readily explicable. We develop this theme in Chapter 8.

The evolution of dispersal

8.1 Introduction

Terrestrial plants are excellent subjects for the study of dispersal because, in nearly all cases, they have just one dispersal event during their lifetime, moving from their parent to a place where they take root. Yet, as we have seen in the previous chapters, there is a huge diversity in the dispersal phase of a plant's life. The movement of a seed, or other propagule, away from a parent may be a chance event governed by the laws of probability and best described by a statistical distribution, but that distribution is affected by plant phenology, plant and seed morphology, and the mode(s) of transport during the dispersal period. Plants exhibit an enormous range of dispersal strategies and there is clearly no perfect strategy. So, how can we explain such diversity? We know that what we observe is the result of evolution and we agree strongly with Dobzhansky's maxim 'Nothing in biology makes sense except in the light of evolution' (Dobzhansky 1973). In this chapter we provide both a flavour of, and some insight into, the evolutionary forces at work shaping the amazing diversity of dispersal strategies that we see.

Evolution by natural selection works in a very simple way. It states that characters present in lineages that leave the most descendents will increase in frequency. It requires variation between individuals in characters that are heritable and a selective force generating differential fitness. One important point is that success cannot be simply the number of seeds produced, or even the number of offspring that germinate successfully. For example, two lineages in competition have different strategies. One lineage has very high numbers of small seeds and the other fewer larger seeds. It would appear that the small-seeded strategy will lead to the higher number of descendents and be favoured. However, if

there is competition between the lineages then the energy stored in larger seeds may give individuals a huge advantage in early growth (e.g. Telenius and Torstensson 1999). This can, in a few generations, lead to the less fecund lineage leaving the most descendents by deploying this more competitive strategy. Livnat et al. (2005) agree that investment in offspring should be seen as an investment in future descendants, but they also advise caution as through time the relatedness to descendants will erode through sex, mutation, and recombination. This erosion of relatedness is often missing from simple analytical models of evolution, but it could be argued that it is incorporated by default into evolutionary individual-based models (Grimm and Railsback 2005).

There will be similar scenarios in the evolution of dispersal strategies. Indeed, if we go back to our two hypothetical lineages it is clear that, on average, large seeds (or other propagules) will tend to disperse less well than small seeds. Large seeds and small seeds will have overlapping dispersal kernels but mean and median distances will be greater for small seeds (see Chapter 5). There will be some situations where lineages leaving more, small, dispersive seeds will be favoured over lineages with large, less dispersive seeds and a dispersive strategy will therefore increase in frequency in a population (Smith and Fretwell 1974). There may also be some circumstances where both strategies remain in a mixed strategy or dispersal polymorphism. We consider some of those scenarios below.

8.1.1 Selective forces acting on dispersal

The observed diversity in dispersal strategies must be the result of different selective forces acting on lineages in the past. Some selective forces will act to increase dispersal rates and dispersal distances while others act to reduce them. We have discussed some of these selective forces in earlier chapters in the context of spatial distributions and inter-specific competition (Chapter 7). What is considered here is the situation where there are two, or more, dispersal or life-history strategies within a species. Which strategy will leave the most descendents in the long term? That is the one that will be favoured through natural selection.

As we have already stated, for evolution by natural selection to work there must be some variation in the character, inheritance of the character, and some selective force acting on that character. Variation in dispersal characteristics is widespread in plants. One example is *Centaurea corymbosa* (Asteraceae), a well-studied, cliff-dwelling species with a very small range and on the verge of extinction, which has both genetic variability (Colas et al. 1997) and considerable variation in its dispersal capabilities (Riba et al. 2005).

What follows in this chapter is a bestiary of the selective forces acting on dispersal distance. First we consider those forces acting against dispersal, and acting to keep evolved dispersal distances short. We then consider those forces that favour dispersal, tending to increase rates of dispersal and dispersal distances. Selection will almost certainly be operating simultaneously both for and against increased dispersal, so the strategy we see now is the result of a complex array of these selective forces operating through the history of the lineage, albeit within the constraints of what it is possible to evolve.

8.1.2 Pollen and seed dispersal

It is important to realize that movement of genes can be achieved in two ways—dispersal of seeds and dispersal of pollen. In population genetics movement of seeds and pollen can have similar effects as both have the effect of moving genetic information. However, in some cases, such as range expansions, the movement of pollen is far from equivalent to the movement of seeds as pollen can never establish a new population. While many of the same forces operate on the two methods, it is clear that when pollen dispersal is easier and less costly than seed

dispersal a plant may exhibit a lower seed dispersal rate than when pollen dispersal is more costly. The effect of altering the relative costs of pollen and seed dispersal is explored by Ravigné et al. (2006) using an analytical modelling approach.

Pollen limitation is often cited as a problem in small or fragmented populations. In *Centaurea corymbosa*, for example, small populations have lower seed set and higher abortion rates even when populations are quite dense (Colas et al. 2001) with many fertilization events occuring between full siblings and attributed to the shortage of pollen (Hardy et al. 2004).

In the remainder of the chapter we don't consider the selective forces acting on pollen dispersal explicitly but concentrate on propagule dispersal. Some of the selective forces we describe act on both pollen and propagule dispersal in the same way, while others will affect only the dispersal of propagules.

8.2 Processes acting against dispersal

8.2.1 Cost of dispersal structures

Organisms always have a finite amount of resource available. Resources are acquired by converting CO_2 and water to sugars through photosynthesis. This pool of available resource can be deployed in a variety of ways: towards growth, storage, and reproduction. Within the allocation to reproduction there is production of seeds and pollen along with an array of associated structures such as flowers and fruits. This is a rich area of research with the study of resource allocations usually termed 'life history theory' (Roff 1992; Stearns 1992). Here we concentrate on the allocations of resources within the reproductive structures, and particularly the seed, in an attempt to understand the extraordinary diversity of dispersal strategies documented in the first two parts of this book.

One of the costs to dispersal is that structures which increase the dispersal capacity of a seed, such as a pappus to catch the wind or a barb to allow it to attach to an animal, take some resource to produce and therefore have a cost. Seeds transported within the gut of an animal or by ants will often have a huge cost associated with the production of a fleshy fruit or the elaiosome (the calorie-rich structure that attracts ants). Slingsby and Bond (1985) showed the elaiosome of *Leucospermum conocarpodendron* (Proteaceae) had no effect on germination in controlled conditions, but removal of the structure completely prevented germination after a fire as survival was dependent on burial of the seed by ants. Structures such as elaiosomes and fleshy fruit may be slightly less costly than first appears, as they are often rich in carbon (i.e. sugars) that some plants have in relative abundance, but that are attractive to the dispersal agents. Edwards et al. (2006) studied over 200 Australian plant species and showed from the relative size of the seed and the elaiosome that ants need increasing rewards to encourage them to remove larger seeds.

8.2.2 Cost in sacrifice of viability and germination

For any given shape, there is a simple trade-off between the size of the seed and its associated structures, and the distance it might travel. Seeds are often provisioned with energy reserves. These reserves allow the seedling to survive for a period after germinating without the requirement for acquiring water and generating its own energy. A large reserve allows the seedling to remain viable for longer and, after germination, to grow more quickly, taller, and more robust than a seed without a reserve, and is therefore a clear benefit in a competitive environment. However, provisioning a seed carries several costs. The first is that fewer seeds will be produced for the same quantity of resource; second is that seeds will be more attractive to granivores (animals predating on seeds); and third is that heavy seeds will have a much lower dispersal capacity, especially if wind-dispersed. From the perspective of the adult plant both the number of seeds and the dispersal capacity of each seed will have an effect on the distance that offspring are likely to travel from the adult. So there is a clear penalty to be paid for being dispersive. Ruderal and fugitive species will tend to have small, plentiful seeds, but be outcompeted by species with fewer, larger seeds. It has long been hypothesized that a trade-off between dispersal ability and competitive ability (e.g. Tilman 1994; page 168) can explain the coexistence of many plant species on the same limiting resource,

although evidence to support this hypothesis has been relatively sparse.

8.2.3 Movement away from areas of local adaptation

Highly dispersive lineages will tend to have individuals scattered across many different environments. It can be hypothesized that such lineages are, therefore, not under strong selection to adapt to any particular environment. Lineages with highly restricted dispersal will experience much more similar environments. This will allow selection to act on the lineage to produce local adaptation, that is, characteristics which are of particular benefit in that environment. Dispersal away from this habitat will often result in an individual growing in a habitat to which it is poorly suited. So one of the costs of dispersal is that the capacity of a lineage to achieve competitive advantage through local adaptation is lost. Garcia-Ramos and Rodriguez (2002) showed how dispersal and local adaptation can interact during rapid range expansion, with high dispersal actually slowing invasion rates in some cases because individuals move too far away from the habitats to which they are adapted. This effect can also occur if genetic material arrives in a population through pollen. Any decline in fitness shown by individuals that result from matings between parents from different populations, is called out-breeding depression.

Kisdi (2002) used a simple model with two habitat types to analyse the relative importance of risk spreading through dispersal and local adaptation. The model showed that it was possible to have two resource specialists with low dispersal, a generalist with high dispersal, or a polymorphism in dispersal strategy depending on the assumptions of environmental stochasticity and the cost of dispersal. Generally, if dispersal has a high cost then local adaptation should be favoured.

8.2.4 Habitat heterogeneity—pattern and variation

The environment is always heterogeneous. Imagine a simple world where there is a dichotomy of habitats: some are high quality and some low.

Individuals in the high quality habitats are able to produce more offspring than those in the low quality habitats. So, if originally individuals were spread at random across the world, the result is that the average individual will be found in a high quality habitat. As individuals benefit from being in a high quality habitat it will obviously be advantageous for them to retain their offspring in the same habitat patch, that is, not engage in any dispersal that risks leaving offspring in the low quality patches (Travis and Dytham 1999).

More generally, habitat heterogeneity is much more subtle and complex than a simple dichotomy. But spatial variation in environment means that many individuals find themselves in habitats or conditions that are different from those experienced by their parents.

8.2.5 Marginal and island populations

The idea of habitat heterogeneity can be stretched further to consider situations where the bad habitats are fatal for dispersers. This is so for island populations where dispersers are lost at sea (Stefan 1984). This provides very strong selection pressure against dispersal. Gros et al. (2006) show how in an individual-based model in static landscapes the populations at the margins will always develop reduced dispersal because the patch edge makes the cost of dispersal higher for them. There is some evidence of reduced dispersal in plant populations on islands (Cody and Overton 1996). However, there is clearly a paradox here. To arrive at an island usually requires good dispersal ability, but for a lineage to maintain a population on an island may require poor dispersal ability and self compatibility (e.g. Baker 1955) and because of the genetic bottleneck caused by a small founding population there will be little variation in the founding population from which reduced dispersal can arise. In the context of animal dispersal this has been termed the paradox of Rockall (Johannesson 1988). It is so called because of the seven British species of *Littorina* (intertidal snails) only the species with the most restricted dispersal is present on the isolated islet of Rockall in the north Atlantic, which is more than 300 km from the nearest land and can only be colonized by a very long-distance dispersal event. The pattern of habitat availability across the whole landscape will affect both the evolved dispersal strategies and the achieved range size as Dytham (2003) showed using a range of simulated archipelagoes.

8.2.6 Mutualistic interaction—mycorrhizae, rhizobia, facilitation, and pollination

A large proportion of plant species exist in intimate symbioses with mycorrhizal fungi and the highly successful family, legumes (Fabaceae), can fix atmospheric nitrogen thanks to a mutualistic interaction with a bacterium (*Rhizobium*) living within root nodules. Around 80% of terrestrial plants form associations with mycorrhizal fungi and most plants probably acquire their fungi by extension along roots rather than from spores in the soil (Buwalda et al. 1982). In many species, (e.g. achlorophyllous orchids that lack the ability to photosynthesize), the relationship is obligate as plant cannot survive without its fungal partner (Bidartondo 2005; Leake 2005). In other cases the relationship is more diffuse, with species benefiting from the association but not requiring it. In arid environments many species benefit from an intra- or inter-specific nurse-protégé relationship, where germination and establishment are strongly favoured in the shade of another plant because water management becomes easier if some of the sunlight is intercepted. Pollination is another mutualistic, interspecific interaction and many plants are obligate out-breeders that require an insect pollinator. If this mutualistic interaction is disrupted it can be catastrophic for both partners and be one of the main causes of plant extinction (Bond 1994).

Whatever the precise form of the relationship, it is clear that species in mutualistic relationships will have an additional cost to dispersal in that they may need to disperse together, and dispersal carries the additional cost of arrival at a location without the mutualist. This additional complication to dispersal is certain to make long-distance dispersal events less likely to be successful, especially if the long-term evolutionary stability of the association is thought to be partly attributed to pseudo-vertical transmission of fungi (Keirs and van der Heijden 2006). Whether this really has a strong effect when one partner in the association persists as a spore in the soil is not clear. In this case, it is almost certain that a growing seedling will encounter its fungal mutualistic partner, as single plants typically have several fungal partners and a single fungus interacts with several plants (Bidartondo 2005).

8.2.7 Perilous dispersal

The process of dispersal carries inherent risks. It seems obvious that the longer the dispersal period the more likely those risks are to lead to mortality. Indeed 'it is by no means obvious what advantage an individual organism gains by undertaking a perilous dispersal movement instead of staying back to compete more safely in the locality where it was reared' (Comins et al. 1980). The dangers of dispersal led many authors, prior to Hamilton and May (1977), to suppose that dispersal in saturated habitats would always be selected against (see Fig. 8.1). The logic is as follows: imagine a situation where there are two habitats, both supporting populations that have a fraction of their offspring dispersing and the rest remaining. At equilibrium both populations support the same number of individuals, so those dispersing are replaced by immigrants at the same rate. Assuming a deterministic world, dispersal in this situation is entirely neutral. However, if there is any increased mortality, or other cost, from dispersing, as would seem likely, then dispersal will be selected against. As Cohen (1967) states, when describing this situation, 'organisms never gain any advantage by changing their locations'. Although this argument is

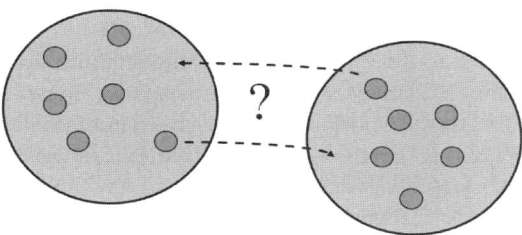

Figure 8.1 Dispersal is a neutral character in a saturated population. This was the world view that prevailed in the 1960s. It was assumed that if populations were in two patches of equal density and the same proportion dispersed from each then, at best, dispersal would be neutral as there would be nothing gained from dispersing. Once mortality during dispersal (indicated by the '?') is considered then dispersal is clearly a poor strategy.

flawed, as we will see below, it led later authors to search for explanations for the evolution of dispersal that operated in circumstances other than saturated populations.

The argument that dispersal is, at best, neutral in saturated environments turned out to be false, but the idea that the peril of dispersal will select against dispersive strategies is real enough. As dispersal becomes costly or difficult, that will always act as a selective force against dispersal rate or distance.

8.3 Processes acting to favour dispersal

Ultimately there is a simple process underpinning the evolution of dispersal strategies. It follows from the simple logic of natural selection. Generally, dispersal will be favoured when, by dispersing, an individual will leave more descendants than an individual that does not disperse and remains a resident. We now consider situations where fitness is increased by dispersing.

8.3.1 Group selection

With so many reasons not to disperse, and yet faced with an enormous range of plant structures and mechanisms to enhance dispersal, it was important to find a mechanism or process that would enhance dispersal. Many authors turned to the idea of group selection to provide an answer (see Van Valen 1971). The idea is that some individuals will act against their own interest, or at least in a sub-optimal way, for the benefit of the population or species as a whole—that is, selection is acting on the group rather than the individual. Unfortunately, although it is clear that there are selective forces operating at levels higher than the individual, this concept has been shown to be generally unhelpful to the understanding of evolutionary ecology (see Mayhew 2006 for discussion).

While we talk here about evolution acting on the individual, this is merely convenience, as it is the indivisible unit of selection. It is more correct to consider the individual genes as the basic units under selection, with natural selection favouring those genes that leave most descendants. While this is not the place for such a discussion, there are many levels of selection that operate between the gene and the individual, such as the chromosome (discussed in Michod 1999, Okasha 2006).

In the area of dispersal evolution, it is posited that species will generate dispersal strategies that maximize total population size, or overall site occupancy, rather than that which maximizes the number of descendants left by an individual. This will tend to enhance dispersal propensity, as individuals are 'sent' to found new populations even if it would be in their interest to remain in the natal site. Furthermore there is a group selection argument to altruistic dispersal as a mechanism to avoid overcrowding, although again it might be in the better interest of an individual to remain. Den Boer (1968) showed that in a variable environment it could be best for individuals to adopt a bet-hedging strategy, sending some individuals to apparently lower quality habitats, as the cost to the lineage of a site failing was so high. The possible benefits of such apparently near suicidal dispersal decisions, without invoking group, selection are discussed below.

While the idea of group selection is now very widely rejected, suggestions of group selection are still found in the literature from time to time. We simply urge caution and remind you that apparently altruistic strategies are not selected for because of benefit to the group, but rather because that strategy increases the inclusive fitness of a lineage, as we will see in the next section.

8.3.2 Kin selection

One of the key components in the understanding of dispersal evolution is the idea of inclusive fitness that was suggested by Hamilton (1964). Inclusive fitness extends the Darwinian notion of fitness being determined by the number of descendants an individual leaves, to consider the success of its relatives. So the fitness of a focal individual is comprised of both its own descendants and the descendants of its relatives scaled by their relatedness to the focal individual. Therefore two plants sharing the same mother, and having been fertilized by pollen from the same plant, will be about 50% related (i.e. they will have approximately half of their genetic material in common), and if they had been fertilized by pollen from different plants they would be less related. Hamilton's rule shows that some behaviours

or strategies that carry a large cost to an individual can be favoured by selection if they give sufficient benefit to related individuals:

$$rB - C > 0$$

where r is the relatedness of the beneficiary to the focal individual, B is the benefit gained by the relative, and C is the cost to the focal individual. Of course this is the fitness effect for just one pair-wise interaction when the overall inclusive fitness should be considered by summing up the rB consequences. Clearly, restricted dispersal will lead to more intense competition between relatives, as has been shown in annual grasses (Cheplick 1993) and this will inevitably lead to a reduction in inclusive fitness.

Models that ignore kin selection show that, in stable populations, lower dispersal rates (or zero dispersal rates) evolve if there is any cost to dispersal (e.g. Balkau and Feldman 1973; Asmussen 1983). Hamilton and May (1977) showed for the first time that inclusive fitness can help to drive evolution of dispersal strategies even when dispersal is very risky. Their model was very simple (see Fig. 8.2). They considered a set of identical patches each able to support one adult plant that was replaced by a seed from its patch. Clearly a strategy of dispersing seeds between patches beats a non-dispersive strategy even when the cost of dispersal is quite high. A dispersive strategy beats a non-dispersing one because dispersing individuals avoid competing with their close relatives and instead compete with unrelated individuals.

Since Hamilton and May, there have been many confirmations of the power of kin selection in dispersal models. In their review, Levin et al. (2003) still cite Hamilton and May (1977) as a key step in the understanding of the evolution of dispersal. However, there are many subtleties that their simple model overlooked. For example it is true that increasing the population size in a patch (i.e. a unit where there is complete mixing of all individuals) will generally decrease the effect of kin competition both in analytical models (Comins et al. 1980; Gandon and Rousset 1999) and in simulation models (Travis and Dytham 1998). This occurs because the relatedness of interacting individuals is, inevitably, reduced in a large patch. Additionally, as the number of patches that dispersers can move to declines, it becomes more

Offspring			x			x	
(after	o	o	o		o	o	o
dispersion)	o	o	o		o	o	o
	o	o	o		o	o	o
	o	o	o		o	o	o
	o	o	o	x	o	o	o
Adults	O	O	O	X	O	O	O
Sites	a	b	c	d	e	f	g

Figure 8.2 A demonstration by Hamilton and May (1977) that dispersal can be favoured in saturated, static landscapes even with high mortality of dispersers (reproduced courtesy of *Nature*). Here there are seven sites (columns labelled a–g). The resident strategy is to not disperse, depicted as 'O' in the bottom row. Adults produce five offspring that remain in their home patch, depicted as 'o'. One of these will go on to replace the adult. However, strategy 'O' is clearly not evolutionary stable as it can be invaded by a dispersive strategy, depicted as 'X'. Assuming that dispersive adults also produce five viable offspring, ('x') they can send four of them away from the home patch and still be certain of retaining site 'd' in the next generation. This can still be an effective strategy even with 50% mortality associated with dispersal, as shown here, as at sites 'c' and 'f' the disperser has a 1/6 chance of replacing the current strategy as five of the competing individuals in those sites are are non-dispersing 'o' and one is a dispersing 'x'.

likely that dispersers will still be competing with kin (e.g. the offspring of the previous generation's dispersers) and this will tend to decrease the evolved dispersal strategy.

There has been much debate about the methods for including inclusive fitness into models. The analytical approaches can be either based on whether the effect is moving from a single actor to recipients or from a single recipient to a group of actors. These contrasting approaches have been unified by Taylor et al. (2007) by making a few simple assumptions about the way selection is operating on the population. Using individual-based simulation models (IBMs) the effects of kin selection are apparent without having to calculate relatedness explicitly. Allowing the evolution of dispersal strategy in IBMs will almost inevitably be driven by inclusive fitness as there will be many different strategies in the population and each will be favoured by the reduction in competition between kin (Travis and Dytham 1998, 1999; Bach et al. 2006).

There are many models of dispersal that do incorporate kin selection. In an analytical model, Rousset and Gandon (2002) showed that kin selection can lead to long distance dispersal even when the cost of dispersal is very high and the probability of a long-distance disperser arriving at a suitable habitat is low. Comins (1982) made the general assumption that relatedness would simply fall with distance dispersed, although nearly all models have a simple assumption of individuals either 'staying' (highly related) or 'leaving' (totally unrelated) and there being no effect of the actual distance moved. There may also be different dispersal decisions made by the parent and the offspring (see Box 8.1).

It has been difficult to establish the relative contribution of kin competition in the evolution of dispersal, as complex models have many effects operating together. Gandon and Michalakis (2001), building on the method of Frank (1986), Taylor (1988), and Taylor and Frank (1996), tried to incorporate the relatedness of individuals within a patch or between patches into an analytical modelling framework. One novel approach taken by Poethke et al. (2007) is to use an individual-based model in a metapopulation context which allows the evolution of dispersal to be driven by a variety of forces. To extract the relative importance of kin selection they ran two parallel versions of the model: one with complete random shuffling (simply exchanging the locations of many pairs of individuals) to break all kin relationships while retaining the spatial structure, and the other without any shuffling of individuals to retain the kin and spatial structure. They found that kin selection has a huge impact on the evolution of dispersal. In some scenarios they found dispersal rates to be ten times higher in unshuffled than shuffled populations. They estimated that up to 30% of the fitness of

Box 8.1 Parent–offspring conflict

It is usual to consider the 'decision' to disperse from the point of view of an individual making a decision to produce some dispersive and some non-dispersive offspring (e.g. Motro 1982). Simple models show that there is an evolutionary stable strategy (ESS) for the fraction of offspring dispersing that is affected by the assumptions of the model for costs of dispersal and so on. From the point of view of one of the offspring it can either stay at home with a high probability of germination, but high competition with both kin and non-kin, or it can disperse with a low probability of survival but with the hope of germinating away from competition. At the ESS these two should balance and the parent and offspring would make the same

decision. However, in some circumstances an offspring may not make exactly the same decision as its parent and there may be some parent–offspring conflict (Trivers 1974, Motro 1983).

Johst and Brandl (1997) showed that the timing of dispersal has a big effect on the evolved dispersal strategy. This could be the result of the avoidance of any competition with an earlier generation if dispersal and germination both come after the death of the parents. While in a deterministic age-structured model (Ronce et al. 2000) the conflict between parent and offspring is shown to be lower for older parents. This leads to a lower dispersal rate for offspring born to older, or senescent, parents.

a dispersing individual could come from the increased fitness of relatives left in the natal patch.

8.3.3 Inbreeding depression

One of the costs of low dispersal rates is that most interactions are with close relatives. This includes mating opportunities and matings between close relatives and leads to an increase in homozygosity for individuals and a general loss of diversity in the gene pool. Although this might be seen as a method for purging unwanted genes or rapidly fixing genes for local adaptation ('the cost of out-crossing', Fisher 1941), there is an increased risk of deleterious genes being expressed, or even fixed. Furthermore there is a large body of evidence that suggests heterozygosity is generally of benefit to individuals. The term 'over-dominance' is used to describe the situation when a heterozygote at a single locus has a higher relative fitness than either homozygote. Mix et al. (2006) in a study of fragmented populations of *Hypochaeris radicata* showed that inbreeding reduced the dispersal capacity of offspring, thus making a direct (albeit correlative) link between homozygosity and dispersal.

It has long been appreciated that inbreeding depression might influence the mating system of plants (e.g. Uyenoyama, 1986; Charlesworth and Charlesworth 1987; Jarne and Charlesworth, 1993). Lande and Shemske (1985) used a simple model

to show the link between the extent of inbreeding depression (i.e. the relative fitness of inbred and out-crossed progeny) and the mating system. As might be expected, the bigger the cost of inbreeding depression the less likely a plant is to self-fertilize. However, self-fertilization can be seen as a selective sieve, as selfing is likely to expose deleterious recessive mutations to selection and therefore increase mean fitness if the individual is able to produce an excess of offspring. It has been hypothesized that this selective mechanism may explain the general observation that perennials outcross while related annuals self-fertilize (Morgan 2001).

So far this consideration of inbreeding depression has been in a non-spatial context. There have been attempts to show how dispersal and inbreeding depression might interact. For example, Theodorou and Couvet (2002) used a mathematical model to show that out-crossing is selected for when gene flow is low, but that the degree of selfing is affected by whether novel genetic material enters a patch by seed or by pollen.

Bengtsson (1978) was the first to consider the effect of inbreeding depression on dispersal, explicitly showing that as the impact of inbreeding depression increased the fraction of individuals leaving an area would also increase. This general result was confirmed by Gandon (1999) using a stepping-stone model and estimates of the relatedness of dispersers and non-dispersers to demonstrate that both kin selection and inbreeding depression play a role in the evolution of dispersal. More sophisticated multi-locus models (e.g. Roze and Rousset 2005) have also shown that inbreeding depression can be a major promoter of dispersal.

Populations are described as having a high 'genetic load' (Haldane 1957) if individuals have genotypes that are far from the favoured genotype. Guillaume and Perrin (2006) explore a model where the inbreeding depression is allowed to vary dynamically through mutation, drift and selection. They show that inbreeding depression will evolve jointly with dispersal and therefore the strength of selection for dispersal coming from inbreeding depression will vary.

The implications of inbreeding depression continue to unfold. For example Kokko and Ots (2006) show that breeding with close relatives can sometimes be beneficial as, under some circumstances, mating between siblings will increase the inclusive fitness of the parents. This may be shown in plants by the high prevalence of self-compatibility. The implications of inbreeding depression for conservation are reviewed in Ouborg et al. (2006). They suggest that the effects of isolation and small population size, which both reduce genetic variation and increase inbreeding, are actually quite different when the evolutionary history of a population is considered as the degree of inbreeding in a lineage will itself be variable.

As Roze and Rousset (2005) point out, it is difficult to disentangle the increased dispersal caused by evolution to avoid inbreeding depression from other factors. Individuals that disperse are less likely to have homozygous offspring, but also will have less kin competition than non-dispersing individuals.

What is clear is that the relationship between dispersal and population genetics is a complex one. Models have demonstrated that inbreeding depression and kin competition both have a clear potential to exert a strong influence on the evolution of dispersal strategies. What is much less clear is the relative importance of these forces through the evolutionary history of a lineage.

8.3.4 Habitat turnover

Species of ephemeral habitats or sites prone to disturbance must be able to recolonize habitats or they will become extinct. Southwood (1962) considered this the prime determinant of evolved dispersal capacity in insects. Species can recolonize by either returning from the seed bank (see below) or by dispersal. Travis and Dytham (1999) used a cellular IBM to demonstrate a very strong, although non-linear, relationship between evolved dispersal rate and the persistence of a patch, with dispersal propensity increasing as patches became more ephemeral, up to a point. A slightly different model by Poethke et al. (2003) confirmed that if patches were destroyed randomly this would tend to increase evolved dispersal rates. They also warned that local extinction rate could not be used to measure the dispersal rate in the population, as local extinction caused by demographic stochasticity, including losses due to dispersal, would have a more ambiguous relationship. Kallimanis et al. (2006) also used an IBM to show that habitat disturbance had a bigger effect on dispersal distance than the pattern of habitat and that if the disturbance affected larger areas this increased the evolved dispersal distance.

8.3.5 Population dynamics

From the early models of Gadgil (1971), it was clear that population ecology will have a strong influence on evolution of dispersal strategy. The size of the patch (previously discussed under kin-selection), the population growth rate, type of competition, and the variability of patches) all affect the evolution of dispersal.

It is generally true that growth rate and dispersal rate are linked. Increasing the growth rate of a species will increase the number of individuals dispersing and therefore increase its rate of spread during an invasion even when the mean dispersal distance is unchanged (see Chapter 6). In the IBM model of Travis and Dytham (1998) evolved dispersal strategy increased with growth rate. This increase is partly attributed to an increase in the intensity of kin competition with increasing birth rate, as individuals that don't disperse very far will be competing with more close relatives. However, it is also true that with increasing growth rate there is a tendency for more temporal variation in population size within a patch. It is generally true that as sites become more variable in space, dispersal is favoured. McPeek and Holt (1992), Ruxton (1996), and Holt and McPeek (1996) showed this in the extreme that when populations have a very high growth rate and become chaotic, this favours the evolution of dispersal.

Part of the selective pressure for dispersal from temporal and spatial variability comes from the fact that the average individual will be in a dense patch and therefore the mean density, as perceived by the individual, is higher than the population mean density (Lloyd 1967). Imagine two patches, one has a single individual, the other has nine, so the mean density is five individuals per patch. However, there are nine individuals experiencing a density of nine and only one experiencing a density of one, so the mean density experienced by individuals is much higher (mean perceived density of 8.2 in this case rather than 5). The effect of this mean crowding is that, on average, an individual will gain by moving as it will be most likely to be moving to a lower density patch. It is for this reason that models show the evolved dispersal strategies increase with more scramble competition and high growth rates as these are both more likely to increase the variability in density between patches (Holt and McPeek 1996, Travis and Dytham 1998).

Even in the absence of spatial heterogeneity birth and death events are inherently stochastic (random) and this will lead to differences in density between patches that are identical (Fig 8.3). The stochasticity of births and deaths will therefore inevitably result in the average individual being in a more dense

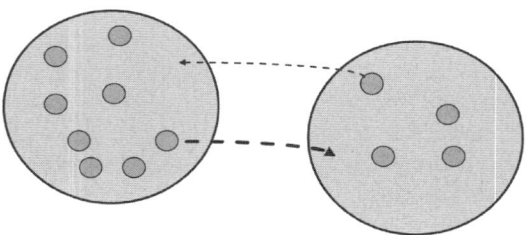

Figure 8.3 Through stochasticity (e.g. random birth and death events) the two identical patches depicted in Fig. 8.1 now have different population sizes. Now an individual moving from the more crowded to the less crowded patch will be at an advantage (thicker line) as it will gain from dispersing by arriving in a lower density patch. Individuals from the low density patch will now suffer increased competition from dispersing. As there are more individuals in the dense patch the average strategy across the population will favour dispersal.

patch and this will apply some selective pressure to disperse because dispersal will tend to move an individual to a patch with a lower density.

8.3.6 Expansion to new areas

As we saw in Chapter 6, the rate of spread during an invasion is influenced strongly by population growth rate and by dispersal kernels. During an invasion into a pristine habitat there is clearly an advantage to dispersing long distances. Some of the earliest theoretical treatments of dispersal considered this situation. Skellam's (1951) prediction of an expanding concentric circle was based on the situation where individuals were introduced into a pristine available habitat and they moved at random within that space. This is the case for very few species. What has been observed in most invasions is that range expansion is not at a constant rate predicted by the Fisher-Skellam model, but is often accelerating (Williamson 1997; Shigesada and Kawasaki 1997), or proceeds at a much faster rate than would be predicted from the mean dispersal distances achieved. It has long been known that post-glacial spread of trees is remarkably fast (Reid 1899), hundreds of metres per year, or several kilometres per generation (Delacourt and Delacourt 1987). This phenomenon is now known as Reid's paradox (Clark et al. 1998a). The explanation for Reid's paradox is that Skellam's assumption of normal distributions of seeds about an adult is incorrect and a 'fat-tailed' dispersal kernel, or a combination of dispersal kernels, allows occasional long distance dispersal events and it is these rare events that are pushing the range forward (Clark 1998; Higgins and Richardson 1999; page 129).

Discarding Skellam's assumption of a normal dispersal kernel might predict the rapid range expansions, but it is less clear how the observed change in rates of expansion during an invasion occurs. It is obvious that if there is variation for dispersal distance within a population then a rapid range expansion will lead to selection, by assortment alone, for a more rapid expansion at the margins (Travis and Dytham 2002; Fig. 8.4). Indeed, if a species can evolve increasingly dispersive propagules then the rate of expansion will continue to rise. However, there will always be a limit as the new area becomes

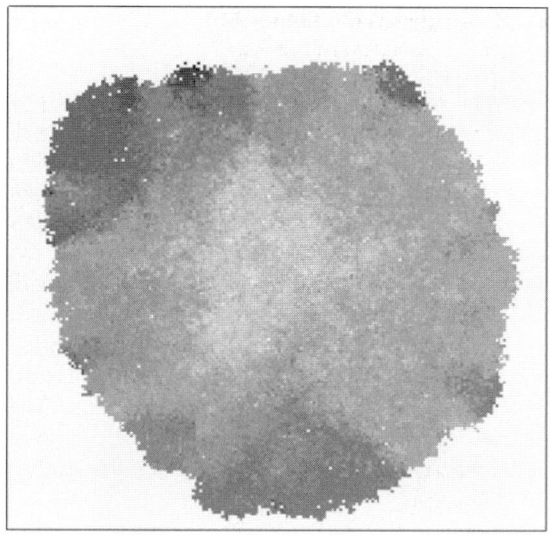

Figure 8.4 Distribution of dispersal strategies across a range when a propagule of individuals is dropped into the middle of a pristine environment. Darker shading indicates longer dispersal distances. Clearly dispersal is favoured at the range margin and the origination of novel dispersal strategies through mutation has led to more rapid range expansion in some areas (Travis and Dytham 2002, reproduced courtesy of the authors).

filled and the highly dispersive individuals start to pay a heavy cost for dispersal. Higgins and Cain (2002) and Kisdi and Geritz (2003) both suggest that a competition-colonization trade off exists, with the more dispersive individuals at the range margins being competitively inferior to those from the range core.

There is some evidence for the increase in dispersal at range margins, but in plants this evidence comes indirectly from clinal variation in morphology. The lodgepole pine (*Pinus contorta*) in the Rocky Mountains of North America shows morphological variation across its range with younger, northern populations having smaller seeds with longer wings than southern populations (Cwynar and MacDonald 1987). This observation fits well with the age of the populations: northern populations are younger than southern ones. However, the evidence is indirect, as other factors, such as climate, also have an effect on morphology. In animals there are some more direct observations of the phenomenon. For example, in cane toads expanding rapidly across Australia there are both morphological and behavioural differences between populations at the expanding edge and those from the core (Phillips et al. 2006). The cane toad appears to be showing an exponential increase in dispersal strategy at the population margin, with the incredible $50\,\mathrm{km\,yr^{-1}}$ movement of the range front between 2000 and 2005

attributed to both morphological (longer legs) and behavioural (increased activity) changes.

Ronce et al. (2005) used a deterministic model to explore the evolution of plasticity in dispersal strategies during periods of range expansion. They predict that there will be both genetic and plastic differences between marginal and core populations, with those populations recently colonized having a higher investment in seed production and dispersal.

When invasions occur into a huge area of suitable habitat then there is clearly a strong advantage to dispersal in any direction and therefore anything which increases dispersal rate or distance will be selected for. However, ranges cannot expand without limit, so even in large continents the invasion will reach limits. Range expansions in response to climate change may also behave differently as very long-distance dispersal is still likely to leave a propagule out of suitable habitat.

8.3.7 Avoidance of predators and pathogens

As has already been discussed earlier in this book (section 7.6.4 on page 154), the presence of specific pests or pathogens could have a profound effect on the distribution of successful offspring. The hypothesis that this could enhance the diversity of tropical forests was proposed independently by Janzen (1970) and Connell (1971). The Janzen–Connell effect can be extended to consider the evolutionary effects of host-specific pathogens and pests. A population will be under strong selective pressure to avoid the mortality associated with short distance dispersal and therefore evolve more long-distance dispersal, or a dispersal distribution with a fatter tail. Of course this will only free an individual from host-specific pests if the species is sufficiently rare for dispersal to allow individuals to escape them. We can hypothesize that as a species becomes more common in a landscape the selection for increased dispersal caused by the Janzen–Connell effect will diminish. Therefore to maintain tropical diversity it should be the case that rare species have better recruitment and survival than common ones. Evidence from seven large scale tropical sites with long-term monitoring shows that this might indeed be the case (Wills et al. 2006).

One interesting example of a dynamic Janzen–Connell effect during range expansion might be the post-glacial expansion of hemlock (*Tsuga canadensis*) and beech (*Fagus grandifolia*) in North America (Moorcroft et al. 2006). They noted that the rapid expansion of hemlock was as would be expected. The two species seem to have very similar habitat requirements, so the presence of hemlock should slow the later expansion of beech. However, the expansion of beech was almost as rapid as that of hemlock. So how could beech expand rapidly into areas already dominated by another species? One possible answer comes from the observation that there may have been a large outbreak of pathogens in the Holocene (Allison et al. 1986). So, using a model of the two tree species and some pathogens, they showed that the advancing wave of hemlock would be followed by a wave of pathogens and this would reduce its competitive ability sufficiently to allow the advancing wave of beech, also without pathogens at the range front, to expand rapidly. This mechanism, if true, is a transient Janzen–Connell effect.

The interaction of the Janzen–Connell effect with the evolution of dispersal could be quite complex. Clearly the presence of pathogens will drive up the dispersal distance, but the optimum distance travelled will depend on many factors, including the rarity of a species in the landscape, the presence of other species, and the habitat heterogeneity.

8.3.8 Avoidance of competition

Will competition lead to evolution for increased dispersal rates and distances? The idea that some species avoid competitive exclusion (Hardin 1960) by running from trouble, with high dispersal rates, is an old one. There are many studies that show with a trade-off of dispersal ability for competitive ability many species can coexist (e.g. Tilman 1994). However, the evidence for such a trade-off is not good, and it could be said that investment in dispersal at the expense of competitive ability is an adaptation to exploit new or disturbed habitats (a ruderal strategy), not a strategy to avoid competition per se. The relative importance of competition in the evolution of dispersal remains a topic worthy of more study.

8.4 Other aspects to the evolution of dispersal

8.4.1 Dispersal and mating system

Plants have a range of mating systems from fully clonal (e.g. *Elodea* spp.: Hydrocharitaceae), through sexual self-fertile, sexual out-breeding hermaphrodite, to fully dioecious. The mating system may have a strong relationship with dispersal strategy. Some species have a range of strategies; for example the Canada goldenrod (*Solidago canadensis*) a long-lived plant from North America, is spreading very rapidly in China through a combination of seed dispersal and clonal, vegetative dispersal (Dong et al. 2006). Soursob (*Oxalis pes-caprae*: Oxalidaceae) is a sterile pentaploid producing bulbils and has spread throughout southern Australia, as has skeleton weed (*Chondrilla juncea*: Asteraceae) an obligate apomict.

Adopting a clonal reproductive strategy usually has the effect of reducing dispersal distance. This might be an advantage in a very patchy habitat when filling up the small patch of good quality habitat is needed (e.g. Stocklin and Winkler 2004). Clonal reproduction may also be an advantage at range margins where suitable habitat is patchy and long distance dispersal is likely to result in arrival in very low quality habitat, although if habitat quality is high, rare long-distance dispersal will allow a species with clonal growth or asexual reproduction to overcome the Allee effect. The alpine *Geum reptans*

(Rosaceae) may show this pattern: an altitudinal transect showed that seed reproduction was most common at intermediate altitudes and clonal reproduction at the lowest and highest altitudes (Weppler and Stocklin 2005).

8.4.2 Dispersal in time

One common feature of plants is that seeds can delay germination and remain viable for a long time (e.g. Fernández et al. 2002). There are costs in reproductive capacity associated with delayed germination, but seed dormancy can function as a bet-hedging strategy (see Seger and Brockmann 1987) in variable environments where the predictability of the next good year is poor, so there is a selective advantage to spread germination across more than one year (e.g. Cohen 1967, Brown and Venable 1986). More generally, delayed germination can be seen as another method of spreading of risk in a heterogeneous landscape (Den Boer 1968, Baskin and Baskin 2000).

Another advantage of dormancy could be to avoid competition with siblings and thus decrease the kin competition without increasing dispersal distance. However, as Kobayashi and Yamamura (2000) point out, delaying germination for a year as an annual grass leads to an individual being in competition with the offspring of its siblings instead. These will be less related kin, but there may be far more of them, so the selective pressure to adopt a strategy of dormancy to avoid kin competition might not be very strong.

Dormancy is another area where parent-offspring conflict could be apparent. The decision to germinate could reside partly with the seed, and partly with the parent plant, through structures external to the seed. Some seeds are retained in woody cones on the mother plant for many years (serotiny). Kobayashi and Yamamura (2000) used an inclusive fitness model to show that the evolutionary stable degree of dormancy is different if under control of the seed rather than the parent.

8.4.3 Conditional dispersal

It has long been realized that adjusting dispersal strategy in response to local conditions would be a good strategy (i.e. to have some phenotypic plasticity in reproductive effort, seed morphology, or seed number). Gadgil (1971) suggested that plants should disperse a greater proportion of their offspring away from high density populations than low density ones.

However, most studies and models of dispersal have assumed that the proportion of individuals leaving a population and the mean distance they travel is constant, regardless of conditions. Ruxton (1996) and Saether et al. (1999) both showed that the inclusion of density-dependent dispersal behaviour into metapopulation models would have a large effect on the behaviour of those models, while Ruxton and Rohani (1999) didn't just rely on density but measured the condition of the individual making the decision to disperse using a measure of individual fitness. In an individual-based model Hovestadt and Poethke (2006) showed that density-dependent dispersal is far superior to density independent dispersal in a range of circumstances.

The evolution of density-dependent dispersal was explored in a landscape of homogeneous patches by Travis et al. (1999). They allowed dispersal probability to vary between individuals and with each individual's strategy described by a linear relationship between density and dispersal. Individuals therefore carry two genes describing the slope and intercept of the relationship between dispersal propensity and density (see Fig. 8.5). They showed that a negative intercept and positive slope always evolved. This means that at low densities no individuals will disperse while at and above the equilibrium density there is a high rate of dispersal. Poethke and Hovestadt (2002) also used a very simple, single gene, system to describe dispersal strategy, with the single gene describing the point at which dispersal is zero and dispersal probability then rising with local density (see Fig 8.5). Kun and Schuring (2006) and Bach et al. (2007) used more complex descriptions of the form of density-dependence, but both confirm that density-dependent dispersal will be strongly favoured under a range of variation in demographic and environmental conditions

Other conditions could be used to determine dispersal strategy. Kisdi (2004) allowed dispersal to evolve based on brood size. She used a threshold number above which all dispersed and below which

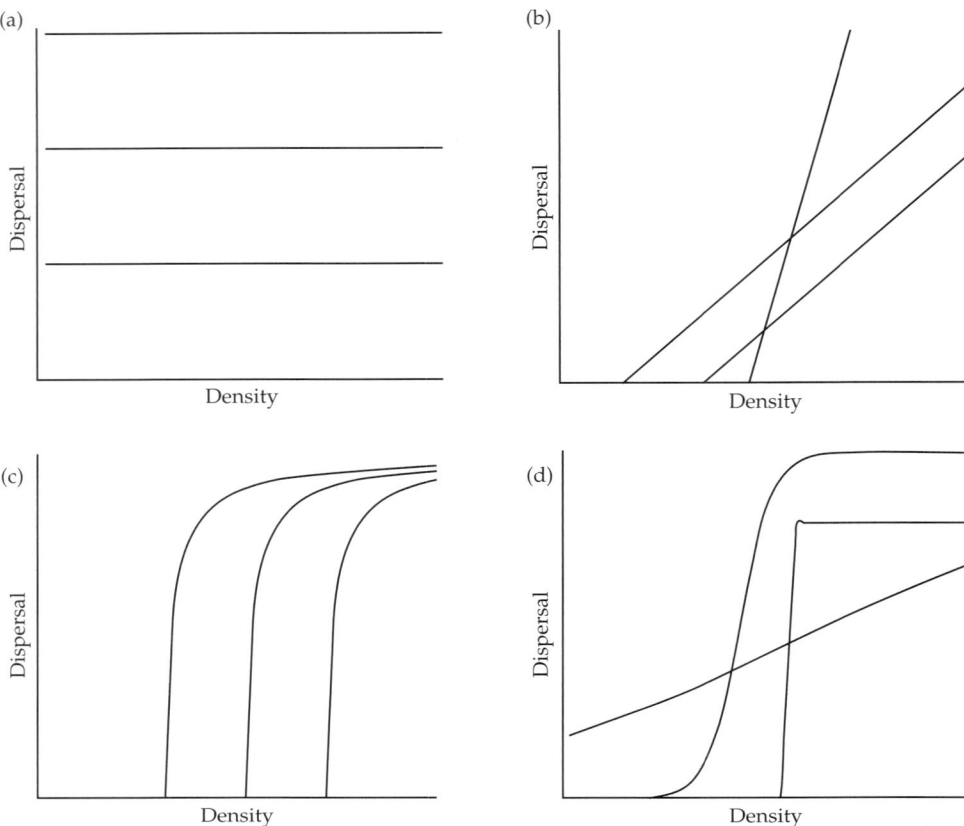

Figure 8.5 Here are four different ways to describe density-dependent dispersal strategies. In each panel we portray the dispersal strategy as a function of the density of individuals. Dispersal can be interpreted as distance dispersed, proportion of propagules leaving the patch, or the mean of a pdf. In a) there are three different dispersal strategies that do not vary with density. This density-independent strategy is a common assumption of models, but will be easily out-competed by almost any density-dependent strategy. In b), following Travis et al. 1999, the dispersal strategy is described by a straight line which needs two parameters to describe the slope and the intercept. There are three example strategies shown, each has a negative intercept and positive slope, so the strategy becomes more dispersive at higher density and there is no dispersal at low density. In c), following, Poethke and Hovestadt (2002), a single parameter describes the dispersal strategy defining the threshold density above which dispersal occurs. Again, three examples are given. Above the threshold density the dispersal strategy increases towards an asymptote of 100% dispersal. In d), following Kun and Schuring (2006) the dispersal strategy is a more complicated function of the density. They use three parameters to describe the function and this can produce a range of responses, as indicated in the three examples shown. The three parameters relate to the maximum possible strategy, the inflection point of the relationship and the steepness of the relationship. Although this allows a hugely flexible description of the relationship between density and dispersal it is difficult to see how the three parameters required can be justified biologically.

none dispersed and found that in large broods there was both a higher number of dispersers and a higher proportion of individuals dispersing. This pattern of a brood-size threshold in dispersal was attributed to the reduction in kin competition experienced by dispersing individuals.

Given the strength of condition-dependent dispersal in several different modelling approaches it is surprising how little empirical data there is available. However, it is difficult to collect such information as disentangling the genotypic effects from the phenotypic plasticity requires seed collected from several sites and then carefully-controlled experiments. Mazer and Lowry (2003) reported one such experiment on the heteromorphic *Spergularia marina*, which produces winged and unwinged seeds. They found that differences in seed morphology were entirely genetic with no evidence of phenotypic plasticity in seed morphology. Imbert and Ronce (2001) showed that both environmental stress and simulated herbivory affected the proportion of propagules with wings in a controlled experiment on the heteromorphic species, *Crepis sancta* (Asteraceae).

It is very difficult to account for all the interactions between dispersal and environmental conditions in either a model or an experiment. For instance, in poor, competitive, or stressful conditions adult plants are likely to have different morphologies from plants in good environments without competitors. As we have already seen in Part A of this book, the distance a propagule disperses can be hugely affected by the morphology of the parent plant.

8.4.4 Dispersal polymorphisms

We have been assuming that there is one optimal dispersal strategy and evolution will tend to move towards that. Simple models, such as that of Roff (1975) predicted that dispersal polymorphisms should be common and Venable (1985) showed that the shape of the fitness curve could lead to a heteromorphism for seed size if it were sufficiently concave (i.e. the fitness of both large seeds and small seeds being higher than the fitness of intermediate sized seeds).

Confirming Venable's prediction, dispersal strategy dimorphism appears to arise quite easily in evolutionary models. Travis and Dytham (1999) found that habitat aggregation led to a divergence in dispersal strategies with both high and low dispersal rates favoured over intermediate. Mathias et al. (2001) used a model with both temporal and spatial variability and showed divergence in the evolution of dispersal strategies which they suggested could be stable enough to provide a first step towards speciation. Hiebeler (2004) showed that it is possible for two dispersal strategies to coexist within a species even without a trade-off of dispersal and competitive ability if the landscape is adjusted appropriately.

Dispersal strategies do vary in natural populations. In some cases individual plants produce seeds with strikingly different dispersal morphologies while in others there are distinctly different dispersal strategies associated with different lineages within the same species (see Chapter 3, page 31). *Spergularia marina* (Caryophyllaceae) is one well-studied species where individuals can produce both winged and unwinged seeds (Salisbury 1958).

8.4.5 Timescales: long, medium, or short

Most ecology operates on short timescales, within the lifetime of an individual, while evolution will tend to operate over longer timescales. When faced with a result from an ecological experiment that operates over one or a few generations it is tempting to extrapolate to longer timescales. However, occasional extreme events may have much more influence on evolution than more 'normal' conditions. Consider two areas in the fire-prone Fynbos region of South Africa. Both are populated with many species of serotinous plants (i.e. those where the seeds are encased in cones until exposed to a fire). There are two life-history strategies available to serotinous Proteaceae: to grow quickly, reproduce and die, or grow slower, allocating resources underground and be able to survive the fire and 'resprout'. If a competition experiment shows that the fast-growing strategy will dominate and that short distance dispersal of seeds allows seedlings to exploit the recently burned ground, we might predict the reseeding strategy to exclude the resprouting one. However, the frequency of fires is a key determinant in the long-term competitive outcome and evolved dispersal strategy as if fires come too frequently the resprouting, slow-growing strategy will be dominant and the reseeding fast growing species can only exploit an area after a fire by much longer distance dispersal. As the model of Yeaton and Bond (1991) indicated, the long-term coexistence of the two strategies is difficult to achieve without invoking occasional disruptive events that occur much less frequently than the normal span of an ecological study.

8.4.6 The shape of the frequency distribution of dispersal distances

The mean or median dispersal distance only tells part of the story. No matter how well propagule morphology is suited to dispersal, a large proportion of propagules fall very close to their parent and a small proportion travel a long way. Dispersal is thus characterized by a frequency distribution of distances rather than the simple dichotomy of short and long distances assumed by many evolutionary models. As we have already discussed in previous chapters, there is a huge body of literature indicating that what is really important for many processes (e.g. range expansion, rescue effects, or arrival on islands) is the shape of the far tail of a dispersal frequency distribution (see Chapter 5). The long distance dispersal events, those which are particularly difficult to detect or study, have a disproportionate influence on both ecological and evolutionary processes (see Chapter 6 for the effect on invasions). Le Corre et al. (1997) and Austerlitz and Garnier-Géré (2003) both showed how important occasional long distance dispersal events can be in evolutionary models of range expansion, while Ibrahim et al. (1996) showed how the shape of the dispersal kernel can have a large effect on the spatial arrangements of genes.

But what frequency distributions could evolve? MacArthur and Wilson (1967) argued that wind-dispersed seeds should have a probability of stopping that is constant and this would generate a dispersal distance distribution that declines exponentially with distance

Hovestadt et al. (2001) used a simulation model without a statistical description of the frequency distribution of dispersal distances, rather allowing the frequency of dispersers in each of the distance 'bins' of a histogram to evolve until a distribution emerged. Each individual held 30 'genes' with a total value of 1 which corresponded to 30 distance categories. Individuals inherited dispersal strategies with occasional mutations. For each mutation two genes were selected at random, one raised slightly and the other lowered by the same amount. In this way they could evolve a dispersal distribution that was independent of any of the usual assumptions about describing a statistical probability density function. The results showed that in spatially autocorrelated landscapes the evolved distribution was much more 'fat tailed' than a negative exponential and was better described by a power function.

It has been suggested that one of the reasons for the widespread reliance on ants as dispersal agents for seeds is the frequency distribution of distances that ants transport seeds (Gomez and Espadaler 1998). Not only do ants bury seeds in sites suitable for germination and away from rodent predation (Manzaneda et al. 2005) but also it is suggested that the distances moved are sufficient to avoid kin competition while remaining in suitable habitat (Giladi 2006).

8.4.7 A complex relationship with granivores

Granivores eat seeds. An eaten seed is a dead seed irrespective of how far the granivore travels before depositing the remnants. Plants may, therefore, be expected to have evolved many defences against granivores—and there are many such defences, both chemical and physical. Even masting— high inter-annual variation in numbers of seeds—is often claimed to be a defence against granivory through satiation of the predators. However, many granivores may actually be agents of dispersal particularly through the habit of seed caching ('scatter-hoarding' and 'larder-hoarding') seen in mammals (e.g. squirrels), birds (e.g. jays), and insects (e.g. beetles and ants, see p. 67). The plant needs the granivore to be inefficient enough to generate dispersal of viable seeds. So, although it appears that the granivore is simply predating the plant's offspring, there is, actually, a mutualism between the granivore and the plant. If seeds were completely impervious to attack or lethally poisonous, the granivores would not visit that species and there would be no dispersal through seed caching.

8.4.8 Directed dispersal

Directed dispersal is a term used to describe a situation when seeds arrive disproportionately at some sites (Howe and Smallwood 1982). This phenomenon is generally thought to be quite rare (Howe 1986) although Wenny (2001) considers that it may actually occur quite widely, with a wide range of species benefiting from increased arrival probabilities in favourable sites. If directed dispersal is present, then this will be certain to increase the probability of dispersal when compared to random dispersal. However, it should be noted that while some directed dispersal can be achieved through abiotic mechanisms, the usual agents of directed dispersal are biotic movement vectors, including vertebrates and ants (Hanzawa et al. 1988), and this means there is mutualism between the plant and the vector that itself may negatively affect dispersal evolution.

8.4.9 Caveats and confusions: phylogenetic independence

One of the perennial problems in plant ecology is in comparative research. It is tempting to analyse a whole mass of species to determine the relationship between a morphological factor and an environmental one. A pattern or relationship is found, say between seed size and levels of disturbance, and this is a strong statistical signal amongst the range of species looked at. Conclusions are made on the basis of analysis where each plant contributes one datum to the analysis and each is treated independently. This is clearly a nonsense if half of the species surveyed come from the same small-seeded family. Accounting for the true independence of species in such analysis is fraught with difficulties and often requires that the evolutionary relatedness of lineages is known exactly (see Felsenstein 1985, Grafen 1989, or review by Harvey and Pagel 1991) as a starting point. As this is rarely, if ever, the case, there will be problems when general patterns are interpreted using species as independent points, although phylogenetic data are accumulating very rapidly.

A good example of the effect of including phylogeny is Kelly and Purvis (1993) who showed that the oft-cited link between seed size and shading (Foster and Janson 1985) disappears entirely once the phylogeny is included in the analysis.

8.5 Conclusions

There are many reasons why dispersal should be selected against. After all, why should an individual that has made it to maturity and been able to produce offspring risk casting those offspring into the unknown? It has long been acknowledged that dispersal must be a huge advantage during range expansion and invasions as well as the colonization of ephemeral habitats. Dispersal is required to take advantage of the empty habitats available and there must, therefore, be strong selection for the development of strategies that aid or enhance dispersal. So, it was believed that perhaps the habit of dispersing propgules is retained solely for the possibility of reaching new habitats. Indeed, as we have shown in this chapter, until the late 1970s there was not seen to be any advantage to dispersal in saturated landscapes. It was the application of inclusive fitness to models of dispersal to account for the reduction of competition between close relatives that is the result of dispersing, combined with the idea that inbreeding depression would reduce fitness, which revealed the true complexity of the evolutionary forces acting on dispersal strategies.

What we have done in this chapter is to outline how the selective forces might increase or decrease evolved dispersal rates or distances. The dispersal strategy adopted by a plant species is clearly shaped by a history of interacting and competing forces. The relative strengths of these forces will wax and wane and the interactions between them shift.

So, although we might not be able to predict the evolutionary path of the dispersal strategy of a lineage with any certainty, we are certainly in a better position to know which forces are likely to be the most important. What is clear from Chapters 6 and 7 is that dispersal strategy is a key part of the life-history of plants and a major determinant of their success and spread. In a world changing rapidly thanks to human activities, it is certain that many species will have dispersal strategies that are inappropriate to their surroundings. These species will need to be identified and translocated if they are to survive.

CHAPTER 9

Concluding remarks

9.1 Introduction

There is no perfect time to write a book to take stock of a discipline. If, as in the case of dispersal, the research is in full flood, by the time the book is published it may be out of date: major new studies may have appeared and new concepts introduced. However, taking stock does help. New researchers will see the historical development of the discipline, the way different components fit together, and good directions to develop interests and ideas. Established researchers like ourselves will see the connections between different parts of the subject, for instance, how measurements of dispersal in the field link to the modelling population dynamics and evolution.

While putting the book together, we made some of the connections ourselves and formed some opinions about the priorities and opportunities for future dispersal research. We could see gaps that, from a population perspective, need to be filled; areas in which some confusion seems to exist; and issues over which care needs to be taken. In this short concluding chapter, we first summarize our take-home messages from the three sections of the book, and then list some recommendations for future research.

9.2 Overview

9.2.1 Part A: Dispersal of individual propagules

An understanding of the mechanisms involved in the dispersal of individual propagules from their point of origin is essential if we are to explain and predict variation in dispersal among environments and over time. This was the focus of Part A of the book. Although various traits of propagule and parent are genetically programmed, dispersal distances are, to a major extent, context-specific. Starting positions for dispersal trajectories are affected by the growth conditions of the parent, the speeds and directions of propagules in motion depends on the availability and 'strength' of the various dispersal vectors, and deposition locations are affected by the vegetation, surface characteristics, and topography of the environment. Propagules vary within and between parents in their mass, aerodynamic properties, attractiveness to animals, ability to survive passage through an animal gut, length of time that they remain buoyant in water, and so on, resulting in variation in their movement. By defining and quantifying the component processes involved in dispersal, we are able to raise and test hypotheses, formulate models, and predict the behaviour of the seed population as a whole (see later chapters). Take, for example, the observations of seeds found high up on buildings (Horn et al. 2001). By understanding the theory of air currents, turbulence, and flight characteristics, it becomes possible to understand the ways in which such heights are achieved and the importance of abnormal wind events for long-distance dispersal (Nathan et al. 2001, 2002).

In taking a mechanistic approach to dispersal, it becomes clear that the maternal plant plays a considerable part in determining the final positions to which propagules move (Chapter 2). A plant is a distributed population of propagule release sites, with spatial coordinates extended vertically and/or horizontally through its growth. In large trees and clonal plants, the release points of many propagules may be metres from where the maternal plant first established. Different positions will make them differentially available to dispersal vectors. However, as yet little is known about the distributions of propagules within plant canopies. The timing of reproductive maturity of the plant will determine the match between dispersal and both vector activity and the habitat conditions for establishment post-dispersal. The pattern of maturation, along with the species' anatomical traits, will determine the threshold force required for propagule separation by dispersal vectors (although this is often ignored in mechanistic models), while in some species maternal structures launch propagules into the air and away from the plant.

The maternal plant also contributes the structures on the propagule that enhance the impact of dispersal vectors (Chapter 3). The shapes of the propagules that have adapted to wind dispersal, for example, are fascinating and naturally draw the interest of researchers. Their potential for long-distance dispersal is indisputable and has been studied in detail. Dispersal by animals is complicated by the huge diversity of animal behaviour, but the traits of propagules that enhance their dispersal have been closely studied in many animal groups and habitats, although there are still issues open to debate. Yet there are many species that lack obvious dispersal traits and these, too, must disperse. The dispersal of 'non-adapted' species by agricultural machinery has been studied in considerable detail. Janzen (1984) argued that herb species may have evolved small seed size, lack of appendages, and other traits in response to grazing from large herbivores. In the absence of animal influence, they are often detached and moved by wind (or from movement of plants by wind) and may be moved on the ground by water. How much dispersal do they achieve?

A focus on adaptations that enhance dispersal naturally draws the attention to *potential* dispersal distances. Although some sea beans *can* travel between continents and orchid seeds *may* be blown across an ocean to an isolated island, such events are extremely rare. Most propagules travel very much shorter distances (Chapter 4), with many failing to be moved at all by the vectors to which they are clearly adapted. The frequency distributions of dispersal distance from a point source under a given set of conditions is usually peaked at short distances and skewed towards longer distances. This

is the case for dispersal by air, water, animals, and farm machinery. Advances in our understanding of air currents, in particular turbulence, have enabled a greater understanding of the causes of long-distance dispersal. While we know, in a general way, the causes of long-distance dispersal by other vectors, our knowledge is based largely on context-specific phenomenological observations. In the case of dispersal by animals, there is a major challenge ahead to predict dispersal, over both short- and long-distances, under conditions other than those in which observations are made.

9.2.2 Part B: Patterns of dispersal from entire plants

Part B considered two types of data that are collected by ecologists to describe dispersal around entire plants: locations of individuals and counts in traps. These naturally lead to two different, but mathematically related, probability density functions, which we have referred to as the distance pdf and the density pdf respectively. They have very different shapes and, when comparing the results from different studies or using data for modelling population dynamics, there is great potential for confusion. We make a suggestion about how to avoid confusion in the future in Chapter 9.3.

Evidence from a range of sources—field data on individual distances, density data converted to frequencies, and models of propagules dispersing from point sources on plant canopies—shows that distance pdfs have low values close to the source, rise to one or more peaks, and decline to a shallow tail. Thus, the shapes are often similar in general to those for individual propagules dispersing from point sources (Chapter 4), but the parameter values may be very different. The exact positions of the key features of the distance pdf depend on the characteristics of the species, the dimensions of the plant (which may be modified by the habitat and competing species), the characteristics of the dispersal vector/s, and the habitat. Multiple peaks may occur through single vectors eliciting multiple outcomes, the existence of discrete features within the habitat (especially with respect to animals), and the summed actions of multiple vectors.

Density pdfs usually have their highest values close to the source and decline in a long tail. The shape is usually (but not always) steepest at the source and is leptokurtic (i.e. having greater kurtosis than a normal distribution). Maps of density around a source plant and transects in different compass directions commonly show that dispersal is anisotropic, with greater dispersal distances and numbers in certain directions. These conclusions, however, are only subjective: a meta-analysis of data sets, supported by a free exchange of raw data among researchers, would provide a helpful baseline for making informed statements on what we have discovered to date.

Despite a great many studies of dispersal in the field, empirical knowledge of dispersal is still limited and potentially biased. Studies usually focus on species likely to be dispersed very long distances by wind or birds and not on species with more limited dispersal. Directly observed distance pdfs are mostly for herbs, while a few are for trees with heavy seeds; density pdfs are usually for species dispersed by wind and animals and there is perhaps a bias towards trees. Most older studies have traps at too few distances to enable us to describe the shapes of their underlying density pdf, either at short or long distances.

Considerable attention has been given in the literature to the selection of mathematical functions to fit to density and distance pdfs. There is no single 'correct' function: different dispersal mechanisms are likely to result in patterns that are best described by different equations. Equations are available with thick or thin tails and (for density pdfs) with origins that decline steeply, are flat, or increase steeply to a peak. Multiple dispersal vectors can be modelled by combining several equations ('mixture models'), but the greater complexity and limited data means that parameters may be poorly estimated. Recent trends in study design, including more traps at very short and very long distances and multiple traps at each distance, should improve our ability to distinguish between alternative shapes of density pdf. Even with good data, however, it is likely to remain difficult to distinguish between distributions with thick or thin tails at long distances. Models of dispersal vectors, in particular wind and animals, have allowed better predictions of what these

tails *might* be. But integrating the effects of multiple dispersers over whole seasons remains a major challenge.

9.2.3 Part C: Dispersal in population dynamics and evolution

Part C took us away from single propagules and the propagules of single plants, to the dynamics of populations and communities. All successful species must, at some stage in their history, spread their ranges. In Chapter 6 we saw how the spread of plants can be predicted from knowledge of the pdf of distances that propagules travel and the number of propagules. The rate of spread during an invasion is indeed, largely determined by rare, long-distance dispersal events, which are very difficult to observe, but depend on the fecundity of individual plants, the number of plants, and the pdf of dispersal distances. Most consideration of range expansion assumes that interactions with resident species are relatively unimportant. However, the state of communities into which species spread could matter; for instance competition may reduce rates of spread by reducing fecundities. Quantifying such effects of competitive interactions during invasions might improve the predictions about range expansion more than attempts to observe very rare long-distance dispersal events.

Chapter 7 moved from expanding populations to events within local populations and showed how dispersal is an important component of local population dynamics. Inferences about dispersal distance from the spatial arrangement of individuals are frustratingly difficult because dispersal is just one process affecting the spatial pattern of individuals we observe. In the context of local spatial dynamics we see some of the advantages and disadvantages of dispersal. Short-distance dispersal inevitably leads to aggregation in the landscape and intense competition for resources. In a homogeneous landscape dispersal over longer distances thus provides some escape from this local competition. However, in an inhomogeneous landscape, there is a cost to longer distance dispersal because propagules can end up in habitats to which they are poorly suited. These within-species processes are themselves typically embedded in a multispecies community containing other species with their own spatial dynamics. The emergent features of multispecies spatial dynamics are largely unknown, but they are potentially important, as can be seen from the Janzen–Connell hypothesis.

It is clear that local conditions such as competitors and habitat have the potential to generate a selective force on dispersal. In Chapter 8 we describe a range of selective forces acting to both increase and decrease dispersal rates and distances. The main forces acting to increase dispersal distance are the exploitation of virgin habitat, often during invasions, the avoidance of kin competition and inbreeding depression, and escape from pathogens. Against these forces, the cost of dispersal, either in development of structures to aid dispersal or in the peril associated with long distance movements, and the risk of arriving in habitat that is unsuitable, are pre-eminent. The dispersal strategies that we see must have arisen from the historical balance of these forces. Exquisite adaptations to aid dispersal, such as barbs and pappi, so carefully described by natural historians, must be considered in this evolutionary context if their contribution to the ecology of plant species is to be appreciated.

9.3 Priorities for dispersal research

Clearly, research directions are very much a matter of personal choice. However, some of the outstanding issues which have emerged from writing this book are below. Some points are potential research areas, and others are points to bear in mind in the course of research on propagule dispersal. The list is not in order of priority and largely follows the order of the chapters.

● **Patterns of dispersal close to the parent.** Although there have been, and continue to be, strong reasons for measuring long distance dispersal, the spatial pattern of seeds immediately around the parent plays an important part in the structure and dynamics of populations and communities. Research in this area will require development of sampling protocols able to describe short distance dispersal.

● **Importance of species with limited dispersal potential.** Not surprisingly, the focus has been on species clearly adapted to enhance dispersal by particular vectors. These can spread rapidly and often achieve remarkable distances. Many species, however, have very restricted dispersal and have major problems in moving about in landscapes without the involvement of people. These species will be severely challenged under climate change and the need to spread through fragmented landscapes.

● **Designs for sampling propagule deposition.** It has been shown clearly that in order to understand the shapes of dispersal pdfs more sampling distances are needed. Debates about kernel shape are unlikely to be resolved by fitting equations to existing data or through the use of maximum likelihood methods on multiple source plant situations. It is also important to know the uncertainty of estimates, and this means some replication of parent plants. (This is not always present in published results, perhaps because there is so much work in getting information on dispersal even from a single parent.)

● **Careful use of terms.** There is much potential for confusion in the use of terms like dispersal curve and kernel. To avoid confusion there would be some merit in always using the density pdf, since most population models use this as a starting point and it can easily be converted to the distance pdf if necessary. Using the term 'kernel' without a precise definition of the function involved would be better avoided.

● **Anisotropic dispersal kernels and the density-dependence of dispersal.** Field studies show repeatedly that dispersal is anisotropic, yet models usually assume that dispersal is equal in all directions. While most models consider that reproduction varies with density, they assume that the shape and scale of a dispersal frequency distribution does not. Many data sets used to parameterize models are from plants grown in the open or at the edge of a stand; dispersal from plants in a stand could be different. How important such effects are on population and community dynamics remains to be seen.

● **Use of dispersal data in population dynamic models.** Some models are in one dimension whereas most field situations are two dimensional. To predict the likely population dynamics of the species (which might result from the shape of its tail), the appropriate conversion needs to be made between one and two dimensions.

• **Integrating dispersal distributions over seasons and vectors.** Many studies, even the most impressive pieces of work, consider only one vector (wind; water; one animal species) and are thus measuring the *vector's* seed shadow rather than that of the plant. Many vectors may act to disperse propagules and the season may be weeks or even months, during which the strength of the vector and the dispersal potential of propagules vary.

• **Attributes of parent plants and artificial propagule releases in estimating dispersal distributions.** Artificial releases usually do not allow for the threshold force required to liberate a seed from the parent; they may thus seriously underestimate the distances dispersed by propagules under realistic conditions. Also, most plants are not point sources. Care is needed to take into account these more subtle features of parents when using artificial releases.

• **Mechanistic dispersal models and estimating dispersal by animal vectors**. Most current approaches for animal vectors are context-specific. The great advances in understanding of dispersal by wind have come from a more mechanistic approach to the behaviour of the vector, allowing predictions of dispersal under conditions other than those in the particular field situation. Taking an approach of this kind to dispersal by animals is perhaps one of the greatest challenges for dispersal at the present time.

• **Interpretation of 'realized' dispersal data.** It is often argued that the distribution of successful offspring (seedlings or other plant stages) is of more direct relevance to population dynamics than the distribution of propagules. It is also often easier to measure. Realized dispersal aggregates a number of biological processes, including dispersal per se, and the resultant distribution will often differ in shape and scale from the distribution of propagules. Where spatial population models do not build in those other processes, it is clearly logical to use realized dispersal distributions and not propagule dispersal distributions. However, if we want to understand *dispersal* and its implications, realized dispersal is not the appropriate quantity to measure.

• **Dispersal and multispecies spatial dynamics.** Going from the spatial dynamics of one plant species to that of two species introduces many complications, yet this is just a small step in the direction of multispecies communities, such as the rainforest at Pasoh discussed in Chapter 7. There is a major open challenge of spatial dynamics in communities, in which the dispersal properties of the constituent species is bound to play a central role.

• **Heritability of dispersal strategies**. Plants are highly plastic and can adopt a huge variety of phenotypes from a single genotype. In quantitative models of evolution of dispersal traits it is important to know how much of the variation in the trait is inherited and how much is due to environmental causes. The variation among individuals has its own interest, and a genetic component may be maintained by selection.

References

Ackerman, J. D. (2002). Diffusivity in a marine macrophyte canopy: implications for submarine pollination and dispersal. *American Journal of Botany*, **89**, 1119–27.

Acosta, F. J., Delgado, J. A., Lopez, F., and Serrano, J. M. (1997). Functional features and ontogenetic changes in reproductive allocation and partitioning strategies of plant modules. *Plant Ecology*, **132**, 71–76.

Adler, F. R. and Mosquera, J. (2000). Is space necessary? Interference competition and limits to biodiversity. *Ecology*, **81**, 3226–32.

Adler, F. R. and Muller-Landau, H. C. (2005). When do localized natural enemies increase species richness? *Ecology Letters*, **8**, 438–47.

Aguiar, M. R. and Sala, O. E. (1997). Seed distribution constrains the dynamics of the Patagonian steppe. *Ecology*, **78**, 93–100.

Akaike, H. (1973). Information theory as an extension of the maximum likelihood principle. In B. N. Petrov and F. Csaksi, eds. *2nd International Symposium on Information Theory*, p. 267–81. Akademiai Kiado, Budapest, Hungary.

Allen, L. J. S., Allen, E. J., and Ponweera, S. (1996). A mathematical model for weed dispersal and control. *Bulletin of Mathematical Biology*, **58**, 815–34.

Allen, R. and Wardrop, A. B. (1964). The opening and shedding mechanism of the female cones of *Pinus radiata*. *Australian Journal of Botany*, **12**, 125–34.

Allison, T. D., Moeller, R. E., and Davis, M. B. (1986). Pollen in laminated sediments provides evidence for a mid-holocene forest pathogen outbreak. *Ecology*, **67**, 1101–05.

Alonso, D., Etienne, R. S., and McKane, A. J. (2006). The merits of neutral theory. *Trends in Ecology and Evolution*, **21**, 451–57.

Andersen, A. N. (1988). Dispersal distance as a benefit of myrmecochory. *Oecologia (Berlin)*, **75**, 507–11.

Andersen, A. N. and Morrison, S. C. (1998). Myrmecochory in Australia's seasonal tropics: Effects of disturbance on distance dispersal. *Australian Journal of Ecology*, **23**, 483–91.

Andersson, E., Nilsson, C., and Johansson, M. E. (2000). Plant dispersal in boreal rivers and its relation to the diversity of riparian flora. *Journal of Biogeography*, **27**, 1095–1106.

Appanah, S. and Rasol, A. M. M. (1995). Dipterocarp fruit dispersal and seedling distribution. *Journal of Tropical Forest Science*, **82**, 258–63.

Archibald, S. and Bond, W. J. (2003). Growing tall vs growing wide: tree architecture and allometry of *Acacia karoo* in forest, savanna, and arid environments. *Oikos*, **102**, 3–14.

Arditti, J. and Ghani. A. K. A. (2000). Numerical and physical properties of orchid seeds and their biological implications. *New Phytologist*, **145**, 367–421.

Askew, A. P., Corker, D., Hodkinson, D. J., and Thompson, K. (1997). A new apparatus to measure the rate of fall of seeds. *Functional Ecology*, **11**, 121–25.

Asmussen, A. M. (1983). Evolution of dispersal in density regulated populations: a haploid model. *Theoretical Population Biology*, **23**, 281–99.

Assunção, R. and Jacobi, C. M. (1996). Optimal sampling for studies of gene flow from a point source using marker genes or marked individuals. *Evolution*, **50**, 918–23.

Augspurger, C. and Kitajima, K. (1992). Experimental studies of seedling recruitment from contrasting seed distributions. *Ecology*, **73**, 1270–84.

Augspurger, C. K. (1983). Offspring recruitment around tropical trees: changes in cohort distance with time. *Oikos*, **20**, 189–96.

Augspurger, C. K. (1986). Morphology and dispersal potential of wind-dispersed diaspores of neotropical trees. *American Journal of Botany*, **73**, 353–63.

Augspurger, C. K. and Franson, S. E. (1987). Wind dispersal of artificial fruits varying in mass, area and morphology. *Ecology*, **68**, 27–42.

Augspurger, C. K. and Hogan, K. P. (1983). Wind dispersal of fruits with variable seed number in a tropical tree (*Lonchocarpus pentaphyllus*: Leguminosae). *American Journal of Botany*, **70**, 1031–37.

Auld, B. A., Menz, K. M., and Tisdell, C. A. (1987). *Weed Control Economics*. Sydney, Academic Press, Sydney.

Austerlitz, F. and Garnier-Gere, P. H. (2003). Modelling the impact of colonisation on genetic diversity and differentiation of forest trees: interaction of life cycle, pollen flow and seed long-distance dispersal. *Heredity,* **90**, 282–90.

Austin, R. B. (1972). Effects of environment before harvesting on viability In E. H. Roberts, ed. *Viability of Seeds*, pp. 114–49. Chapman and Hall, London.

Azuma, A. and Okuno, Y. (1987). Flight of a samara, *Alsomitra macrocarpa. Journal of Theoretical Biology,* **129**, 263–74.

Bach, L. A., Thomsen, R., Pertoldi, C., and Loeschcke, V. (2006). Kin competition and the evolution of dispersal in an individual-based model. *Ecological Modelling,* **192**, 658–66.

Bach, L. A., Ripa, J., and Lundberg, P. (2007). On the evolution of conditional dispersal under environmental and demographic stochasticity. *Evolutionary Ecology Research,* In press.

Baddeley, A. J., Møller, J., and Waagepetersen, R. (2000). Non- and semi-parametric estimation of interaction in inhomogeneous point patterns. *Statistica Neerlandica,* **54**, 329–50.

Baker, G. A. and O'Dowd, D. J. (1982). Effects of parent plant density on the production of achene types in the annual *Hypochoeris glabra. Journal of Ecology,* **70**, 201–15.

Baker, H. G. (1955). Self-compatibility and establishment after 'long-distance' dispersal. *Evolution,* **9**, 347–49.

Baker, H. G. (1974). The evolution of weeds. *Annual Review of Ecology and Systematics,* **5**, 1–24.

Balkau, B. J. and Feldman, M. W. (1973). Selection for migration modification. *Genetics,* **74**, 171–74.

Ballaré, C. L., Scopel, A. L., Ghersa, C. M., and Sanchez, R. A. (1987). The demography of *Datura ferox* (L.) in soybean crops. *Weed Research,* **27**, 91–102.

Barnett, R. J. (1977). The effect of burial by squirrels on germination and survival of oak and hickory nuts. *American Midland Naturalist,* **98**, 319–30.

Barroso, J., Navarrete, L., Sánchez del Arco, M. J., Fernandez-Quintanilla, C., Lutman, P. J. W., Perry, N. H., and Hull, R. I. (2006). Dispersal of *Avena fatua* and *Avena sterilis* patches by natural dissemination, soil tillage and combine harvesters. *Weed Research,* **46**, 118–28.

Barry, R. G. and Chorley, R. J. (1998). *Atmosphere, Weather and Climate,* 7th edition. Routledge, London.

Baskin, C. C. and Baskin, J. M. (2000). Ecology and evolution of specialized seed dispersal, dormancy and germination strategies. *Plant Species Biology,* **15**, 95–96.

Bastow Wilson, B. and Lee, W. G. (1989). Infiltration invasion. *Functional Ecology,* **3**, 379–80.

Batschelet, E. (1981). *Circular Statistics in Biology.* Academic Press, London.

Beer, T. and Swaine, M. D. (1977). On the theory of explosively dispersed seeds. *New Phytologist,* **78**, 681–94.

Begon, M., Harper, J. L., and Townsend, C. R. (1996a). *Ecology: Individuals, Populations and Communities.* 3rd edition. Blackwell, Oxford.

Begon, M., Mortimer, M., and Thompson, D. J. (1996b) *Population Ecology: A Unified Study of Plants and Animals,* 3rd edition. Blackwell Science, Oxford.

Bell, A. D. (1974). Rhizome organization in relation to vegetative spread in *Medeola virginiana. Journal of the Arnold Arboretum,* **55**, 458–68.

Bell, A. D. (1979). The hexagonal branching pattern of rhizomes of *Alpinia speciosa* L. (Zingiberaceae). *Annals of Botany,* **43**, 209–23.

Bengtsson, B. O. (1978). Avoiding inbreeding: at what cost? *Journal of Theoretical Biology,* **73**, 439–44.

Beverton, R. J. H. and Holt, S. J. (1957). On the dynamics of exploited fish populations. *Fisheries Investigations Series 2(19).* Ministry of Agriculture, Fisheries and Food, London.

Bewley, J. D. and Black, M. (1994). *Seeds: Physiology of Development and Germination,* 2nd edition. Plenum Press, New York.

Bialozyt, R., Ziegenhagen, B., and Petit, R. J. (2006). Contrasting effects of long distance seed dispersal on genetic diversity during range expansion. *Journal of Evolutionary Biology,* **19**, 12–20.

Bidartondo, M. I. (2005). The evolutionary ecology of mycoheterotrophy. *New Phytologist,* **167**, 335–52.

Bithell, M. and Macmillan, W. D. (2007). Escape from the cell: Spatially explicit modelling with and without grids. *Ecological Modelling,* **200**, 59–78.

Blanco-Moreno, J. M., Chamorro, L., Masalles. R. M., and Recasens, J. (2004). Spatial distribution of *Lolium rigidum* seedlings following seed dispersal by combine harvesters. *Weed Research,* **44**, 375–87.

Blundell, A. G. and Peart, D. R. (2004). Density-dependent population dynamics of a dominant rain forest canopy tree. *Ecology,* **85**, 704–15.

Boedeltje, G., Bakker, J. P., Bekker, R. M., van Groenendael, J. M., and Soesbergen, M. (2003). Plant dispersal in a lowland stream in relation to occurrence and three specific life-history traits of the species in the species pool. *Journal of Ecology,* **91**, 855–66.

Boedeltje, G., Bakker, J. P., Brinke, A. T., van Groenendael, J. M., and Soesbergen, M. (2004). Dispersal phenology of hydrochorous plants in relation to discharge, seed release time and buoyancy of seeds: the flood pulse concept supported. *Journal of Ecology,* **92**, 786–796.

Bolker, B. M. (2003). Combining endogenous and exogenous spatial variability in analytical population models. *Theoretical Population Biology,* **64**, 255–270.

Bolker, B. M. and Pacala, S. W. (1997). Using moment equations to understand stochastically driven spatial pattern formation in ecological systems. *Theoretical Population Biology*, **52**, 179–97.

Bolker, B. M. and Pacala, S. W. (1999). Spatial moment equations for plant competition: understanding spatial strategies and the advantages of short dispersal. *American Naturalist*, **153**, 575–602.

Bolker, B. M., Pacala, S. W., and Neuhauser, C (2003). Spatial dynamics in model plant communities: what do we really know? *American Naturalist*, **162**,135–48.

Bonaccorso, F. J., Glanz, W. E., and Sanford, C. M. (1980). Feeding assemblages of mammals at fruiting *Dipteryx panamensis* trees in Panama: seed predation, dispersal, and parasitism. *Revista de Biologia Tropical*, **28**, 61–72.

Bond, W. J. (1994). Do mutualisms matter – assessing the impact of pollinator and disperser disruption on plant extinction. *Philosophical Transactions of the Royal Society B: Biological Sciences*, **344**, 83–90.

Borchert, R. (1983). Phenology and control of flowering in tropical trees. *Biotropica*, **15**, 81–89.

Borchert, R., Renner, S. S., Calle, Z., Navarrete, D., Tye, A., Gautier, L., Spichiger, R., and von Hildebrand, P. (2005). Photoperiodic induction of flowering near the Equator. *Nature*, **3259**, 1–3.

Boucher, D. H. (1981). Seed predation by mammals and forest dominance by *Quercus oleoides*, a tropical lowland oak. *Oecologia*, **49**, 409–14.

Bradbury, I. K. (1981). Dynamics, structure and performance of shoot populations of the rhizomatous herb *Solidago canadensis* L. in abandoned pastures. *Oecologia* (Berlin), **48**, 271–76.

Bradley, J. (1971). *Bush Regeneration: The Practical Way to Eliminate Exotic Plants from Natural Reserves*. The Mosman Parklands and Ashton Park Association, Sydney.

Brain, P. and Marshall, E. J. P. (1999). Modeling cultivation effects using fast Fourier transforms. *Journal of Agricultural, Biological and Environmental Statistics*, **4**, 276–89.

Brewer, S. J., Rand, T., Levine, J. M., and Bertness, M. D. (1998). Biomass allocation, clonal dispersal, and competitive success in three salt marsh plants. *Oikos*, **82**, 347–53.

Brewer, S. W. (2001). Predation and dispersal of large and small seeds of a tropical palm. *Oikos*, **92**, 245–55.

Brooker, M. I. H. and Kleinig, D. A. (2004). *Field Guide to Eucalypts Volume 3 Northern Australia*, 2nd edition. Bloomings Books, Melbourne.

Brown, J. S. and Venable, D. L. (1986). Evolutionary ecology of seed-bank annuals in temporally varying environments. *American Naturalist*, **127**, 31–47.

Bruun, H, H. and Poschlod, P. (2006). Why are small seeds dispersed through animal guts: large numbers or seed size *per se*? *Oikos*, **113**, 402–11.

Buckley, Y. M., Brockerhoff, E., Langer, L., Ledgard, N., North, H., and Rees, M. (2005). Slowing down a pine invasion despite uncertainty in demography and dispersal. *Journal of Applied Ecology*, **42**, 1020–30.

Bullock, J. M. and Clarke, R. T. (2000). Long distance seed dispersal by wind: measuring and modelling the tail of the curve. *Oecologia*, **124**, 506–521.

Bullock, J. M. and Moy, I. L. (2004). Plants as seed traps: inter-specific interference with dispersal. *Acta Oecologia*, **25**, 35–41.

Bullock, J. M., Moy, I. L., Coulson, S. J., and Clarke, R. T. (2003). Habitat-specific dispersal: environmental effects on the mechanisms and patterns of seed movement in a grassland herb *Rhinanthus minor*. *Ecography*, **26**, 692–704.

Bullock, J. M., Shea, K., and Skarpaas, O. (2006). Measuring plant dispersal: An introduction to field methods and experimental design. *Plant Ecology*, **186**, 217–34.

Bullock, S. H. and Primack, R. B. (1977). Comparative experimental study of seed dispersal on animals. *Ecology*, **58**, 681–86.

Burrows, F. M. (1973). Calculation of the primary trajectories of plumed seeds in steady winds with variable convection. *New Phytologist*, **72**, 647–64.

Burrows, F. M. (1986). Aerial motion. In D. R. Murray, ed. *Seed Dispersal*, pp. 1–47. Academic Press, Sydney.

Burton, M. G. (2000). Effects of soil and landscape characteristics on the population dynamics of wild *Helianthus annuus* L. PhD Thesis, University of Nebraska, Lincoln.

Buwalda, J. G., Ross, G. J. S., Stribley, D. P., and Tinker, P. B. (1982). The developments of endomycorrhizal root systems. III. The mathematical representation of the spread of vesicular-arbuscular mycorrhizal infection in root systems. *New Phytologist*, **91**, 669–82.

Cain, M. L., Damman, H., and Muir, A. (1998). Seed dispersal and the Holocene migration of woodland herbs. *Ecological Monographs*, **68**, 325–47.

Calder, W. A. (1984). *Size, function, and life history*. Harvard University Press, Cambridge, Mass.

Canham, C. D. and Uriarte, M. (2006). Analysis of neighbourhood dynamics of forest ecosystems using likelihood methods and modeling. *Ecological Applications*, **16**, 62–73.

Cantrell, R. S. and Cosner, C. (1991). The effects of spatial heterogeneity in population dynamics. *Journal of Mathematical Biology*, **29**, 315–38.

Cantrell, R. S. and Cosner, C. (2001). Spatial heterogeneity and critical patch size: Area effects via diffusion in closed environments. *Journal of Theoretical Biology*, **209**, 161–71.

Carey, P. D. and Watkinson, A. R. (1993). The dispersal and fates of seeds of the winter annual grass *Vulpia ciliata*. *Journal of Ecology*, **81**, 759–67.

Carlo, T. A. (2005). Interspecific neighbours change seed dispersal pattern of an avian-dispersed plant. *Ecology*, **86**, 2440–49.

Caswell, H., Lensink, R., and Neubert, M. G. (2003). Demography and dispersal: Life table response experiments for invasion speed. *Ecology*, **84**, 1968–78.

Chambers, J. C. (2000). Seed movements and seedling fates in disturbed sagebrush steppe ecosystems: implications for restoration. *Ecological Applications*, **10**, 1400–13.

Chapman, D. S., Dytham, C., and Oxford, G. S. (2007). Modelling population redistribution in a leaf beetle: an evaluation of alternative dispersal functions. *Journal of Animal Ecology*, **76**, 36–44.

Charlesworth, D. and Charlesworth, B. (1987). Inbreeding depression and its evolutionary consequences. *Annual Review of Ecology and Systematics*, **18**, 237–68.

Chase, J. M., Amarasekare, P., Cottenie, K., Gonzalez, A., Holt, R. D., Holyoak, M., Hoopes, M. F., Leibold, M. A., Loreau, M., Mouquet, N., Shurin, J. B., and Tilman, D. (2005). Competing theories for competitive metacommunities. In M. Holyoak, M. A. Leibold and R. D. Holt, eds. *Metacommunities: spatial dynamics and ecological communities*, pp. 335–54. Chicago University Press, Chicago.

Chave, J., Muller-Landau, H. C., and Levin, S. A. (2002). Comparing classical community models: theoretical consequences for patterns of diversity. *American Naturalist*, **159**, 1–23.

Cheam, A. H. (1986). Seed production and seed dormancy in wild radish (*Raphanus raphanistrum* L.) and some possibilities for increasing control. *Weed Research*, **26**, 405–13.

Cheplick, G. P. (1993). Sibling competition is a consequence of restricted dispersal in an annual cleistogamous grass. *Ecology*, **74**, 2161–64.

Chettleburgh, M. R. (1952). Observations on the collection and burial of acorns by jays in Hainault Forest. *British Birds*, **45**, 359–64.

Cici, S-Z-H., Adkins, S., Sindel, B., and Hanan, J. (2006). Canopy interference of chickpea with sowthistle – model development. In C. Preston, J. H. Watts, and N. D. Crossman, eds. *Papers and Proceedings of the 15th Australian Weeds Conference 'Managing Weeds in a Changing Climate'*, p. 426. Weed Management Society of South Australia, Adelaide.

Cipollini, M. L. and Levey, D. J. (1997a). Why are some fruits toxic? Glycoalkaloids in *Solanum* and fruit choice by vertebrates. *Ecology*, **78**, 782–98.

Cipollini, M. L. and Levey, D. J. (1997b). Secondary metabolites of fleshy vertebrate-dispersed fruits: Adaptive hypotheses and implications for seed dispersal. *American Naturalist*, **150**, 346–72.

Clark, C. J., Poulsen, J. R., Bolker, B. M., Connor, E. F., and Parker, V. T. (2005). Comparative seed shadows of bird-, monkey- and wind-dispersed trees. *Ecology*, **86**, 2684–94.

Clark, J. S. (1998). Why trees migrate so fast: Confronting theory with dispersal biology and the paleorecord. *American Naturalist*, **152**: 204–24.

Clark, J. S., Lewis, M., and Horvath, L. (2001). Invasion by extremes: Population spread with variation in dispersal and reproduction. *American Naturalist*, **157**: 537–54.

Clark, J. S., Lewis, M., McLachlan, J. S., and HilleRis-Lambers, J. (2003). Estimating population spread: What can we forecast and how well? *Ecology*, **84**, 1979–88.

Clark, J. S., Fastie, C., Hurtt, G., Jackson, S. T., Johnson, C., King, G. A., Lewis, M., Lynch, J., Pacala, S., Prentice, C., Schupp, E. W., Webb, T., and Wyckoff, P. (1998a). Reid's paradox of rapid plant migration: Dispersal theory and interpretation of paleoecological records. *Bioscience*, **48**, 13–24.

Clark, J. S., Macklin, E., and Wood, L. (1998b). Stages and spatial scales of recruitment limitation in southern Appalachian forests. *Ecological Monographs*, **68**, 213–35.

Clark, J. S., Silman, M., Kern, R., Macklin, E., and HilleRisLambers, J. (1999). Seed dispersal near and far: patterns across temperate and tropical forests. *Ecology*, **80**, 1475–94.

Cobby, J. M., Chapman, P. F., and Pike, D. J. (1986). Design of experiments for estimating inverse quadratic polynomial responses. *Biometrics*, **42**, 659–664.

Cody, M. L. and Overton, J. McC. (1996). Short-term evolution of reduced dispersal in island plant populations. *Journal of Ecology*, **84**, 53–61.

Cohen, D. (1967). Optimizing reproduction in a randomly varying environment when a correlation may exist between the conditions at the time a choice has to be made and the subsequent outcome. *Journal of Theoretical Biology*, **16**, 1–14.

Colas, B., Olivieri, I., and Riba, M. (1997). *Centaurea corymbosa*, a cliff-dwelling species tottering on the brink of extinction: a demographic and genetic study. *Proceedings of the National Academy of Science*, **97**, 3471–76.

Colas, B., Olivieri, I., and Riba, M. (2001). Spatio-temporal variation of reproductive success and conservation of narrow-endemic *Centaurea corymbosa* (Asteraceae). *Biological Conservation*, **99**, 375–86.

Collingham, Y. C. and Huntley, B. (2000). Impacts of habitat fragmentation and patch size upon migration rates. *Ecological Applications*, **10**, 131–44.

Collingham, Y. C., Hill, M. O., and Huntley, B. (1996). The migration of sessile organisms: a simulation model with measurable parameters. *Journal of Vegetation Science*, **7**, 831–46.

Comins, H. N. (1982). Evolutionary stable strategies for localise dispersal in two dimensions. *Journal of Theoretical Biology*, **94**, 57—606.

Comins, H. N., Hamilton, W. D., and May, R. M. (1980). Evolutionary stable dispersal strategies. *Journal of Theoretical Biology*, **82**, 205–30.

Condit, R., Ashton, P. S., Baker, P., Bunyavejchewin, S., Gunatilleke, S., Gunatilleke, N., Hubbell, S. P., Foster, R. B., Itoh, A., LaFrankie, J. V., Seng Lee, H., Losos, E., Manokaran, N., Sukumar, R., and Yamakura, T. (2000). Spatial patterns in the distribution of tropical tree species. *Science*, **288**, 1414–18.

Condit, R., Pitman, N., Leigh, E. G., Chave, J., Terborgh, J., Foster, R. B., Núñez, P., Aguilar, S., Valencia, R., Villa, G., Muller-Landau, H. C., Losos, E., and Hubbell, S. P. (2002). Beta-diversity in tropical forest trees. *Science*, **295**, 666–69.

Connell, J. H. (1971). On the role of natural enemies in preventing competitive exclusion in some marine animals and in rain forest trees. In P. J. den Boer and G. R. Gradwell, eds. *Dynamics of Populations*, pp. 298–310. Center for Agricultural Publishing and Documentation, Wageningen.

Corlett, R. T. (1996). Characteristics of vertebrate-dispersed fruits in Hong Kong. *Journal of Tropical Ecology*, **12**, 819–33.

Cory, V. L. (1927). Activities of livestock on the range. *Texas Agricultural Experiment Station Bulletin*, No. **367**, 47 pp.

Cousens, R. and Mortimer, M. (1995). *Dynamics of Weed Populations*. Cambridge University Press, Cambridge.

Cousens, R. and Moss, S. R. (1990). A model of the effects of cultivation on the vertical distribution of weed seeds within the soil. *Weed Research*, **30**, 61–70.

Cousens, R., Doyle, C. J., Wilson, B. J., and Cussans, G. W. (1986). Modelling the economics of controlling *Avena fatua* in winter wheat. *Pesticide Science*, **17**, 1–12.

Cousens, R. D. and Rawlinson, A. A. (2001). When will plant morphology affect the shape of a seed dispersal kernel? *Journal of Theoretical Biology*, **211**, 229–38.

Cousens, R. D., Dale, M. R. T., Taylor, J., Law, R., Moerkerk, M., and Kembel, S. W. (2006). Interpretation of causes of pattern in communities where environmental change is rapid and species longevity is short. *Journal of Vegetation Science*, **17**, 509–24.

Creel, S., Winnie, J. A., Maxwell, B., Hamlin, K., and Creel, M. (2005). Elk alter habitat selection as an antipredator response to wolves. *Ecology*, **86**, 3387–97.

Cremer, K. W. (1966). Dissemination of seed from *Eucalyptus regnans*. *Australian Forestry*, **30**, 33–37.

Cussans, G. W. and Marshall, E. J. P. (1990). Weed ecology and management. In J. Palmer and J. Abbott, eds. *Institute of Arable Crops Research Report for 1990*, pp.51–56, Institute of Arable Crops Research Institute, Rothamsted.

Cwynar, L. C. and MacDonald, G. M. (1987). Geographical variation of lodgepole pine in relation to population history. *American Naturalist*, **129**, 463–69.

Dammer, P. (1892). Polygonaceenstudien. I. Die Verbreitungsausrustung. *Englers Botanische Jahrbuch*, **15**, 260–85.

Darley-Hill, S. and Johnson, W. C. (1981). Acorn dispersal by blue jays (*Cyanocitta cristata*). *Oecologia*, **50**, 231–32.

Darwin, C. (1859). *On the Origin of Species by Means of Natural Selection*. Murray, London.

Debski, I., Burslem, D. F. R. P., Palmiotto, P. A., LaFrankie, J. V., Lee, H. S., and Manokaran, N. (2002). Habitat preferences of *Aporosa* in two Malaysian forests: implications for abundance and coexistence. *Ecology*, **83**, 2005–18.

Debussche, M. and Isenmann, P. (1989). Fleshy fruit characters and the choices of bird and mammal seed dispersers in a Mediterranean region. *Oikos*, **56**, 327–38.

Debussche, M., Cortez, J., and Rimbault, I. (1987). Variation in fleshy fruit composition in the Mediterranean region: The importance of ripening season, life-form, fruit type and geographical distribution. *Oikos*, **49**, 244–252.

Delacourt, P. and Delacourt, H. (1987). *Long-term Forest Dynamics of the Termperate Zone*. Springer, New York.

Den Boer, P. J. (1968). Spreading of risk and stabilization of animal numbers. *Acta Biotheoretica*, **18**, 165–94.

Dennis, A. J. and Westcott, D. A. (2007). Estimating dispersal kernels produced by a diverse community of vertebrates. In A. J. Dennis, E. W. Schupp, R. J. Green, and D. A. Westcott, eds. *Seed Dispersal: Theory and Its Application in a Changing World*, CAB International, Wallingford.

Dieckmann, U. and Law, R. (2000). Relaxation projections and the method of moments. In U. Dieckmann, R. Law, and J. A. J. Metz *The Geometry of Ecological Interactions: Simplifying Spatial Complexity*, pp. 412–55. Cambridge University Press, Cambridge.

Diggle, P. J. (2003). *Statistical Analysis of Spatial Point Patterns*. Arnold, London.

Dinerstein, E. (1986). Reproductive ecology of fruit bats and the seasonality of fruit production in a Costa Rican cloud forest. *Biotropica*, **18**, 307–18.

Dingler, H. (1889). *Die Bewegung der pflanzlichen flugorgane*. Theodor Ackerman, Munich.

Dobzhansky, T. (1973). Nothing in biology makes sense except in the light of evolution. *The American Biology Teacher*, **35**, 125–29.

Dong, M., Lu, J. Z., Yhang, W. J., Chen, J. K., and Li, B. (2006). Canada goldenrod (*Solidago canadensis*) an invasive alien weed spreading rapidly in China. *Acta Phytotaxonomica Sinica*, **44**, 72–85.

Donohue, K. (1998). Maternal determinants of seed dispersal in *Cakile edentula*: fruit, plant, and site traits. *Ecology*, **79**, 2771–88.

Donohue, K. (1999). Seed dispersal as a maternally influenced character: Mechanistic basis of maternal effects and selection on maternal characters in an annual plant. *American Naturalist*, **154**, 674–89.

Dow, B. D. and Ashley, M. V. (1996). Microsatellite analysis of seed dispersal and parentage of saplings in bur oak, *Quercus macrophylla*. *Molecular Ecology*, **5**, 615–27.

Dytham, C. (2003). How landscapes affect the evolution of dispersal behaviour in reef fishes: results from an individual-based model. *Journal of Fish Biology*, **63**, 213–25.

Edwards, W., Dunlop, M., and Rodgerson, L. (2006). The evolution of rewards: seed dispersal, seed size and elaiosome size. *Journal of Ecology*, **94**, 687–94.

Ehrlén, J. and Eriksson, O. (2003). Large-scale spatial dynamics of plants: a response to Freckleton and Watkinson. *Journal of Ecology*, **91**, 316–20.

Elbaum, R., Zalzman, L., Burget, I., and Fratzl, P. (2007). The role of wheat awns in the seed dispersal unit. *Science*, **316**, 884–86.

Ellner, S. P. (2001). Pair approximation for lattice models with multiple interaction scales. *Journal of Theoretical Biology*, **210**, 435–47.

Estrada, A. and Coates-Estrada, R. (1986). Frugivory by howling monkeys (*Alouatta palliata*) at Los Tuxtlas, Mexico: Dispersal and fate of seeds. In A. Estrada and T. H. Fleming, eds. *Frugivores and Seed Dispersal*, pp. 93–104. Junk, Dordrecht.

Estrada, A. and Coates-Estrada, R. (2001). Species composition and reproductive phenology of bats in a tropical landscape at Los Tuxtlas, Mexico. *Journal of Tropical Ecology*, **17**, 627–46.

Etheridge, A. M. (2004). Survival and extinction in a locally regulated population. *Annals of Applied Probability*, **16**, 188–214.

Felsenstein, J. (1985). Phylogenies and the comparative method. *American Naturalist*, **125**, 1–15.

Fenner, M. and Thompson, K. (2005). *The Ecology of Seeds*. Cambridge University Press, Cambridge.

Fernandez, R. J., Golluscio, R. A., Bisigato, A. J., and Soriano, A. (2002). Gap colonization in the Patagonian semidesert: seed bank and diaspore morphology. *Ecography*, **25**, 336–44.

Figuerola, J. and Green, A. J. (2002). Dispersal of aquatic organisms by waterbirds: a review of past research and priorities for future studies. *Freshwater Biology*, **47**, 483–94.

Finnigan, J. J. and Brunet, Y. (1995). Turbulent airflow in forests on flat and hilly terrain. In M. P. Coutts and J. Grace, eds, *Wind and Trees*, pp. 3–40. Cambridge University Press, Cambridge.

Fischer, S. F., Poschlod, P., and Beinlich, B. (1996). Experimental studies on the dispersal of plants and animals on sheep in calcareous grasslands. *Journal of Applied Ecology*, **33**, 1206–22.

Fisher, R. A. (1937). The wave of advance of advantageous genes. *Annals of Eugenics*, **7**, 355–369.

Fisher, R. A. (1941). Average excess and average effect of a gene substitution. *Annals of Eugenics*, **11**, 53–63.

Ford, H. (1985). Life history strategies in two co-existing agamospecies of dandelion. *Biological Journal of the Linnean Society*, **25**, 169–186.

Ford, R. H., Sharik, T. L., and Feret, P. P. (1983). Seed dispersal of the endangered Virginia round-leaf birch (*Betula uber*). *Forest Ecology and Management*, **6**, 115–128.

Fort, J. (2007). Fronts from complex two-dimensional dispersal kernels: Theory and application to Reid's paradox. *Journal of Applied Physics*, **101**, 094701, 7 pp.

Foster, B. L. (2001). Constraints on colonization and species richness along a grassland productivity gradient: the role of propagule availability. *Ecology Letters*, **4**, 530–35.

Foster, B. L. and Tilman, D. (2003). Seed limitation and the regulation of community structure in oak savanna grassland. *Journal of Ecology*, **91**, 999–1007.

Foster, S. A. and Janson, C. H. (1985). The relationship between seed size and establishment conditions in tropical woody plants. *Ecology*, **66**, 773–80.

Fox, J. F. (1982). Adaptation of gray squirrel behaviour to autumn germination by white oak acorns. *Evolution*, **36**, 800–809.

Frank, S. A. (1986). Dispersal polymorphisms in subdivided populations. *Journal of Theoretical Biology*, **122**, 303–9.

Frankie, G. W., Baker, H. G., and Opler, P. A. (1974). Comparative phenological studies of trees in tropical wet and dry forests in the lowlands of Costa Rica. *Journal of Ecology*, **62**, 881–919.

Freckleton, R. P. and Watkinson, A. R. (2003). Are all plant populations metapopulations? *Journal of Ecology*, **91**, 321–24.

French, K. (1992). Phenology of fleshy fruits in a wet sclerophyll forest in southeastern Australia: are birds an important influence? *Oecologia*, **90**, 366–73.

French, K. (1996). The gut passage rate of silvereyes and its effect on seed viability. *Corella*, **20**, 16–19.

French, K., O'Dowd, D. J., and Lill, A. (1992). Fruit removal of *Coprosma quadrifida* (Rubiaceae) by birds in southeastern Australia. *Australian Journal of Ecology*, **17**, 35–42.

Fuentes, M. (2004). Slight differences among individuals and the unified neutral theory of biodiversity. *Theoretical Population Biology*, **66**, 199–203.

Gadgil, M. (1971). Dispersal: population consequences and evolution. *Ecology*, **52**, 253–61.

Gage, E. A. and Cooper, D. J. (2005). Patterns of willow seed dispersal, seed entrapment, and seedling establishment in a heavily browsed montane riparian ecosystem. *Canadian Journal of Botany*, **83**, 678–87.

Gandon, S. (1999). Kin competition, the cost of inbreeding and the evolution of dispersal. *Journal of Theoretical Biology,* **200**, 345–64.

Gandon, S. and Michalakis, Y. (2001). Multiple causes of the evolution of dispersal. In J. Clobert, E. Danchin, A. A. Dhondt, and J. D. Nichols, eds. *Dispersal*, pp.155–167. Oxford University Press, Oxford.

Gandon, S. and Rousset, F, (1999). Evolution of stepping stone dispersal rates. *Proceedings of the Royal Society B,* **266**, 2507–13.

Garcia-Ramos, G. and Rodriguez, D. (2002). Evolutionary speed of species invasions. *Evolution*, **56**, 661–68.

Gardener, C. J., McIvor, J. G., and Jansen, A. (1993). Passage of legume and grass seeds through the digestive tract of cattle and their survival in faeces. *Journal of Applied Ecology*, **30**, 63–74.

Gardner, R. H., Milne, B. T., Turner, M. G., and O'Neill, R. V. (1987). Neutral models for the analysis of broad-scale landscape patterns. *Landscape Ecology*, **1**, 19–28.

Garrison, W. J., Miller, G. L., and Raspet, R. (2000). Ballistic seed projection in two herbaceous species. *American Journal of Botany*, **87**, 1257–64.

Gautier-Hion, A., Duplantier, J–M., Quris, R., Feer, F., Sourd. C., Decoux, J–P., Dubost, G., Emmons, L., Erard, C., Hecketsweiler, P., Moungazi, A., Roussilhon, C., and Thiollay, J–M. (1985). Fruit characters as a basis of fruit choice and seed dispersal in a tropical forest vertebrate community. *Oecologia* (Berlin), **65**, 324–37.

Geber, M. A. (1989). Interplay of morphology and development on size inequality: a Polygonum greenhouse study. *Ecological Monographs,* **59**, 267–88.

Geertsema, W. (2002). *Plant survival in dynamic habitat networks in agricultural landscapes*. PhD thesis, Wageningen University.

Geng, S. and Hills, F. J. (1978). A procedure for determining numbers of experimental and sampling units. *Agronomy Journal*, **70**, 441–44.

Giladi, I. (2006). Choosing benefits or partners: a review of the evidence for the evolution of myrmecochory. *Oikos*, **112**, 481–92.

Gill, G. S., Cousens, R. D., and Allan, M. R. (1996). Germination, growth, and development of herbicide resistant and susceptible populations of rigid ryegrass (*Lolium rigidum*). *Weed Science*, **44**, 252–56.

Gillespie, D. T. (1976). A general method for numerically simulating stochastic time evolution of coupled chemical reactions. *Journal of Computational Physics*, **22**, 403–34.

Gleeson, S. K. and Tilman, D. (1990). Allocation and the transient dynamics of succession on poor soils. *Ecology,* **71**, 1144–55.

Godoy, J. A. and Jordano, P. (2001). Seed dispersal by animals: exact identification of source trees with endocarp DNA microsatellites. *Molecular Ecology*, **10**, 2275–2283.

Gómez, C. and Espadaler, X. (1998). Myrmecochorous dispersal distances: a world survey. *Journal of Biogeography*, **25**, 573–80.

González-Andujar, J. L. and Perry, J. N. (1995). Models for the herbicidal control of the seed bank of *Avena sterilis*: the effects of spatial and temporal heterogeneity and of dispersal. *Journal of Applied Ecology*, **32**, 578–87.

Goodall, J. (1986). *The Chimpanzees of Gombe: Patterns of Behaviour*. Harvard University Press, Cambridge.

Gorb, S. N., Gorb, E. V., and Punttila, P. (2000). Effects of redispersal of seeds by ants on the vegetation pattern in a deciduous forest: a case study. *Acta Oecologica*, **21**, 293–301.

Gosper, C. R., Stansbury, C. D., and Vivian-Smith, G. (2005). Seed dispersal of fleshy-fruited invasive plants by birds: contributing factors and management options. *Diversity and Distributions*, **11**, 549–58.

Graae, B. J. (2002). The role of epizoochorous seed dispersal of forest plant species in a fragmented landscape. *Seed Science Research*, **12**, 113–21.

Grafen, A. (1989). The phylogenetic regression. *Philosophical Transactions of the Royal Society of London B. Biological Sciences*, **326**, 119–57.

Green, D. G. (1994). Connectivity and complexity in ecological systems. *Pacific Conservation Biology*, **1**, 194–200.

Greene, D. F. and Calogeropoulos, C. (2002). Measuring and modeling seed dispersal of terrestrial plants. In J. M. Bullock, R. E. Kenward, and R. S. Hails, eds. *Dispersal Ecology*, pp. 3–23. Blackwell Science, Oxford.

Greene, D. F. and Johnson, E. A. (1989). A model of wind dispersal of winged or plumed seeds. *Ecology*, **70**, 339–47.

Greene, D. F. and Johnson, E. A. (1992). Fruit abscission in *Acer saccharum* with reference to seed dispersal. *Canadian Journal of Botany*, **70,** 2277–83.

Greene, D. F. and Johnson, E. A. (1995). Long-distance wind dispersal of tree seeds. *Canadian Journal of Botany*, **73**, 1036–45.

Greene, D. F. and Johnson, E. A. (1996). Wind dispersal of seeds from a forest into a clearing. *Ecology*, **77**, 595–609.

Greene, D. F. and Johnson, E. A. (1997). Secondary dispersal of tree seeds on snow. *Journal of Ecology*, **85**, 329–40.

Greene, D. G., Canham, C., Coates, K. D., and Lepage, P. T. (2004). An evaluation of alternative dispersal functions for trees. *Journal of Ecology*, **92**, 75—66.

Gregory, P. H. (1973). *The Microbiology of the Atmosphere*. Leonard Hill, Aylesbury.

Grimm, V. and Railsback, S. F. (2005) *Individual-Based Modelling and Ecology*. Princeton University Press, Princeton.

Gros, A., Poethke, H. J., and Hovestadt, T. (2006). Evolution of local adaptations in dispersal strategies. *Oikos*, **114**, 544–52.

Groves, R. H. (1998). *Recent Incursions of Weeds to Australia 1971–1995*. Cooperative Research Centre for Weed Management Systems, Technical Series No. 3. Adelaide, South Australia.

Guillaume, F. and Perrin, N. (2006). Joint evolution of dispersal and inbreeding load. *Genetics*, **173**, 497–509.

Hadeler, K. P. and Rothe, F. (1975) Travelling fronts in nonlinear diffusion equations. *Journal of Mathematical Biology*, **2**, 251–63.

Hadj-Chikh, L. Z., Steele, M. A., and Smallwood, P. D. (1996). Caching decisions by grey squirrels: a test of the handling time and perishability hypotheses. *Animal Behaviour*, **52**, 941–48.

Haldane, J. B. S. (1957). The cost of natural selection. *Journal of Genetics*, **55**, 511–24.

Hallé, F., Oldeman, R. A. A., and Tomlinson, P. B. (1978). *Tropical Trees and Forests: An Architectural Analysis*. Springer-Verlag, Berlin.

Hallwachs, W. (1986). Agoutis (*Dasyprocta punctata*): The inheritors of guapinol (*Hymenaea courbaril*: Leguminosae). In A. Estrada and T. H. Fleming, eds. *Frugivores and Seed Dispersal*, pp. 285–304. Junk, Dordrecht.

Hamilton, W. D. (1964). The genetical evolution of social behaviour. I. *Journal of Theoretical Biology*, **7**, 1–16.

Hamilton, W. D. and May, R. M. (1977). Dispersal in stable habitats. *Nature*, **269**, 578–81.

Hampe, A. (2004). Extensive hydrochory uncouples spatiotemporal patterns of seedfall and seedling recruitment in a 'bird-dispersed' riparian tree. *Journal of Ecology*, **92**, 797–807.

Hampe, A. and Bairlein, F. (2000). Modified dispersal-related traits in disjunct populations of bird-dispersed *Frangula alnus* (Rhamnaceae): a result of its Quaternary distribution shifts? *Ecography*, **23**, 603–13.

Hancock, J. (1953). Grazing behaviour of cattle. *Animal Breeding Abstracts*, **21**, 1–13.

Hanski, I. A. and Gilpin, M. E. (1997). *Metapopulation Biology*. Academic Press, San Diego.

Hanzawa, F. M., Beattie, A. J., and Culver, D. C. (1988). Directed dispersal – demographic analysis of an ant-seed mutualism. *American Naturalist*, **131**, 1–13.

Hard, J. S. (1964). Vertical distribution of cones in red pine. *Research Note LS-51, Lake States Forest Experiment Station, U. S. Department of Agriculture*.

Hardin, G. (1960). The competitive exclusion principle. *Science*, **131**, 1292–97.

Hardy, O. J., Gonzalez-Martinez, S. C., Colas, B., Freville, H., Mignot, A., and Olivieri, I. (2004). Fine-scale genetic structure and gene dispersal in *Centaurea corymbosa* (Asteraceae) II. Correlated paternity within and among sibships. *Genetics*, **168**, 1601–14.

Harlow, W. M., Cote, W. A., and Day, A. C. (1964). The opening mechanism of pine cone scales. *Journal of Forestry*, **62**, 538–40.

Harms, K. E., Wright, S. J., Calderón, O., Hernández, A., and Herre, E. A. (2000). Pervasive density-dependent recruitment enhances seedling diversity in a tropical forest. *Nature*, **404**, 493–95.

Harper, J. L. (1977). *Population Biology of Plants*. Academic Press, London.

Harrison, S. (1997). How natural habitat patchiness affects the distribution of diversity in Californian serpentine chaparral. *Ecology*, **78**, 1898–906.

Harvey, P. H. and Pagel, M. D. (1991). *The Comparative Method in Evolutionary Biology*. Oxford series in Ecology and Evolution, Volume 1. Oxford University Press, Oxford.

Hassell, M. P. (1975). Density-dependence in single-species populations. *Journal of Animal Ecology*, **44**, 283–95.

Hastings, A. (1980). Disturbance, coexistence, history and competition for space. *Theoretical Population Biology*, **163**, 491–504.

He, T. H., Krauss, S. L., Lamont, B. B., Miller, B. P., and Enright, N. J. (2004). Long-distance see dispersal in a metapopulation of *Banksia hookeriana* inferred from a population allocation analysis of amplified fragment length polymorphism data. *Molecular Ecology*, **13**, 1099–109.

Hegde, S. G., Shaanker, R.U., and Ganeshaiah, K. N. (1991). Evolution of seed size in the bird-dispersed tree *Santalum album* L.: A trade off between seedling establishment and dispersal efficiency. *Evolutionary Trends in Plants*, **5**, 131–35.

Hengeveld, R. (1989). *Dynamics of Biological Invasions*. Chapman and Hall, London.

Hensen, I. and Müller, C. (1997). Experimental and structural investigations of anemochorous dispersal. *Plant Ecology*, **133**, 169–80.

Herrera, C. M. (1984). A study of avian frugivores, bird-dispersed plants, and their interaction in Mediterranean scrublands. *Ecological Monographs*, **54**, 1–23.

Herrera, C. M. (1987). Vertebrate-dispersed plants of the Iberian peninsula: A study of fruit characteristics. *Ecological Monographs*, **57**, 305–31.

Hiebeler, D. (2000). Populations on fragmented landscapes with spatially structured heterogeneities: landscape generation and local dispersal. *Ecology*, **81**, 1629–41.

Hiebeler, D. (2004). Competition between near and far dispersers in spatially-structured habitats. *Theoretical Population Biology*, **66**, 205–18.

Higgins, S. I. and Cain, M. L. (2002). Spatially realistic plant metapopulation models and the colonization-competition trade-off. *Journal of Ecology*, **90**, 616—26.

Higgins, S. I. and Richardson, D. M. (1999). Predicting plant migration rates in a changing world: the role of long-distance dispersal. *American Naturalist*, **153**, 464–75.

Higgins, S. I., Richardson, D. M., and Cowling, R. M. (1996). Modeling invasive plant spread: the role of plant-environment interactions and model structure. *Ecology*, **77**, 2043–54.

Higgins, S. I., Clark, J. S., Nathan, R., Hovestadt, T., Schurr, F., Fragoso, J. M. V., Aguiar, M. R., Ribbens, E., and Lavorel, S. (2003a). Forecasting plant migration rates: managing uncertainty for risk assessment. *Journal of Ecology*, **91**, 341–47.

Higgins, S. I., Nathan, R., and Cain, M. L. (2003b). Are long-distance dispersal events in plants usually caused by nonstandard means of dispersal? *Ecology*, **84**, 1945–56.

Higgins, S. I., Lavorel, S., and Revilla, E. (2003c). Estimating plant migration rates under habitat loss and fragmentation. *Oikos*, **101**, 354–66.

Hilty, S. L. (1980). Flowering and fruiting periodicity in a premontane rain forest in Pacific Colombia. *Biotropica*, **12**, 292–306.

Hinds, T. E., Hawksworth, F. G., and McGinnies, W. J. (1963). Seed discharge in *Arceuthobium*: A photographic study. *Science*, **140**, 1236–38.

Hinds, T. E., Hawksworth, F. G., and McGinnies, W. J. (1965). Seed dispersal velocity in four dwarfmistletoes. *Science*, **148**, 517–19.

Hocking, P. J. and Liddle, M. J. (1986). The biology of Australian weeds: 15. *Xanthium occidentale* Bertol. complex and *Xanthium spinosum* L. *Journal of the Australian Institute of Agricultural Science*, **52**, 191–221.

Hodgson, J. G. and Grime, J. P. (1990).The role of dispersal mechanisms, regenerative strategies and seed banks in the vegetation dynamics of the British landscape. In R. G. H. Bunce and D. C. Howard, eds., *Species Dispersal in Agricultural Habitats*, pp. 65–81. Belhaven Press, London.

Holbrook, K. M. and Smith, T. B. (2000). Seed dispersal and movement patterns in two species of *Ceratogymna* hornbills in a West African tropical lowland forest. *Oecologia*, **125**, 249–57.

Holmes, E. E. (1993). Are diffusion models too simple? A comparison with telegraph models of invasion. *American Naturalist*, **142**, 779–95.

Holmes, E. E. and Wilson, H. B. (1998). Running from trouble: long-distance dispersal and the competitive coexistence of inferior species. *American Naturalist*, **151**, 578–86.

Holmes, E. E., Lewis, M. A., Banks, J, E., and Veit, R. R. (1994). Partial differential equations in ecology: spatial interactions and population dynamics. *Ecology*, **75**, 17–29.

Holt, R. D. and McPeek, M. A. (1996). Chaotic population dynamics favors the evolution of dispersal. *American Naturalist*,**148**, 709–18.

Holthuijzen, A. M. A., Sharik, T. L., and Fraser, J. D. (1987). Dispersal of eastern red cedar (*Juniperus virginiana*) into pastures: an overview. *Canadian Journal of Botany*, **65**, 1092–95.

Holyoak, M., Leibold, M. A., and Holt, R. D. (eds) (2005) *Metacommunities: Spatial Dynamics and Ecological Communities.* Chicago University Press, Chicago.

Hood, L. A., Swaine, M. D., and Mason, P. A. (2004). The influence of spatial patterns of damping-off disease and arbuscular mycorrhizal colonization on tree seedling establishment in Ghanaian tropical forest soil. *Journal of Ecology*, **92**, 816–23.

Hoppes, W. G. (1988). Seedfall pattern of several species of bird-dispersed plants in an Illinois woodland. *Ecology*, **69**, 320–29.

Horn, H. S., Nathan, R., and Kaplan, S. R. (2001). Long-distance dispersal of tree seeds by wind. *Ecological Research*, **16**, 877–85.

Hovestadt, T. and Poethke, H. J. (2006). The control of emigration and its consequences for the survival of populations. *Ecological Modelling*, **190**, 443–53.

Hovestadt, T., Messner, S., and Poethke, H. J. (2001). Evolution of reduced dispersal mortality and 'fat tailed' dispersal kernels in autocorrelated landscapes. *Proceedings of the Royal Society B*, **268**, 385–91.

Howard, C. L., Mortimer, A. M., Gould, P., Putwain, P. D., Cousens, R., and Cussans, G. W. (1991). The dispersal of weeds: seed movement in arable agriculture. *Proceedings of the Brighton Crop Protection Conference – Weeds – 1991*, 821–28, British Crop Protection Council, Farnham.

Howe, H. F. (1982). Fruit production and animal activity in two trees. In E. G. Leigh, A. S. Rand and D. M. Windsor, eds. *The Ecology of a Tropical Forest: Seasonal Rhythms and Long-term Changes*, pp. 189–200. Smithsonian Institution Press, Washington D.C.

Howe, H. F. (1986). Seed dispersal by fruit-eating birds and mammals. In D. R. Murray ed. *Seed Dispersal*, pp. 123–89. Academic Press, Sydney.

Howe, H. F. and Smallwood, J. (1982). Ecology of seed dispersal. *Annual Reviews of Ecology and Systematics*, **13**, 201–28.

Howe, H. F., Schupp, E. W., and Westley, L. C. (1985). Early consequences of seed dispersal for a neotropical tree (*Virola surinamensis*). *Ecology*, **66**, 781–91.

Hubbell, S. P. (1979). Tree dispersion, abundance, and diversity in a tropical dry forest. *Science*, **203**, 1299–309.

Hubbell, S. P. (1980). Seed predation and the coexistence of tree species in tropical forests. *Oikos*, **35**, 214–29.

Hubbell, S. P. (2001). *The Unified Neutral Theory of Biodiversity and Biogeography*. Princeton University Press, Princeton.

Hughes, L. and Westoby, M. (1992). Effect of diaspore characteristics on removal of seeds adapted for dispersal by ants. *Ecology*, **73**, 1300–12.

Hughes, L., Dunlop, M., French, K., Leishman, M. R., Rice, B., Rodgerson, L., and Westoby, M. (1994). Predicting dispersal spectra: a minimal set of hypotheses based on plant attributes. *Journal of Ecology*, **82**, 933–50.

Humston, R., Mortensen, D. A., and Bjornstad, O. N. (2005). Anthropogenic forcing on the spatial dynamics of an agricultural weed: the case of the common sunflower. *Journal of Applied Ecology*, **42**, 863–72.

Hutchings, M. J. and Bradbury, I. K. (1986). Ecological perspectives on clonal perennial herbs: When does it pay for individual clone modules to be functionally integrated or independent? *BioScience*, **36**, 178–82.

Hutchings, M. J. and Wijesinghe, D. K. (1997). Patchy habitats, division of labour and growth dividends in clonal plants. *Trends in Ecology and Evolution*, **12**, 390–94.

Ibrahim, K. M., Nichols, R. A., and Hewitt, G. M. (1996) Spatial patterns of genetic variation generated by different forms of dispersal during range expansion. *Heredity*, **77**, 282–91.

Imbert, E. and Ronce, O. (2001). Phenotypic plasticity for dispersal ability in the seed heteromorphic *Crepis sancta* (Asteraceae). *Oikos*, **93**, 126–34.

Jacobs, L. F. (1992). The effect of handling time on the decision to cache by grey squirrels. *Animal Behaviour*, **43**, 522–24.

Janson, C, H. (1983). Adaptation of fruit morphology to dispersal agents in a neotropical forest. *Science*, **219**, 187–89.

Janzen, D. H. (1970). Herbivores and the number of tree species in tropical forests. *American Naturalist*, **104**, 501–28.

Janzen, D. H. (1984). Dispersal of small seeds by big herbivores: foliage is the fruit. *American Naturalist*, **123**, 338–53.

Janzen, D. H. and Martin, P. S. (1982). Neotropical anachronisms: The fruits the gomphotheres ate. *Science*, **215**, 19–27.

Jarne, P. and Charlesworth, D. (1993). The evolution of the selfing rate in functionally hermaphrodite plants and animals. *Annual Review of Ecology and Systematics*, **24**, 441–46.

Johannesson, K. (1988). The paradox of Rockall: why is a brooding gastropod (*Littorina saxatilis*) more widespread than one having planktonic larval dispersal stage (*L. littorea*)? *Marine Biology*, **99**, 507–23.

Johnson, C. K. and West, N. E. (1988). Laboratory comparisons of five seed-trap designs for dry, windy environments. *Canadian Journal of Botany*, **66**, 346–48.

Johnson, E. A. and Fryer, G. I. (1992). Physical characterization of seed microsites – movement on the ground. *Journal of Ecology*, **80**, 823–36.

Johst, K. and Brandl, R. (1997). Evolution of dispersal: the importance of temporal order of reproduction and dispersal. *Proceedings of the Royal Society B*, **264**, 23–30.

Jones, F. A., Chen, J., Weng, G-J., and Hubbell, S. P. (2005). A genetic evatuation of seed dispersal in the neotropical tree *Jacaranda copaia* (Bignoniaceae). *American Naturalist*, **166**, 543–55.

Jongejans, E. and Schippers, P. (1999). Modeling seed dispersal by wind in herbaceous species. *Oikos*, **87**, 362–72.

Jongejans, E. and Telenius, A. (2001). Field experiments on seed dispersal by wind in ten umbelliferous species (Apiaceae). *Plant Ecology*, **152**, 67–78.

Jordano, P. (1992). Fruits and frugivory. In M. Fenner ed. *Seeds: The Ecology of Regeneration in Plant Communities*, pp. 105–56. CAB International, Wallingford.

Jordano, P. (1995). Frugivore-mediated selection on fruit and seed size: birds and St. Lucie's cherry, *Prunus mahaleb*. *Ecology*, **76**, 2627–39.

Jordano, P. and Godoy, J. A. (2002). Frugivore-generated seed shadows: a landscape view of demographic and genetic effects. In D. J. Levey, W. R. Silva, and M. Galetti, eds. *Seed Dispersal and Frugivory: Ecology, Evolution and Conservation*, pp. 305–21. CAB International, Wallingford.

Jordano, P. and Schupp, E. W. (2000). Seed disperser effectiveness: the quantity component and patterns of seed rain for *Prunus mahaleb*. *Ecological Monographs*, **70**, 591–615.

Jordano, P., Garcia, C., Godoy, J. A., and Garcia-Castano, J. L. (2007). Differential contribution of frugivores to complex seed dispersal patterns. *PNAS*, **104**, 3278–82.

Kallimanis, A. S., Kunin, W. E., Halley, J. M., and Sgardelis, S. P. (2006). Patchy disturbance favours longer dispersal distance. *Evolutionary Ecology Research*, **8**, 529–41.

Kaplin, B. A. and Lambert, J. E. (2002). Effectiveness of seed dispersal by *Cercopithecus* monkeys: Implications for seed input into degraded areas. In D. J. Levey, W, R, Silva, and M. Galetti, eds. *Seed Dispersal and Frugivory: Ecology, Evolution and Conservation*, pp. 351–64. CAB International, Wallingford.

Kareiva, P. M. and Shigesada, N. (1983). Analyzing insect movement as a correlated random walk. *Oecologia*, **56**, 234–38.

Katul, G. G., Porporato, A., Nathan, R., Siqueira, M., Soons, M. B., Poggi, D., Horn, H. S. and Levin, S, A. (2005). Mechanistic analytical models for long-distance seed dispersal by wind. *American Naturalist*, **166**, 368–81.

Keddy, P. A. (1980). Population ecology in an environmental mosaic: *Cakile edentula* on a gravel bar. *Canadian Journal of Botany*, **58**, 1095–100.

Kelley, A. D. and Bruns, V. F. (1975). Dissemination of weed seeds by irrigation water. *Weed Science*, **23**, 486–93.

Kelly, C. K. and Purvis A (1993). Seed size and establishment conditions in tropical trees: on the use of taxonomic relatedness in determining ecological patterns. *Oecologia*, **94**, 356–60.

Kiers, E. T. and van der Heijden, M. G. A. (2006). Mutualistic stability in the arbuscular mycorrhizal symbiosis: exploring hypotheses of evolutionary cooperation. *Ecology*, **87**, 1627–36.

Kinezaki, N., Kawasaki, K., Takasu, F., and Shigesada, N. (2003). Modeling biological invasions into periodically fragmented environments. *Theoretical Population Biology*, **64**, 291–302.

Kinlan, B. P. and Gaines, S. D. (2003). Propagule dispersal in marine and terrestrial environments: A community perspective. *Ecology*, **84**, 2007–20.

Kinlan, B. P., Gaines, S. D., and Lester, S. E. (2005). Propagule dispersal and the scales of marine process. *Diversity and Distributions*, **11**, 139–48.

Kira, T., Ogawa, H., and Shinozaki, K. (1953). Intraspecific competition among higher plants. 1. Competition-density-yield inter-relationships in regularly dispersed populations. *Journal of the Polytechnic Institute, Osaka City University D*, **4**, 1–16.

Kisdi, É. (2002). Dispersal: risk spreading versus local adaptation. *American Naturalist*, **159**, 579–96.

Kisdi, É (2004). Conditional dispersal under kin competition: extension of the Hamilton/May model to brood size-dependent dispersal. *Theoretical Population Biology*, **66**, 369–80.

Kisdi, É. and Geritz, S. A. H (2003). On the coexistence of perennial plants by the competition-colonization trade-off. *American Naturalist*, **161**, 350—54.

Knight, R. S. and Siegfried, W. R. (1983). Inter-relationships between type, size and colour of fruits and dispersal in southern African trees. *Oecologia*, **56**, 405–12.

Ko, I. W. P., Corlett, R. T., and Xu, R-J. (1998). Sugar composition of wild fruits in Hong Kong, China. *Journal of Tropical Ecology*, **14**, 381–87.

Kobayashi, Y. and Yamamura, N. (2000). Evolution of seed dormancy due to sib competition: effect of dispersal and inbreeding. *Journal of Theoretical Biology*, **202**, 11–24.

Kokko, H. and Ots, I. (2006). When not to avoid inbreeding. *Evolution*, **60**, 467–75.

Kolmogoroff, A., Petrovsky, I., and Piscounoff, N. (1937). Étude de l'équation de la diffusion avec croissance de la quantité de matière et son application à une problème biologique. *Moscow University Bulletin Series International Section A*, **1**, 1–25.

Kot, M. (1992). Discrete-time travelling waves: Ecological examples. *Journal of Mathematical Biology*, **30**, 413–36.

Kot, M. and Schaffer, W. M. (1986). Discrete-time growth-dispersal models. *Mathematical Biosciences*, **80**, 109–36.

Kot, M., Lewis, M. A., and van den Driessche, P. (1996). Dispersal data and the spread of invading organisms. *Ecology*, **77**, 2027–42.

Koyama, H. and Kira, T. (1956). Intraspecific competition among higher plants VIII. Frequency distribution of individual plant weight as affected by the interaction between plants. *Osaka Shiritsu Daigaku, Institute of Polytechnics, Journal series D*, **7**, 73–94.

Krebs, C. J. (1985). *Ecology: The Experimental Analysis of Distribution and Abundance.* 3rd Edition. Harper and Row, New York.

Kubitzki, K. and Ziburski, A. (1994). Seed dispersal in flood plain forests of Amazonia. *Biotropica*, **26**, 30–43.

Kun, A. and Scheuring, I. (2006). The evolution of density-dependent dispersal in a noisy spatial population model. *Oikos*, **115**, 308–20.

Lacap, C. D. A., Vermaat, J. E., Rollon, R. N., and Nacorda, H. M. (2002). Propagule dispersal of the SE Asian seagrasses *Enhalus acoroides* and *Thalassia hemprichii. Marine Ecology Progress Series*, **235**, 75–80.

Lande, R. and Schemske, W. (1985). The evolution of self-fertilization and inbreeding depression in plants. I. Genetic models. *Evolution*, **39**, 25–40.

Lang, D. C. (1987). *The Complete Book of British Berries.* Threshold Books, London.

Lanza, J., Schmitt, M. A., and Awad, A. B. (1992). Comparative chemistry of elaiosomes of three species of *Trillium. Journal of Chemical Ecology*, **18**, 209–21.

Latimer, A. M., Silander, J. A., and Cowling, R. M. (2005). Neutral ecological theory reveals isolation and rapid speciation in a biodiversity hot spot. *Science*, **309**, 1722–25.

Lavergne, S. and Molofsky, J. (2007). Increased genetic variation and evolutionary potential drive the success of an invasive grass. *Proceedings of the National Academy of Science*, **104**, 3883–88.

Law, R. and Dieckmann, U. (2000). A dynamical system for neighbourhoods in plant communities. *Ecology*, **81**, 2137–48.

Law, R., Murrell, D. J., and Dieckmann, U. (2003). Population growth in space and time: spatial logistic equations. *Ecology*, **84**, 252–62.

Le Corre, V., Machon, N., Petit, R. J., and Kremer, A. (1997). Colonization with long-distance seed dispersal and genetic structure of maternally inherited genes in forest trees: a simulation study. *Genetical Research, Cambridge*, **69**, 117–25.

Leake, J. R. (2005). Plants parasitic on fungi: unearthing the fungi in myco-heterotrophs and debunking the 'saprophytic' plant myth. *Mycologist*, **19**, 113–22.

Levey, D. J. (1986). Methods of seed processing by birds and seed deposition patterns. In A. Estrada and T. H. Fleming, eds. *Frugivores and Seed Dispersal*, pp. 147–58. Junk, Dordrecht.

Levey, D. J. (1987). Seed size and fruit-handling techniques of avian frugivores. *American Naturalist*, **129**, 471–85.

Levey, D. J. and Cipollini, M. L. (1998). A glycoalkaloid in ripe fruit deters consumption by Cedar Waxwings. *The Auk*, **115**, 359–67.

Levey, D. J. and del Rio, C. M. (2001). It takes guts (and more) to eat fruit: lessons from avian nutritional ecology. *The Auk*, **118**, 819–31.

Levin, D. A. and Kerster, H. W. (1968). Local gene dispersal in *Phlox*. *Evolution*, **22**, 130–39.

Levin, D. A. and Kerster, H. W. (1969). Density-dependent gene dispersal in *Liatris*. *American Naturalist*, **103**, 61–74.

Levin, S. A., Muller-Landau, H. C., Nathan, R., and Chave, J. (2003). The ecology and evolution of seed dispersal: A theoretical perspective. *Annual Reviews of Ecology, Evolution and Systematics*, **34**, 575–604.

Levine, J. M. and Murrell, D. J. (2003). The community-level consequences of seed dispersal patterns. *Annual Review of Ecology Evolution and Systematics*, **34**, 549–74.

Levins, R, (1969). Some demographic and genetic consequences of environmental heterogeneity for biological control. *Bulletin of the Entomological Society of America*, **15**, 237–40.

Lewis, J. R. (1964). *The Eology of Rocky Shores*. English Universities Press, London.

Lewis, M. A. (1997). Variability, patchiness, and jump dispersal in the spread of an invading population. In D. Tilman and P. Kareiva, eds. *Spatial Ecology: The Role of Space in Population Dynamics and Interspecific Interactions*, pp. 46–69. Princeton University Press, Princeton, New Jersey.

Lewis, M. A. and Kareiva, P. (1993). Allee dynamics and the spread of invading organisms. *Theoretical Population Biology*, **43**, 141–58.

Lewis, M. A. and Pacala, S. (2000). Modeling and analysis of stochastic invasion processes. *Journal of Mathematical Biology*, **41**, 387–429.

Lewis, M. A., Neubert, M. G., Caswell, H., Clark, J., and Shea, K. (2006) A guide to calculating discrete-time invasion rates from data. In M. W. Cadotte, S. M. McMahon, and T. Fukami, eds. *Conceptual Ecology and Invasion Biology: Reciprocal Approaches to Nature*, pp. 169–92. Springer, The Netherlands.

Liddle, M. J. and Elgar, M. A. (1984). Multiple pathways in diaspore dispersal, exemplified by studies of Noogoora Burr (*Xanthium occidentale* Bertol., Compositae). *Botanical Journal of the Linnean Society*, **88**, 303–15.

Livnat, A., Pacala, S. W., and Levin, S. A. (2005). The evolution of intergenerational discounting in offspring quality. *American Naturalist*, **165**, 311–21.

Lloyd, M. (1967). Mean crowding. *Journal of Animal Ecology*, **36**, 1–30.

Lonsdale, W. M. (1993). Rates of spread of an invading species – *Mimosa pigra* in northern Australia. *Journal of Ecology*, **81**, 513–21.

Lonsdale, W. M. and Lane, A. M. (1994). Tourist vehicles as vectors of weed seeds in Kakadu National Park, northern Australia. *Biological Conservation*, **69**, 277–83.

Lopez, O. R. (2001). Seed flotation and postflooding germination in tropical terra firme and seasonally flooded forest species. *Functional Ecology*, **15**, 763–71.

Lord, J. M., Markey, A. S., and Marshall, J. (2002). Have frugivores influenced the evolution of fruit traits in New Zealand? In D. J. Levey, W. R. Silva, and M. Galetti, eds. *Seed Dispersal and Frugivory: Ecology, Evolution and Conservation*, pp. 55–68. CAB International, Wallingford.

Loskutov, I. G. (2001). Influence of vernalization and photoperiod to the vegetation period of wild species of oats (*Avena* spp.). *Euphytica*, **117**, 125–31.

Mabry, C. M. (2004). The number and size of seeds in common versus restricted woodland herbaceous species in central Iowa, USA. *Oikos*, **107**, 497–504.

MacArthur, R. and Wilson, E. O. (1967). *The theory of island biogeography*. Princeton University Press, Princeton.

Mack, R. N. (1985). Invading plants: Their potential contribution to population biology. In J. White, ed. *Studies on Plant Demography: A Festschrift for John L. Harper*, pp. 127–42. Academic Press, New York.

Mack, R. N. and Lonsdale, W. M. (2001). Humans as global plant dispersers: Getting more than we bargained for. *BioScience*, **51**, 95–102.

Mahdi, A. and Law, R. (1987). On the spatial organization of plant species in a limestone grassland community. *Journal of Ecology*, **75**, 259–76.

Maier, A., Emig, W., and Leins, P. (1999). Dispersal patterns of some *Phyteuma* species (Campanulaceae). *Plant Biology*, **1**, 408–17.

Manzaneda, A. J., Fedriani, J. M., and Rey, P. J. (2005). Adaptive advantages of myrmecochory: the predator-avoidance hypothesis tested over a wide geographic range. *Ecography*, **28**, 583–92.

Mark, S. and Olesen, J. M. (1996). Importance of elaiosome size to removal of ant-dispersed seeds. *Oecologia*, **107**, 95–101.

Marshall, E. J. P. (1990). Interference between sown grasses and the growth of rhizome of *Elymus repens* (couch grass). *Agriculture, Ecosystems and Environment*, **33**, 11–22.

Marshall, E. J. P. and Butler, R. (1991). Seed rain patterns. *Institute of Arable Crops Research Report 1990*, p. 52.

Mathias, A., Kisdi, É., and Olivieri, I. (2001). Divergent evolution of dispersal in a heterogeneous landscape. *Evolution*, **55**, 246–59.

Matlack, G. R. (1987). Diaspore size, shape, and fall behaviour in wind-dispersed plant species. *American Journal of Botany*, **74**, 1150–60.

Matlack, G. R. (1989). Secondary dispersal of seed across snow in *Betula lenta*, a gap-colonizing tree species. *Journal of Ecology*, **77**, 853–69.

Maxwell, B. D. and Ghersa, C. (1992). The influence of weed seed dispersion versus the effect of competition on crop yield. *Weed Technology*, **6**, 196–204.

May, R. M. (1973). On relationships among various types of population models. *American Naturalist*, **107**, 46–57.

Mayhew, P. J. (2006). *Discovering Evolutionary Ecology. Bringing Together Ecology and Evolution*. Oxford University Press, Oxford.

Mazer, S. J. and Lowry, D. E. (2003). Environmental, genetic and seed mass effects on winged seed production in the heteromorphic *Spergularia marina* (Caryophyllaceae). *Functional Ecology*, **17**, 637–50.

McCanny, S. J. (1985). Alternatives in parent-offspring relationships in plants. *Oikos*, **45**, 148–49.

McCartney, H. A. (1990). Dispersal mechanisms through the air. In R. G. H Bunce and D. C. Howard, eds. *Species Dispersal in Agricultural Habitats*, pp. 133–58. Belhaven Press, London.

McConkey, K. R. and Drake, D. R. (2002). Extinct pigeons and declining bat populations: Are large seeds still being dispersed in the tropical Pacific. In D. J. Levey, W. R. Silva, and M. Galetti, eds. *Seed Dispersal and Frugivory: Ecology, Evolution and Conservation*, pp. 381–95. CAB International, Wallingford.

McConkey, K. R. and Drake, D. R. (2006). Flying foxes cease to function as seed dispersers long before they become rare. *Ecology*, **87**, 271–76.

McDonnell, M. J. and Stiles, E. W. (1983). The structural complexity of old field vegetation and the recruitment of bird-dispersed plant species. *Oecologia*, **56**, 109–16.

McEuen, A. B. and Curran, L. M. (2004). Seed dispersal and recruitment limitation across spatial scales in temperate forest fragments. *Ecology*, **85**, 507–18.

McEvoy, P. B. and Cox, C. S. (1987). Wind dispersal distances in dimorphic achenes of ragwort, *Senecio jacobaea*. *Ecology*, **68**, 2006–15.

McInerny, G., Travis, J. M. J., and Dytham, C. (2007). Range shifting on a fragmented landscape. *Ecological Informatics*, **2**, 1–8.

McPeek, M. A. and Holt, R. D. (1992). The evolution of dispersal in spatially and temporally varying environments. *American Naturalist*, **140**, 1010–27.

Mead, A., Grundy, A. C., Brain, P., and Marshall, E J. P. (2003). Development of a model for the joint horizontal and vertical movement of seeds following cultivation. *Aspects of Applied Biology 69, Seedbanks: Determination, Dynamics and Management*, pp. 179–86.

Meagher, T. R. and Thompson, E. (1987). Analysis of parentage for naturally established seedlings of *Chamaelirium luteum* (Liliaceae). *Ecology*, **68**, 803–12.

Mech, R. and Prusinkiewicz, P. (1996). Visual Models of Plants Interacting with Their Environment. Proceedings of SIGGRAPH 96 (New Orleans, Louisiana, August 4-9, 1996). In *Computer Graphics Proceedings, Annual Conference Series*, 1996, ACM SIGGRAPH, pp. 397–410, Addison-Wesley, Reading, Mass.

Medway, L. (1972). Phenology of a tropical rain forest in Malaya. *Biological Journal of the Linnean, Society*, **4**, 117–46.

Merritt, D. M. and Wohl, E. E. (2002). Processes governing hydrochory along rivers: Hydraulics, hydrology and dispersal phenology. *Ecological Applications*, **12**, 1071–87.

Michaels, H. J., Benner, B., Hartgerink, A. P., Lee, T. D., Rice, S., Willson, M. F., and Bertin, R. I. (1988). Seed size variation: magnitude, distribution, and ecological correlates. *Evolutionary Ecology*, **2**, 157–66.

Michaux, B. (1989). Reproductive and vegetative biology of *Cirsium vulgare* (Savi) Ten. (Compositae: Cynareae). *New Zealand Journal of Botany*, **27**, 401–14.

Michod, R. E. (1999). *Darwinian Dynamics. Evolutionary Transitions in Fitness and Individuality*. Princeton University Press, Princeton.

Mikich, S. B., Bianconi, G. V., Maia, B. H. L. N. S., and Teixeira, S. D. (2003). Attraction of the fruit-eating bat *Carollia perspicillata* to *Piper gaudichaudianum* essential oil. *Journal of Chemical Ecology*, **29**, 2379–83.

Milton, K. (1984). The role of food-processing factors in primate food choice. In P. Rodman and J. G. H. Cant, eds. *Adaptations for Foraging in Non-human Primates*, pp. 249–79. Columbia University Press, NY.

Minami, S and Azuma, A (2003). Various flying modes of wind-dispersed seeds. *Journal of Theoretical Biology*, **225**, 1–14.

Mix, C., Pico, F. X., van Groenendael, J. M., and Ouborg, N. J. (2006). Inbreeding and soil conditions affect dispersal and components of performance of two plant species in fragmented landscapes. *Basic and Applied Ecology*, 7, 59–69.

Moerkerk, M. R. (2002). Seed box survey of field crops in Victoria during 1996 and 1997. *Proceedings of the 13th Australian Weeds Conference, Perth*, pp. 55–58, Plant Protection Society of Western Australia, Perth.

Moermond, T. C. and Denslow, J. S. (1983). Fruit choice in neotropical birds: effects of fruit type and accessibility on selectivity. *Journal of Animal Ecology*, 52, 407–20.

Moermond, T. C., Denslow, J. S., Levey, D. J., and Santana, E. (1986). The influence of morphology on fruit choice in neotropical birds. In A, Estrada and T. H. Fleming, eds. *Frugivores and Seed Dispersal* pp. 137–46. Junk, Dordrecht.

Monjardino, M., Diggle, A., and Moore, J. (2004). What is the impact of harvesting technology on the spread of new weeds in cropping systems? In B. M. Sindel and S. B. Johnson, eds. *Papers and Proceedings, 14th Australian Weeds Conference, Weed Management: Balancing People, Planet, Profit*, pp. 580–83. Weed Society of New South Wales, Sydney.

Monteith, J. L. and Unsworth, M. H, (1990). *Principles of Environmental Physics*, 2nd edition. Edward Arnold, New York.

Moody, M. E. and Mack, R. N. (1988). Controlling the spread of plant invasions: The importance of nascent foci. *Journal of Applied Ecology*, 25, 1009–21.

Moorcroft, P. R., Pacala, S. W., and Lewis, M. A. (2006), Potential role of natural enemies during tree range expansions following climate change. *Journal of Theoretical Biology*, 241, 601–16.

Morales, J. M. and Carlo, T. A. (2006). The effects of plant distribution and frugivore density on the scale and shape of dispersal kernels. *Ecology*, 87, 1489–96.

Morgan, M. T. (2001). Consequences of life history for inbreeding depression and mating system evolution in plants. *Proceedings of the Royal Society B*, 268, 1817–24.

Morse, D. H. and Schmitt, J. (1985). Propagule size, dispersal ability, and seedling performance in *Asclepias syriaca*. *Oecologia* (Berlin), 67, 372–79.

Motro, U. (1982). Optimal rates of dispersal I. Haploid populations. *Theoretical Population Biology*, 21, 394–411.

Motro, U. (1983). Optimal rates of dispersal III. Parent-offspring conflict. *Theoretical Population Biology*, 23, 159–68.

Mouissie, A. M., Lengkeek, W., and van Diggelen, R. (2005). Estimating adhesive seed dispersal distances: field experiments and correlated random walks. *Functional Ecology*, 19, 478–86.

Mouquet, N. and Loreau, M. (2002). Coexistence in metacommunities: the regional similarity hypothesis. *American Naturalist*, 159, 420–26.

Mouquet, N. and Loreau, M. (2003). Community patterns in source-sink metacommunities. *American Naturalist*, 162, 544–57.

Murray, D. R. (1986). Seed dispersal by water. In D. R. Murray ed. *Seed Dispersal*, pp. 49–85. Academic Press, Sydney.

Murray, K. G. (1988). Avian seed dispersal of three neotropical gap-dependent plants. *Ecological Monographs*, 58, 271–98.

Murrell, D. J. and Law, R. (2003). Heteromyopia and the spatial coexistence of similar competitors. *Ecology Letters*, 6, 48–59.

Murrell, D. J., Dieckmann, U., and Law, R. (2004). On moment closures for population dynamics in continuous space. *Journal of Theoretical Biology*, 229, 421–32.

Nadeau, L. B. and King, J. R. (1991). Seed dispersal and seedling establishment of *Linaria vulgaris* Mill. *Canadian Journal of Plant Science*, 71, 771–82.

Nathan, R. (2005). Long-distance dispersal research: building a network of yellow brick roads. *Diversity and Distributions*, 11, 125–30.

Nathan, R. and Casagrandi, R. (2004). A simple mechanistic model of seed dispersal, predation and plant establishment: Janzen-Connell and beyond. *Journal of Ecology*, 92, 733–46.

Nathan, R. and Katul, G. G. (2005). Foliage shedding in deciduous forests lifts up long-distance seed dispersal by wind. *Proceedings of the National Academy of Sciences*, 102, 8251–56.

Nathan, R. and Muller-Landau, H. C. (2000). Spatial patterns of seed dispersal, their determinants and consequences for recruitment. *Trends in Ecology and Evolution*, 15, 278–85.

Nathan, R., Katul, G. G., Horn, H. S., Thomas, S. M., Oren, R., Avissar, R., Pacala, S. W., and Levin, S. A. (2002). Mechanisms of long-distance dispersal of seeds by wind. *Nature*, 418, 409–13.

Nathan, R., Safriel, U. N., Noy-Meir, I., and Schiller, G. (2000). Spatiotemporal variation in seed dispersal and recruitment near and far from *Pinus halepensis* trees. *Ecology*, 81, 2156–69.

Nathan, R., Safriel, U. N., and Noy-Meir, I. (2001). Field validation and sensitivity analysis of a mechanistic model for tree seed dispersal by wind. *Ecology*, 82, 374–88.

Neigel, J. E. (2002). Is F_{ST} obsolete? *Conservation Genetics*, 3, 167–73.

Neubert, M. G. and Caswell, H. (2000). Demography and dispersal: calculation and sensitivity analysis of invasion speed for structured populations. *Ecology*, 81, 1613–28.

Neubert, M. G. and Parker, I. M. (2004). Predicting rates of spread for invasive species. *Risk Analysis*, **24**, 817–31.

Niklas, K. J. (1994). *Plant Allometry: The Scaling of Form and Process*. Chicago University Press, Chicago.

Nilsson, C., Andersson, E., Merritt, D. M., and Johansson, M. E. (2002). Differences in riparian flora between riverbanks and river lakeshores explained by dispersal traits. *Ecology*, **83**, 2878–87.

North, A. and Ovaskainen, O. (2007). Interactions between dispersal, competition, and landscape heterogeneity. *Oikos*, **116**, 1106–19.

O'Dowd, D. J. and Gill, A. M. (1986). Seed dispersal syndromes in Australian *Acacia*. In D. R. Murray, ed. *Seed Dispersal*, pp. 87–121. Academic Press, Sydney.

O'Toole, J. J. and Cavers, P. B. (1983). Input to seed banks of proso millet (*Panicum miliaceum*) in southern Ontario. *Canadian Journal of Botany*, **63**, 1023–30.

Okasha, S. (2006). *Evolution and the Levels of Selection*. Oxford University Press, Oxford.

Okubo, A. and Levin, S. A. (1989). A theoretical framework for data analysis of wind dispersal of seeds and pollen. *Ecology*, **70**, 329–38.

Okubo, A., Ackerman, J. D., and Swaney, D. P. (2001). Passive diffusion in ecosystems. In A. Okubo and S. A. Levin, eds. *Diffusion and Ecological Problems: Modern Perspectives*, 2nd edition, pp. 31–106. Springer, New York.

Olivieri, I. and Gouyon, P. H. (1985). Seed dimorphism for dispersal: theory and implications. In J. Haeck and J. W. Woldendrop, eds. *Structure and Functioning of Plant Populations*, pp. 77–90. North Holland Publications, Amsterdam.

Olivieri, I., Swan, M., and Gouyon, P-H. (1983). Reproductive system and colonizing strategy of two species of *Carduus* (Compositae). *Oecologia* (Berlin), **60**, 114–17.

Oppenheimer, P. (1986). Real time design and animation of fractal plants and trees. *Computer Graphics*, **20**, 55–64.

Orth, R. J., Luckenbach, M., and Moore, K. A. (1994). Seed dispersal in a marine macrophyte: implications for colonization and restoration. *Ecology*, **75**, 1927–39.

Osada, N., Takeda, H., Furukawa, A., and Awang, M. (2001). Fruit dispersal of two dipterocarp species in a Malaysian rain forest. *Journal of Tropical Ecology*, **17**, 911–17.

Ouborg, N. J., Piquot, Y., and Van Groenendael, J. M. (1999). Population genetics, molecular markers and the study of dispersal in plants. *Journal of Ecology*, **87**, 551–68.

Ouborg, N. J., Vergeer, P., and Mix, C. (2006). The rough edges of the conservation genetics paradigm for plants. *Journal of Ecology*, **94**, 1233–48.

Ovaskainen, O. and Cornell, S. J. (2006). Space and stochasticity in population dynamics. *Proceedings of the National Academy of Sciences*, **103**, 12781–86.

Overton, J. M. (1996). Spatial autocorrelation and dispersal in mistletoes: field and simulation results. *Vegetatio*, **125**, 83–98.

Owen, M. (1980). *Wild Geese of the World*. Batsford, London.

Palumbi, S. R. (2003). Population genetics, demographic connectivity, and the design of marine reserves. *Ecological Applications*, **13**, S146–58.

Parciak W (2002). Environmental variation in seed number, size, and dispersal of a fleshy-fruited plant. *Ecology*, **83**, 780–93.

Pärtel, M. and Zobel, M. (2007). Dispersal limitation may result in the unimodal productivity-diversity relationship: a new explanation for a general pattern. *Journal of Ecology*, **95**, 90–94.

Peart, D. R. (1985). The quantitative representation of seed and pollen dispersal. *Ecology*, **66**, 1081–83.

Peres, C. A. (1994). Primate responses to phenological changes in the Amazonian terra firme forest. *Biotropica*, **26**, 98–112.

Peroni, P. A. (1994). Seed size and dispersal potential of *Acer rubrum* (Aceraceae) samaras produced by populations in early and late successional environments. *American Journal of Botany*, **81**, 1428–34.

Phillips, B. L., Brown, G. P., Webb, J. K., and Shine R. (2006). Invasion and the evolution of speed in toads. *Nature*, **439**, 803.

Pielaat, A., Lewis, M. A., Lele, S., and de-Camino-Beck, T. (2006). Sequential sampling designs for catching the tail of dispersal kernels. *Ecological Modelling*, **190**, 205–22

Pielou, E. C. (1977). *Mathematical Ecology*. Wiley, New York.

Piggin, C. M. (1978). Dispersal of *Echium plantagineum* L. by sheep. *Weed Research*, **18**, 155–60.

Pike, D. J. and Hasted, A. M. (1987). Experimental design and response surface analysis of pesticide trials. *Pesticide Science*, **19**, 297–307.

Piqueras, J., Klimeš, L., and Redbo-Torstensson, P. (1999). Modelling the morphological response to nutrient availability in the clonal plant *Trientalis europaea* L. *Plant Ecology*, **141**, 117–27.

Pitt, W. C., Box, P. W., and Knowlton, F. F. (2003). An individual-based model of canid populations: modelling territoriality and social structure. *Ecological Modelling*, **166**, 109–21.

Plank, J. E. van der (1963). *Plant Diseases: Epidemics and Control*. Academic Press, New York.

Platt, W. J. and Weis, M. I. (1977). Resource partitioning and competition within a guild of fugitive prairie plants. *American Naturalist*, **111**, 479–513.

Plotkin, J. B., Chave, J., and Ashton, P. S. (2002). Cluster analysis of spatial patterns in Malaysian tree species. *American Naturalist*, **160**, 629–44.

Poethke, H. J. and Hovestadt, T. (2002). Evolution of density and patch-size dependent dispersal rates. *Proceedings of the Royal Society B*, **269**, 637–45.

Poethke, H. J., Hovestadt, T., and Mitesser, O. (2003). Local extinction and the evolution of dispersal rates: causes and correlations. *American Naturalist*, **161**, 631–40.

Poethke, H. J., Pfenning, B., and Hovestadt, T. (2007). The relative contribution of individual and kin selection to the evolution of density-dependent dispersal rates. *Evolutionary Ecology Research*, **9**, 41–50.

Pond, K. R., Ellis, W. C., Matis, J. H., Ferreiro, H. M., and Sutton, J. D. (1988). Compartment models for estimating attributes of digesta flow in cattle. *British Journal of Nutrition*, **60**, 571–95.

Portnoy, S. and Willson, M. F. (1993). Seed dispersal curves: behaviour of the tail of the distribution. *Evolutionary Ecology*, **7**, 25–44.

Potthoff, M., Johst, K., Gutt, J., and Wissel, C. (2006). Clumped dispersal and species coexistence. *Ecological Modelling*, **198**, 247–54.

Poulsen, J. R., Clark, C. J., and Smith, T. B. (2001). Seed dispersal by a diurnal primate community in the Dja Reserve, Cameroon. *Journal of Tropical Ecology*, **17**, 787–808.

Price, M. V. and Jenkins, S. H. (1986). Rodents as seed consumers and dispersers. In D. R. Murray ed. *Seed Dispersal*, pp. 191–235. Academic Press, Sydney.

Prusinkiewicz, P. and Hanan, J. (1990). Visualization of botanical structures and processes using parametric L-systems. In D. Thaimann ed. *Scientific Visualization and Graphics Simulation*, pp. 183–201. John Wiley, New York.

Prusinkiewicz, P. and Lindenmayer, A. (1996). *The Algorithmic Beauty of Plants*. Springer, New York.

Quinn, J. A., Mowrey, D. P., Emanuele, S. M., and Whalley, R. D. B. (1994). The 'Foliage is the Fruit' hypothesis: *Buchloe dactyloides* (Poaceae) and the shortgrass prairie of North America. *American Journal of Botany*, **81**, 1545–54.

Rabinowitz, D., and Rapp, J. K. (1981). Dispersal abilities of seven sparse and common grasses from a Missouri prairie. *American Journal of Botany*, **68**, 616–24.

Raemaekers, J. (1980). Causes of variation between months in the distance traveled daily by gibbons. *Folia Primatologia*, **34**, 46–60.

Ramos-Fernández, G., Mateos, J. L., Miramontes, O., and Cocho, G. (2004) Lévy walk patterns in the foraging movements of spider monkeys (*Ateles geoffroyi*). *Behavioural Ecology and Sociobiology*, **55**, 223–30.

Randall, A. G. (1974). Seed dispersal into two spruce-fir clear-cuts in eastern Maine. *Research in the Life Sciences*, **21**, 1–15.

Ravigné, V., Olivieri, I., González-Martinez, S. C., and Rousset, F. (2006). Selective interactions between short-distance pollen and seed dispersal in self-compatible species. *Evolution*, **60**, 2257–71.

Raybould, A. F., Clarke, R. T., Bond, J. M., Welters, R. E., and Gliddon, C. J. (2002). Inferring patterns of dispersal from allele frequency data. In J. M. Bullock, R. E. Kenward, and R. S. Hails, eds. *Dispersal Ecology*, pp. 89–110. Blackwell Science, Oxford.

Reid, C. (1899). *The Origin of the British Flora*. Dulau, London.

Rejmánek, M. and Richardson, D. M. (1996). What attributes make some plant species more invasive. *Ecology*, **77**, 1655–61.

Renton, M., Hanan, J., and Burrage, K. (2005). Using the canonical modeling approach to simplify the simulation of function in functional-structural plant models. *New Phytologist*, **166**, 845–57.

Rew, L. J., Froud-Williams, R. J., and Boatman, N. D. (1996). Dispersal of *Bromus sterilis* and *Anthriscus sylvestris* seed within arable field margins. *Agriculture, Ecosystems and Environment*, **59**, 107–14.

Riba, M., Mignot, A., Freville, H., Colas, B., Imbert, E., Vile, D., Virevaire, M., and Olivieri, I. (2005). Variation in dispersal traits in a narrow-endemic plant species, *Centaurea corymbosa* (Asteraceae). *Evolutionary Ecology*, **19**, 443–53.

Ribbens, E., Silander, J. A., and Pacala, S. W. (1994). Seedling recruitment in forests: calibrating models to predict patterns of tree seedling dispersion. *Ecology*, **75**, 1794–806.

Richardson, B. A., Klopfenstein, N. B., and Brunsfeld, S. J. (2002). Assessing Clark's nutcracker seed-caching flights using maternally inherited mitochondrial DNA of whitebark pine. *Canadian Journal of Forest Research*, **32**, 1103–07.

Ricker, W. E. (1954). Stock and recruitment. *Journal of the Fisheries Research Board of Canada*, **11**, 559–623.

Ridley, H. N. (1930). *The Dispersal of Plants Throughout the World*. Reeve, Ashford.

Ripley, B. D. (1976). The second-order analysis of stationary point processes. *Journal of Applied Probability*, **13**, 225–66.

Roff, D. (1992). *The Evolution of Life Histories: Theory and Analysis*. Chapman and Hall, New York.

Roff, D. (1975). Population stability and the evolution of dispersal in a heterogeneous environment. *Oecologia*, **19**, 217–37.

Römermann, C., Tackenberg, O., and Poschlod, P. (2005). How to predict attachment potential of seeds to sheep and cattle coat from simple morphological seed traits. *Oikos*, **110**, 219–30.

Ronce, O., Gandon, S., and Rousset, F. (2000). Kin selection and natal dispersal in an age-structured population. *Theoretical Population Biology*, **58**, 143–59.

Ronce, O., Brachet, S., Olivieri, I., Gouyon, P. H., and Clobert, J. (2005). Plastic changes in seed dispersal along ecological succession: theoretical predictions from an evolutionary model. *Journal of Ecology,* **93**, 431–40.

Room, P. M. (1983). 'Falling apart' as a lifestyle: the rhizome architecture and population growth of *Salvinia molesta. Journal of Ecology,* **71**, 349–65.

Roth, J. K. and Vander Wall, S. B. (2005) Primary and secondary seed dispersal of bush chinquapin (Fagaceae) by scatterhoarding rodents. *Ecology,* **86**, 2428–39.

Rousset, F. and Gandon, S. (2002). Evolution of the distribution of dispersal distance under distance-dependent cost of dispersal. *Journal of Evolutionary Biology,* **15**, 515–23.

Roze, D. and Rousset, F. (2005). Inbreeding depression and the evolution of dispersal rates: A multi-locus model. *American Naturalist,* **166**, 708–21.

Rudis, V. A., Ek, A. R., and Balsiger, J. W. (1978). Within-stand seedling dispersal for isolated *Pinus strobus* within hardwood stands. *Canadian Journal of Forest Research,* **8**, 10–13.

Ruiz de Clavijo, E. and Jiménez, M. J. (1998). The influence of achene type and plant density on growth and biomass allocation in the heterocarpic annual *Catananche lutea* (Asteraceae). *International Journal of Plant Sciences,* **159**, 637–47.

Russo, S. E., Portnoy, S., and Augspurger, C. K. (2006). Incorporating animal behavior into seed dispersal models: Implications for seed shadows. *Ecology,* **87**, 3160–74.

Ruxton, G. D. (1996). Density-dependent migration and stability in a system of linked popualtions. *Bulletin of Mathematical Biology,* **58**, 643–60.

Ruxton, G. D. and Rohani, P. (1999). Fitness-dependent dispersal in metapopulations and its consequence for persistence and synchrony. *Journal of Animal Ecology,* **68**, 530–39.

Saether, B. E., Engen, S., and Lande, R. (1999). Finite metapopulation models with density-dependent migration and stochastic local dynamics. *Proceedings of the Royal Society B,* **266**, 113–18.

Sagar, G. R. and Mortimer, A. M. (1976). An approach to the study of the population dynamics of plants with special reference to weeds. *Advances in Applied Biology,* **1**, 1–47.

Salisbury, E. J. (1942). *The Reproductive Capacity of Plants.* Studies in Quantitative Biology. G Bell, London.

Salisbury, E. J. (1958). *Spergularia marina* and *Spergularia marginata* and their heteromorphic seeds. *Kew Bulletin,* **1**, 41–51.

Salisbury, E. J. (1961). *Weeds and Aliens.* Collins, London.

Sallabanks, R. (1992). Fruit fate, frugivory, and fruit characteristics: a study of the hawthorn, *Crataegus monogyna* (Rosaceae). *Oecologia,* **91**, 296–304.

Sallabanks, R. (1993). Hierarchical mechanisms of fruit selection by an avian frugivore. *Ecology,* **74**, 1326–36.

Schaefer, H. M. and Schmidt, V. (2004). Detectability and content as opposing signal characteristics in fruits. *Proceedings of the Royal Society of London B,* **271**, S370–73.

Schaefer, H. M., Schmidt, V., and Winkler, H. (2003). Testing the defence trade-off hypothesis: How contents of nutrents and secondary compounds affect fruit removal. *Oikos,* **102**, 318–28.

Schippers, P., Ter Borg, S. J., Van Groenendael, J. M., and Habekotte, B. (1993). What makes *Cyperus esculentus* (yellow nutsedge) an invasive species? – A spatial model approach. *Proceedings of the Brighton Crop Protection Conference – Weeds – 1993*, pp. 495–504, British Crop Protection Council, Farnham.

Schmidt, V., Schaefer, H. M., and Winkler, H. (2004). Conspicuousness, not colour as foraging cue in plant-animal signaling. *Oikos,* **106**, 551–57.

Schneider, R. L. and Sharitz, R. R. (1988). Hydrochory and regeneration in a bald cypress-water tupelo swamp forest. *Ecology,* **69**, 1055–63.

Schupp, E. W. and Fuentes, M. (1995). Spatial patterns of seed dispersal and the unification of plant population ecology. *Ecoscience,* **2**, 267–75.

Schurr, F. M., Bond, W. J., Midgley, G. F., and Higgins, S. I. (2005). A mechanistic model for secondary seed dispersal by wind and its experimental validation. *Journal of Ecology,* **93**, 1017–28.

Seger, J. and Brockmann, H. J. (1987). What is bet-hedging? *Oxford Surveys in Evolutionary Biology,* **4**, 182–211.

Seidler, T. G. and Plotkin, J, B. (2006). Seed dispersal and spatial pattern in tropical trees. *PLoS Biology* **4**, e344. DOI: 10.1371/journal.pbio.0040344.

Setter, M., Bradford, M., Dorney, B., Lynes, B., Mitchell, J., Setter, S., and Westcott, D. (2002). Pond apple – are the endangered cassowary and feral pig helping this weed to invade Queensland's wet tropics? In H. Spafford Jacob, J. Dodd, J. and J. H. Moore, eds. *Papers and Proceedings, 13th Australian Weeds Conference, 'Weeds: Threats Now, and Forever?'*, pp. 173–76. Plant Protection Society of WA, Perth, Western Australia.

Shaw, M. W. (1995). Simulation of population expansion and spatial pattern when individual dispersal distributions do not decline exponentially with distance. *Proceedings of the Royal Society, London,* **B 259**, 243–48.

Shea, K., Possingham, H. P., Murdoch, W. W., and Roush, R. (2002). Active adaptive management in insect pest and weed control: Intervention with a plan for learning. *Ecological Applications,* **12**, 927–36.

Sheldon, J. C. and Burrows, F. M. (1973). The dispersal effectiveness of the achene-pappus units of selected compositae in steady winds with convection. *New Phytologist,* **72**, 665–75.

Shigesada, N. and Kawasaki, K. (1997). *Biological Invasions: Theory and Practice*. Oxford University Press, Oxford.

Shigesada, N., Kawasaki, K., and Teramoto, E. (1986). Traveling periodic waves in heterogeneous environments. *Theoretical Population Biology*, **30**, 143–60.

Shigesada, N., Kawasaki, K., and Takeda, Y. (1995). Modelling stratified diffusion in biological invasions. *The American Naturalist*, **146**, 229–51.

Shirtliffe, S. J. and Entz, M. H. (2005). Chaff collection reduces seed dispersal of wild oat (*Avena fatua*) by a combine harvester. *Weed Science*, **53**, 465–70.

Silverman, B. W. (1986). *Density Estimation for Statistics and Data Analysis*. Chapman and Hall, London.

Sinoquet, H. and Rivet, P. (1997). Measurement and visualization of the architecture of an adult tree based on a three-dimensional digitising device. *Trees*, **11**, 265–70.

Sintes, T., Marbà, N., Duarte, C. M., and Kendrick, G. A. (2005) Nonlinear processes in seagrass colonisation explained by simple clonal growth rules. *Oikos*, **108**, 165–75.

Skarpaas, O., Stabbetorp, O. E., Rønning, I., and Svennungsen, T. O. (2004). How far can a hawk's beard fly? Measuring and modelling the dispersal of *Crepis praemorsa*. *Journal of Ecology*, **92**, 747–57.

Skarpaas, O., Shea, K., and Bullock, J. M. (2005). Optimizing dispersal study design by Monte Carlo simulation. *Journal of Applied Ecology*, **42**, 731–39.

Skellam, J. G. (1951). Random dispersal in theoretical populations. *Biometrika*, **38**, 196–218.

Skidmore, B. A. and Heithaus, E. R. (1988). Lipid cues for seed-carrying ants in *Hepatica Americana*. *Journal of Chemical Ecology*, **14**, 2185–96.

Slade, A. J. and Hutchings, M. J. (1987). The effects of nutrient availability on foraging in the clonal herb *Glechoma hederacea*. *Journal of Ecology*, **75**, 95–112.

Slingsby, P. and Bond, W. J. (1985). The influence of ants on the dispersal distance and seedling recruitment of *Leucospermum conocarpodendron* (L.) Buek (Proteaceae). *South African Journal of Botany*, **51**, 30–34.

Smallwood, P. D., Steele, M. A., and Faeth, S. H. (2001). The ultimate basis of the caching preferences of rodents, and the oak-dispersal syndrome: tannins, insects, and seed germination. *American Zoologist*, **41**, 840–51.

Smith, C. C. and Follmer, D. (1972) Food preferences of squirrels. *Ecology*, **53**, 82–91.

Smith, C. C. and Fretwell, S. D. (1974). The optimal balance between size and number of offspring. *American Naturalist*, **108**, 499–506.

Smith, C. C. and Reichman, O. J. (1984). The evolution of food caching by birds and mammals. *Annual Reviews of Ecology and Systematics*, **15**, 329–51.

Smith, G. S., Curtis, J. P., and Edwards, C. M. (1992). A method for analysing plant architecture as it relates to fruit quality using three-dimensional computer graphics. *Annals of Botany*, **70**, 265–69.

Smith, R. B. (1973). Factors affecting dispersal of dwarf mistletoe seeds from an overstory western hemlock tree. *Northwest Science*, **47**, 9–19.

Snäll, T., O'Hara, R. B., and Arjas, E. (2007). A mathematical and statistical framework for modelling dispersal. *Oikos*, **116**, 1037–50.

Snyder, R. E. and Chesson, P. (2003). Local dispersal can facilitate coexistence in the presence of permanent spatial heterogeneity. *Ecology Letters*, **6**, 301–309.

Snyder, R. E. and Chesson, P. (2004). How the spatial scales of dispersal, competition, and environmental heterogeneity interact to affect coexistence. *American Naturalist*, **164**, 633–50.

Sokal, R. R. and Rohlf, F. J. (1995). *Biometry: The Principles and Practice of Statistics in Biological Research*, 3rd edition. W. H. Freeman, New York.

Soons, M. B. and Heil, G. W. (2002). Reduced colonization capacity in fragmented populations of wind-dispersed grassland forbs. *Journal of Ecology*, **90**, 1033–43.

Soons, M. B. and Ozinga, W. A. (2005). How important is long-distance seed dispersal for the regional survival of plant species? *Diversity and Distribution*, **11**, 165–72.

Soons, M. B., Heil, G. W., Nathan, R., and Katul, G. G. (2004a). Determinants of long-distance seed dispersal by wind in grasslands. *Ecology*, **85**, 305—68.

Soons, M. B., Nathan, R., and Katul, G. G. (2004b). Human effects on long-distance wind dispersal and colonization by grassland plants. *Ecology*, **85**, 3069–79.

Sorensen, A. E. (1984). Nutrition, energy and passage time: experiments with fruit preference in European blackbirds. *Journal of Animal Ecology*, **53**, 545–57.

Sorensen, A. E. (1986). Seed dispersal by adhesion. *Annual Review of Ecology and Systematics*, **17**, 443–63.

Sork, V. L. (1984). Examination of seed dispersal and survival in red oak, *Quercus rubra* (Fagaceae), using metal-tagged acorns. *Ecology*, **65**, 1020–22.

Southwood, T. R. E. (1962). Migration of terrestrial arthropods in relation to habitat. *Biological Reviews*, **37**, 171–214.

St. John-Sweeting, R. S. and Morris, K. A. (1990). Seed transmission through the digestive tract of the horse. *Proceedings of the 9th Australian Weeds Conference, Adelaide, South Australia*, pp. 137–39, Crop Science Society of South Australia, Adelaide.

Stallings, G. P., Thill, D. C., Mallory-Smith, C. A., and Lass, L. W. (1995). Plant movement and seed dispersal of Russian Thistle (*Salsola iberica*). *Weed Science*, **43**, 63–69.

Stamp, N. E. (1989). Seed dispersal of four sympatric grassland annual species of *Erodium*. *Journal of Ecology*, **77**, 1005–20.

Stamp, N. E. and Lucas, J. R. (1983). Ecological correlates of explosive seed dispersal. *Oecologia* **59**, 272–78.

Stearns, S. (1992). *The Evolution of Life Histories*. Oxford University Press, Oxford.

Steele, M., Wauters, L. A., and Larsen, K. W. (2005). Selection, predation and dispersal of seeds by tree squirrels in temperate and boreal forests: are tree squirrels keystone granivores? In P-M. Forget, J. E. Lambert, P. E. Hilme, and S. B. Vander Wall, eds. *Seed Fate*, pp. 205–21. CAB International, Wallingford.

Steele, M. A. (1998). *Tamiasciurus hudsonicus*. *Mammalian Species*, **586**, 1–9.

Steele, M. A. and Smallwood, P. D. (2001) Acorn dispersal by birds and mammals. In W. J. McShea and W. M. Healy, eds. *Oak Forest Ecosystems: Ecology and Management for Wildlife*, pp. 182–96. Johns Hopkins University Press, Baltimore.

Stefan, A. S. (1984). To fly or not to fly? Colonization of Baltic islands by winged and wingless carabid beetles. *Journal of Biogeography*, **11**, 413–26.

Stergios, B. G. (1976). Achene production, dispersal, seed germination, and seedling establishment in *Hieracium aurantiacum* in an abandoned field community. *Canadian Journal of Botany*, **54**, 1189–97.

Stiles, E. W. (1980). Patterns of fruit presentation and seed dispersal in bird-disseminated woody plants in the eastern deciduous forest. *American Naturalist*, **116**, 670–88.

Stiles, E. W. (1992). Animals as seed dispersers. In M. Fenner, ed. *Seeds: The Ecology of Regeneration in Plant Communities*, pp. 87–104. CAB International, Wallingford.

Stiles, E. W. and White, D. W. (1986). Seed deposition patterns: influence of season, nutrients, and vegetation structure. In A. Estrada and T. H. Fleming, eds. *Frugivores and Seed Dispersal*, pp. 45–54. Junk, Dordrecht.

Stocklin, J. and Winkler, E. (2004). Optimum reproduction and dispersal strategies of a clonal plant in a metapopulation: a simulation study with *Hieracium pilosella*. *Evolutionary Ecology*, **18**, 563–84.

Stoll, P. and Prati, D. (2001). Intraspecific aggregation alters competitive interactions in experimental plant communities. *Ecology*, **82**, 319–27.

Stoyan, D. and Penttinen, A. (2000). Recent applications of point process methods in forestry statistics. *Statistical Science*, **15**, 61–78.

Sun, C., Ives, A. R., Kraeuter, H. J., and Moermond, T. C. (1997) Effectiveness of three turacos as seed dispersers in a tropical montane forest. *Oecologia*, **112**, 94–103.

Swaine, M. D. and Beer, T. (1977). Explosive seed dispersal in *Hura crepitans* L. (Euphorbiaceae). *New Phytologist*, **78**, 695–708.

Swaine, M. D., Dakubu, T., and Beer, T. (1979). On the theory of explosively dispersed seeds: a correction. *New Phytologist*, **82**, 777–81.

Tackenberg, O. (2003). Modeling long-distance dispersal of plant diaspores by wind. *Ecological Monographs*, **73**, 173–89.

Tackenberg, O., Poschlod, P., and Bonn, S. (2003). Assessment of wind dispersal potential in plant species. *Ecological Monographs*, **73**, 191–205.

Tackenberg, O., Römermann, C., Thompson, K., and Poschlod, P. (2006). What does diaspore morphology tell us about external animal dispersal? Evidence from standardized experiments measuring seed retention on animal-coats. *Basic and Applied Ecology*, **7**, 45–58.

Taghizadeh, M. S. (2007) Determinants of Seed Dispersal Distance in the Weed *Raphanus raphanistrum* L. PhD thesis, The University of Melbourne.

Tang, A. M. C., Corlett, R. T., and Hyde, K. D. (2005). The persistence of ripe fleshy fruits in the presence and absence of frugivores. *Oecologia*, **142**, 232–37.

Taylor, C. M. and Hastings, A. (2004). Finding optimal control strategies for invasive species: a density-structured model for *Spartina alterniflora*. *Journal of Applied Ecology*, **41**, 104—57.

Taylor, P. D. (1988). An inclusive fitness model for dispersal of offspring. *Journal of Theoretical Biology*, **130**, 363–78.

Taylor, P. D. and Frank, S. A. (1996). How to make a kin selection model. *Journal of Theoretical Biology*, **180**, 27–37.

Taylor, P. D., Wild, G., and Gardner, A. (2007). Direct fitness or inclusive fitness: How shall we model kin selection? *Journal of Theoretical Biology*, **20**, 301–9.

Telenius, A. and Torstensson, P. (1999). Seed type and seed size variation in the heteromorphic saltmarsh annual *Spergularia salina* along the coast of Sweden. *Plant Biology*, **1**, 585–93.

ter Steege, H. (1994). Flooding and drought tolerance in seeds and seedlings of two *Mora* species segregated along a soil hydrological gradient in the tropical rain forest of Guyana. *Oecologia*, **100**, 356–67.

Terborgh, J. (1986). Keystone plant resources in the tropical forest. In M. E. Soule ed. *Conservation Biology II*, pp. 330–44. Sinauer, Sunderland, Mass.

Theodorou, K. and Couvet, D. (2002). Inbreeding depression and heterosis in a subdivided population: influence of the mating system. *Genetical Research*, **80**, 107–16.

Thiede, D. A. and Augspurger, C. K. (1996). Intraspecific variation in seed dispersion of *Lepidium campestre* (Brassicaceae). *American Journal of Botany*, **83**, 856–66.

Thomas, A. G., Gill, A. M., Moore, P. H. R. and Forcella, F. (1984). Drought feeding and the dispersal of weeds. *Journal of the Australian Institute of Agricultural Science,* **50**, 103–7.

Tilman, D. (1994). Competition and biodiversity in spatially structured habitats. *Ecology,* **75**, 2–16.

Tilman, D. (1997). Community invasibility, recruitment limitation, and grassland biodiversity. *Ecology,* **78**, 81–92.

Tilman, D., Lehman, C. L., and Kareiva, P. (1997) Population dynamics in spatial habitats. In D. Tilman and P. Kareiva, eds. *Spatial Ecology: The Role of Space in Population Dynamics and Interspecific Interactions,* pp. 3–20. Princeton University Press, Princeton.

Topping, C. J., Hansen, T. S., Jensen, T. S., Jepsen, J. U., Nikolaisen, F., and Odderskaer, P. (2003). ALMaSS, an agent-based model for animals in temperate European landscapes. *Ecological Modelling* **167**, 65–82.

Trapp, E. J. (1988). Dispersal of heteromorphic seeds in *Amphicarpaea bracteata* (Fabaceae) *American Journal of Botany* **75**, 1535–39.

Traveset, A. and Verdu, M. (2002). A meta-analysis of the effect of gut treatment on seed germination. In D. J. Levey, W. R. Silva, and M. Galetti, eds. *Seed Dispersal and Frugivory: Ecology, Evolution and Conservation,* pp. 339–50. CAB International, Wallingford.

Travis, J. M. J. and Dytham, C. (1998). The evolution of dispersal in a metapopulation: a spatially explicit, individual-based model. *Proceedings of the Royal Society B,* **265**, 17–23.

Travis, J. M. J. and Dytham, C. (1999). Habitat persistence, habitat availability and the evolution of dispersal. *Proceedings of the Royal Society B,* **266**, 723–28.

Travis, J. M. J. and Dytham, C. (2002). Dispersal evolution during invasions. *Evolutionary Ecology Research,* **4**, 1119–29.

Travis, J. M. J., Murrell, D. J., and Dytham, C. (1999). The evolution of density-dependent dispersal. *Proceedings of the Royal Society B,* **266**, 1837–42.

Tribe, D. E. (1949). Some seasonal observations on the grazing habits of sheep. *Empire Journal of Experimental Agriculture,* **17**, 105–15.

Trivers, R. L. (1974). Parent-offspring conflict. *American Zoologist,* **14**, 249–64.

Tufto, J., Engen, S., and Hindar, K. (1996). Inferring patterns of migration from gene frequencies under equilibrium conditions. *Genetics,* **144**, 1909–19.

Tufto, J., Engen, S., and Hindar, K. (1997). Stochastic dispersal processes in plant populations. *Theoretical Population Biology,* **52**, 16–26.

Tufto, J., Raybould, A. F., Hindar, K., and Engen, S. (1998). Analysis of genetic structure and dispersal patterns in a population of sea beet. *Genetics,* **149**, 1975–85.

Turkington, R. A. and Harper, J. L. (1979). The growth, distribution and neighbour relationships of *Trifolium repens* in a permanent pasture. Ordination, pattern and contact. *Journal of Ecology,* **67**, 201–208.

Tutin, C. E. G., Williamson, E. A., Roger, M. E., and Fernandez, M. (1991). A case study of a plant-animal relationship: *Cola lizae* and lowland gorillas in the Lopé Reserve, Gabon. *Journal of Tropical Ecology,* **7**, 181–99.

Uyenoyama, M. K. (1986). Inbreeding and the cost of meiosis: the evolution of selfing in populations practicing biparental inbreeding. *Evolution,* **40**, 388–404.

van den Bosch, F., Metz, J. A. J., and Diekmann, O. (1990). The velocity of spatial population expansion. *Journal of Mathematical Biology,* **28**, 529–65.

van der Pijl, L. (1982). *Principles of Dispersal in Higher Plants.* Springer-Verlag, Berlin.

van Rheede van Oudtshoorn, K., and van Rooyen, M. W. (1999). *Dispersal Biology of Desert Plants.* Springer, Berlin.

Van Schaik, C. P. (1986). Phenological changes in a Sumatran rain forest. *Journal of Tropical Ecology,* **2**, 327–47.

Van Schaik, C. P., Terborgh, J. W., and Wright, S. J. (1993). The phenology of tropical forests: Adaptive significance and consequences for primary consumers. *Annual Review of Ecology and Systematics,* **24**, 353–77.

Van Valen, L. (1971). Group selection and the evolution of dispersal. *Evolution,* **25**, 591–98.

Vander Wall, S. B. (2001). The evolutionary ecology of nut dispersal. *The Botanical Review,* **67**, 74–117.

Vellend, M., Myers, J. A., Gardescu, S., and Marks, P. L. (2003). Dispersal of *Trillium* seeds by deer: implications for long-distance migration of forest herbs. *Ecology,* **84**, 1067–72.

Venable, D. L. (1985). The evolutionary ecology of seed heteromorphism. *American Naturalist,* **126**, 577–95.

Venable, D. L. and Levin, D. A. (1985). Ecology of achene dimorphism in *Heterotheca latifolia. Journal of Ecology,* **73**, 133–45.

Vibrans, H. (1999). Epianthropochory in Mexican weed communities. *American Journal of Botany,* **86**, 476–81.

Viswanathan, G. M., Afanasyev, V., Buldyrev, S. V., Murphy, E. J., Prince, P. A., and Stanley H. E. (1996). Lévy flight search patterns of wandering albatrosses. *Nature,* **381**, 413–15.

Waagepetersen, R. P. (2007). An estimating function approach to inference for inhomogeneous Neyman Scott processes. *Biometrics,* **63**, 252–58.

Wadsworth, R. A., Collingham, Y C., Willis, S. G., Huntley, B., and Hulme, P. E. (2000). Simulating the spread and management of alien riparian weeds: Are they out of control? *Journal of Applied Ecology,* **37**, 28–38.

Wagner, S. (1997). A model describing the fruit dispersal of ash (*Fraxinus excelsior*) taking into account directionality. *Allgemeine Forst und Jagdzeitung*, **168**, 149–55.

Wahaj, S. A., Levey, D. J., Sanders, A. K., and Cipollini, M. L. (1998). Control of gut retention time by secondary metabolites in ripe *Solanum* fruits. *Ecology*, **79**, 2309–19.

Waide, R. B., Willig, M. R., Steiner, C. F., Mittelbach, G., Gough, L., Dodson, S. I., Juday, G. P., and Parmenter, R. (1999). The relationship between productivity and species richness. *Annual Review of Ecology and Systematics*, **30**, 257–300.

Wallinga, J. (1995). The role of space in plant population dynamics: Annual weeds as an example. *Oikos*, **74**, 377–83.

Wallinga, J., Kropff, M. J., and Rew, L. J. (2002). Patterns of spread of annual weeds. *Basic and Applied Ecology*, **3**, 31–38.

Ward, R. G. and Brookfield, M. (1992). The dispersal of the coconut: Did it float or was it carried to Panama. *Journal of Biogeography*, **19**, 467–80.

Wareing, P. F. (1958). Reproductive development in *Pinus sylvestris*. In K. V. Thimann ed. *The Physiology of Forest Trees*, pp. 643–54. Ronald Press, New York.

Watkinson, A. R. (1978). The demography of a sand dune annual: Vulpia fasciculate III. The dispersal of seeds. *Journal of Ecology*, **66**, 483–98.

Watkinson, A. R. and Powell, J. C. (1997). The life history and population structure of *Cycas armstrongii* in monsoonal northern Australia. *Oecologia*, **111**, 341–49.

Watkinson, A. R. and White, J. (1985). Some life-history consequences of modular construction in plants. *Philosophical Transactions of the Royal Society of London* B, **313**, 31–51.

Watt, A, S. (1947). Pattern and process in the plant community. *Journal of Ecology*, **35**, 1–22.

Wauters, L. A. and Casale, P. (1996). Long-term scatter-hoarding by Eurasian red squirrels (*Sciurus vulgaris*). *Journal of Zoology, London*, **238**, 195–207.

Wauters, L. A. and Lens, L. (1995). Effects of food availability and density on red squirrel (*Sciurus vulgaris*) reproduction. *Ecology*, **76**, 2460–69.

Weiblen, G. D. and Thomson, J. D. (1995). Seed dispersal in *Erythronium grandiflorum* (Liliaceae). *Oecologia*, **102**, 211–19.

Weinberger, H. F. (1982). Long-time behavior of a class of biological models. *SIAM Journal on Mathematical Analysis*, **13**, 353–96.

Weinberger, H. F. (2002). On spreading speeds and traveling waves for growth and migration models in a periodic habitat. *Journal of Mathematical Biology*, **45**, 511–48.

Weiner, J., Berntson, G. M., and Thomas, S. C. (1990). Competition and growth form in a woodland annual. *Journal of Ecology*, **78**, 459–69.

Weiner, J., Martinez, S., Müller-Schärer, H., Stoll, P., and Schmid, B. (1997). How important are environmental maternal effects in plants? A study with *Centaurea maculosa*. *Journal of Ecology*, **85**, 133–42.

Wender, N. J., Polisetti, C. R., and Donohue, K. (2005). Density-dependent processes influencing the evolutionary dynamics of dispersal: A functional analysis of seed dispersal in *Arabidopsis thaliana* (Brassicaceae). *American Journal of Botany*, **92**, 960–71.

Wenny, D. G. (2001). Advantages of seed dispersal: a re-evaluation of directed dispersal. *Evolutionary Ecology Research*, **3**, 51–75.

Wenny, D. G. and Levey, D. J. (1998). Directed seed dispersal by bellbirds in a tropical cloud forest. *Proceedings of the National Academy of Sciences of the U.S.A.*, **95**, 6204–07.

Weppler, T. and Stocklin, J. (2005). Variation of sexual and clonal reproduction in the alpine *Geum reptans* in contrasting altitudes and successional stages. *Basic and Applied Ecology*, **6**, 305–16.

Werner, P. A. and Platt, W. J. (1976). Ecological relationships of co-occurring golden rods (*Solidago*: Compositae). *American Naturalist*, **110**, 959–71.

Westcott, D. A., Bentrupperbäumer, J., Bradford, M. G., and McKeown, A. (2005). Incorporating patterns of disperser behaviour into models of seed dispersal and its effects on estimated dispersal curves. *Oecologia*, **146**, 57–67.

Whalley, R. D. B. and Burfitt, J. M. (1972). Ecotypic variation in *Avena fatua* L., *A. sterilis* L. (*A. ludoviciana*), and *A. barbata* Pott. In New South Wales and southern Queensland. *Australian Journal of Agricultural Research*, **23**, 799–810.

Wheelwright, N. T. (1985). Fruit size gape width and the diets of fruit-eating birds. *Ecology*, **66**, 808–18.

Wheelwright, N. T. (1986). A seven-year study of individual variation in fruit production in tropical bird-dispersed tree species in the family Lauraceae. In A. Estrada and T. H. Fleming, eds. *Frugivores and Seed Dispersal*, pp. 19–35. Junk, Dordrecht.

Whelan, B. (1988). *Reconciling continuous soil variation and crop growth: a study of some implications of field variability for site-specific crop management*. PhD thesis, University of Sydney.

Whitney, K. D. (2005). Linking frugivores to the dynamics of a fruit color polymorphism. *American Journal of Botany*, **92**, 859–67.

Whitney, K. D. and Lister, C. E. (2005). Fruit colour polymorphism in *Acacia ligulata*: seed and seedling performance, clinal patterns, and chemical variation. *Evolutionary Ecology*, **18**, 165–86.

Wiegand, T. and Moloney, K. (2004). Rings, circles and null-models for point pattern analysis in ecology. *Oikos*, **104**, 209–29

Wiens, J. A. (1997). Metapopulation dynamics and landscape ecology. In I. A. Hanski and M. E. Gilpin, eds. *Metapopulation Biology: Ecology, Genetics, and Evolution*, pp. 43–62. Academic Press, San Diego.

Williamson, G. B., Costa, F., and Vera, C. V. M. (1999). Dispersal of Amazonian trees: hydrochory in *Swartzia polyphylla*. *Biotropica*, **31**, 460–65.

Williamson, M. (1997). *Biological Invasions*. Chapman and Hall, London.

Wills, C. and Condit, R. (1999). Similar non-random processes maintain diversity in two tropical rainforests. *Proceedings of the Royal Society London, Series B*, **266**, 1445–52.

Wills, C., Harms, K. E., Condit, R., King, D., Thompson, J., He, F., Muller-Landau, H. C., Ashton, P., Losos, E., Comita, L., Hubbell, S., LaFrankie, J., Bunyavejchewin, S., Dattaraja, H. S., Davies, S., Esufali, S., Foster, R., Gunatilleke, N., Gunatilleke, S., Hall, P., Itoh, A., John, R., Kiratiprayoon, S., de Lao, S. L., Massa, M., Nath, C., Noor, M. N. S., Kassim, A. R., Sukumar, R., Suresh, H. S., Sun, I-F., Tan, S., Yamakura, T., and Zimmerman, J. (2006). Nonrandom processes maintain diversity in tropical forests. *Science*, **311**, 527–31.

Willson, M. F. (1993). Dispersal mode, seed shadows, and colonization patterns. *Vegetatio*, **107/108**, 261–80.

Willson, M. F. (1994). Fruit choices by captive American Robins. *The Condor*, **96**, 494–502.

Wilson, B. J. (1970). Studies of the shedding of seed of *Avena fatua* in various cereal crops and the presence of this seed in the harvested material. *Proceedings of the British Weed Control Conference*, pp. 831–36, British Crop Protection Council, Croydon.

With, K. A. (2002) Using percolation theory to assess landscape connectivity and effects of habitat fragmentation. In K. J. Gutzwiller, ed *Applying Landscape Ecology in Biological Conservation*, pp. 105–30. Springer-Verlag, New York.

With, K. A. (2004). Assessing the risk of invasive spread in fragmented landscapes. *Risk Analysis*, **24**, 803–15.

With, K. A. and King, A. W. (1999). Dispersal success on fractal landscapes: a consequence of lacunarity thresholds. *Landscape Ecology*, **14**, 73–82.

With, K. A., Pavuk, D. M., Worchuck, J. L., Oates, R. K., and Fisher, J. L. (2002). Threshold effects of landscape structure on biological control in agroecosystems. *Ecological Applications*, **12**, 52–65.

Witztum, A. and Schulgasser, K. (1995a). The mechanics of seed expulsion in *Acanthaceae*. *Journal of Theoretical Biology*, **176**, 531–42.

Witztum, A. and Schulgasser, K. (1995b). Seed dispersal ballistics in *Blepharis ciliaris*. *Israel Journal of Plant Sciences*, **43**, 147–50.

Witztum, A., Schulgasser, K., and Vogel, S. (1996). Upwind movement of achenes of *Centauria eriophora* L. on the ground. *Annals of Botany*, **78**, 431–36.

Wolf, B. O., Martinez del Rio, C., and Babson, J. (2002). Stable isotopes reveal that Saguaro fruit provides different resources to two desert dove species. *Ecology*, **83**, 1286–93.

Woolcock, J. L. and Cousens, R. (2000). A mathematical analysis of factors affecting the rate of spread of patches of annual weeds in an arable field. *Weed Science*, **48**, 27–34.

Wright, S. (1951). The genetical structure of populations. *Annals of Eugenics*, **15**, 323–54

Xiao, Z., Zhang, Z., and Wang, Y. (2005). Effects of seed size on dispersal distance in five rodent-dispersed fagaceous species. *Acta Oecologia*, **28**, 221–229.

Yeaton, R. I. and Bond, W. J. (1991). Competition between two shrub species: dispersal differences and fire promote coexistence. *American Naturalist*, **138**, 328–41.

Zasada, J. C. (1986). Natural regeneration of trees and tall shrubs on forest sites in interior Alaska. In K. Cleve, F. S. Chapin, P. W. Flanagan, L. A. Viereck, and C. T. Dyrness, eds. *Forest Ecosystems in the Alaskan Taiga, A Synthesis of Structure and Function* pp. 44–73. Spinger-Verlag, New York.

Zasada, J. C. and Lovig, D. (1983). Observations on primary dispersal of white spruce, *Picea glauca*, seed. *Canadian Field Naturalist*, **97**, 104–06.

Zobel, M., Otsus, M., Liira, J., Moora, M., and Möls, T. (2000). Is small-scale species richness limited by seed availability or microsite availability? *Ecology*, **81**, 3274–82.

Plant species index

Nomenclature follows that given in the source papers: some names may have been revised.

Index